Process Development Division 2014

Core Programming Area at the 2014 AIChE Annual Meeting

Atlanta, Georgia, USA
16-21 November 2014

ISBN: 978-1-5108-1270-3

Printed from e-media with permission by:

Curran Associates, Inc.
57 Morehouse Lane
Red Hook, NY 12571

Printed by Curran Associates, Inc. (2015)

For permission requests, please contact AIChE
at the address below.

AIChE
120 Wall Street, FL 23
New York, NY 10005-4020

Phone: (800) 242-4363
Fax: (203) 775-5177

www.aiche.org

Additional copies of this publication are available from:

Curran Associates, Inc.
57 Morehouse Lane
Red Hook, NY 12571 USA
Phone: 845-758-0400
Fax: 845-758-2634
Email: curran@proceedings.com
Web: www.proceedings.com

TABLE OF CONTENTS

376192 Custom Design and Evaluation of Oldershaw Columns for Pilot Plant Operations

Abstract:

Custom Design and Evaluation of Oldershaw Columns for Pilot Plant Operations

Pilot plant distillation operations are typically challenged to provide realistic feedback on separations capabilities and recycle stream compositions by mimicking commercial scale operations as closely as possible. Matching commercial scale operating conditions to provide information for scale-up can be especially difficult for columns with unique hydraulic loads. Since columns represent a significant capital expenditure and can be long lead time items impacting the project schedule, it is imperative that the column perform as expected at startup to avoid additional expense and lost time due to rework.

The challenge for pilot plant designers is that little data and design methodologies are available for columns at the smaller scale. Design correlations have been studied and published extensively for commercial scale equipment, but these correlations have often been untested or underreported at smaller scales. Oldershaw column manufacturers typically offer off-the-shelf designs that have been proven to work in typical applications over the years, but they offer little assistance in evaluating the columns for a specific application prior to purchase. Standard offerings may also be a poor choice for unique process conditions such as high or low L/V ratios, fouling services, etc.

MATRIC (Mid-Atlantic Technology, Research and Innovation Center) has developed a methodology for evaluating Oldershaw column designs based on process simulations and to custom design Oldershaw trays to meet specific process needs. This methodology has been developed through a combination of extrapolating commercial scale correlations, testing in MATRIC labs, and experienced based observations on the pilot scale.
An outline of the sizing and evaluation methodology is presented along with a case study demonstrating the use of the methodology to size a pilot scale column with a unique L/V ratio and fouling characteristics. The adaptation of design correlations to the pilot scale are discussed and the methodology is followed to yield a custom tray design. Project execution considerations for custom designed columns and validating the tray efficiencies at startup are also discussed briefly.

382382 Novel Municipal Solid Waste to Liquid Transportation Fuels Processes: Mathematical Model of MSW Gasification, Process Synthesis, and Global Optimization Strategies

Abstract:

The United States consumes over 18 million barrels of petroleum-based products per day, with the transportation sector representing the majority (nearly 70%) of this consumption. With growing concerns over expensive crude oil prices and increased scrutiny over high levels of greenhouse gas (GHG) emissions, the U.S. transportation sector faces major challenges that must be addressed through the investigation of novel processes to produce liquid fuels. Three feedstocks that have been touted as alternatives to petroleum include coal, biomass, and natural gas. A recent review has illustrated the key contributions in the production of liquid fuels from single and hybrid combinations of these three feedstocks [1]. This study, however, focuses on a fourth type of feedstock: municipal solid waste (MSW).

The United States generates over 250 million tons of MSW each year [2]. Its competitive energy content, as well as its negative cost, make it an attractive precursor to liquid fuels. Facilities normally receive a tipping fee, which varies between $24-$70/ton in the U.S., for the disposal of municipal solid waste [3]. Thus, we seek to determine whether the production of liquid transportation fuels from municipal solid waste is economically viable.

A stoichiometric municipal solid waste gasifier model [4] that was able to predict the effluent to 8.75% when compared with experimental data is incorporated in a thermochemical based process synthesis superstructure to produce liquid fuels from multiple conversion pathways [5-16]. Municipal solid waste is first processed in a refuse derived fuel (RDF) facility that converts the MSW into a higher-calorific fuel (RDF). The RDF is then gasified and the synthesis gas (syngas) effluent is sent to either the scrubbing system or to a dedicated water-gas-shift reactor. Syngas is then directed to either the Fischer-Tropsch reactor section or to a methanol synthesis reactor. The Fischer-Tropsch hydrocarbons are upgraded to fuel-grade products, while the methanol is converted to either gasoline or olefins. The olefins are subsequently converted into gasoline and distillate. Simultaneous heat and power integration will ensure that waste heat is converted into electricity using a series of heat engines.

The optimal plant topology that can produce liquids at the lowest possible cost is determined using a rigorous deterministic global optimization branch-and-bound strategy. The effect of plant capacity and product distribution on the optimal process topology is investigated. The overall cost of the refinery, the total plant cost, and the break-even oil price will also be illustrated. Additionally, the effect that the tipping fee has on the overall cost of liquid fuels is investigated parametrically.

[1] Floudas, C. A.; Elia, J. A.; Baliban, R. C. Hybrid and single feedstock energy processes for liquid transportation fuels: A critical review. Computers & Chemical Engineering 2012, 41 (6), 24-51.
[2] EPA, Municipal Solid Waste Generation, Recycling and Disposal in the United States: Factsand Figures for 2011. Document Number: EPA530-R-13-001, http://www.epa.gov/osw/nonhaz/municipal/pubs, 2013.
[3] Valkenburg, C.; Walton, C.; Thompson, B.; Gerber, M.; Jones, S.; Stevens, D. Municipal solid waste (MSW) to liquid fuels synthesis, Volume 1: Availability of Feedstock and Technolog. Richland, WA: Pacific Northwest National Laboratory 2009.
[4] Onel, O.; Niziolek, A. M.; Hasan, M.; Floudas, C. A. Municipal solid waste to liquid transportation fuels–Part I: Mathematical modeling of a municipal solid waste gasifier. Computers & Chemical Engineering 2014 DOI:10.1016/j.compchemeng.2014.03.008.
[5] Baliban, R. C., Elia, J. A., Floudas, C. A. Toward novel biomass, coal, and natural gas processes for satisfying current transportation fuel demands, 1: Process alternatives, gasification modeling, process simulation, and economic analysis. Industrial & Engineering Chemistry Research 2010, 49, 7343-7370.
[6] Elia, J. A., Baliban, R. C., Floudas, C. A. Toward novel biomass, coal, and natural gas processes for satisfying current transportation fuel demands, 2: Simultaneous heat and power integration. Industrial & Engineering Chemistry Research 2010, 49, 7371-7388.
[7] Baliban, R. C., Elia, J. A., Floudas, C. A. Optimization framework for the simultaneous process synthesis, heat and power integration of a thermochemical hybrid biomass, coal, and natural gas facility. Comp. Chem. Eng. 2011, 35, 1647-1690.
[8] Baliban, R. C., Elia, J. A., Floudas, C. A. Simultaneous process synthesis, heat, power, and water integration of thermochemical hybrid biomass, coal, and natural gas facilities. Comp. Chem. Eng. 2012, 37, 297-327.
[9] Baliban, R. C., Elia, J. A., Misener, R., Floudas, C. A. Global Optimization of a MINLP Process Synthesis Model for Thermochemical Based Conversion of Hybrid Coal, Biomass, and Natural Gas to Liquid Fuels. Comp. Chem. Eng. 2012, 42, 64-86.
[10] Baliban, R. C., Elia, J. A., Weekman, V., Floudas, C. A. Process Synthesis of Hybrid Coal, Biomass, and Natural Gas to Liquids via Fischer-Tropsch Synthesis, ZSM-5 Catalytic Conversion, Methanol Synthesis, Methanol-to-Gasoline, and Methanol-to-Olefins/Distillate Technologies. Comp. Chem. Eng. 2012, 47 (12), 29-56.
[11] Baliban, R. C.; Elia, J. A.; Floudas, C. A. Biomass to liquid transportation fuels (BTL) systems: process synthesis and global optimization framework. Energy Environ. Sci. 2013, 6 (1), 267-287.
[12] Baliban, R. C.; Elia, J. A.; Floudas, C. A. Novel Natural Gas to Liquids Processes: Process Synthesis and Global Optimization Strategies. AIChE Journal 2013, 59 (2), 505-531.
[13] Baliban, R. C.; Elia, J. A.; Floudas, C. A. Biomass and Natural Gas to Liquid Transportation Fuels: Process Synthesis, Global Optimization, and Topology Analysis. Industrial & Engineering Chemistry Research 2013, 52 (9), 3381-3406.
[14] Baliban, R. C.; Elia, J. A.; Floudas, C. A.; Gurau, B.; Weingarten, M. B.; Klotz, S. D. Hardwood Biomass to Gasoline, Diesel, and Jet Fuel: 1. Process Synthesis and Global Optimization of a Thermochemical Refinery. Energy & Fuels 2013, 27 (8), 4302-4324.
[15] Baliban, R. C.; Elia, J. A.; Floudas, C. A.; Xiao, X.; Zhang, Z.; Li, J.; Cao, H.; Ma, J.; Qiao, Y.; Hu, X. Thermochemical Conversion of Duckweed Biomass to Gasoline, Diesel, and Jet Fuel: Process Synthesis and Global Optimization. Industrial & Engineering Chemistry Research 2013, 52 (33), 11436-11450.
[16] Niziolek, A. M.; Onel, O.; Elia, J. A.; Baliban, R. C.; Xiao, X.; Floudas, C. A. Coal and Biomass to Liquid Transportation Fuels: Process Synthesis and Global Optimization Strategies. Industrial & Engineering Chemistry Research 2014 DOI: 10.1021/ie500505h.

379885 Thermomechanical and Electrical Properties of Ionic Liquid Imbibed Poly(ethylene glycol) Gels Designed for Use in Lithium Batteries

Abstract:

Lithium ion batteries (LIBs) are attractive as energy storage devices because of their higher volumetric and gravimetric energy densities than the nickel metal hydride batteries that are currently used in hybrid vehicles. However, conventional LIB electrolytes are volatile and flammable liquids, and therefore, limit the durability and temperature range of operation of the device. Ionic liquids (ILs) have been found to be promising alternatives because of their low volatility, satisfactory ionic conductivity, and good electrochemical stability [1], but to overcome the problems of hermetic sealing and lithium dendrite formation associated with liquids, there is a significant interest in solid polymer electrolytes for lithium conduction. Completely dry solid polymer electrolytes have rather low room-temperature ionic conductivities, often below 10^{-5} S/cm. Liquid electrolytes (e.g., those based on blends of ethylene carbonate and diethyl carbonate), on the other hand, have at least three orders of magnitude higher conductivity [1].

With the goal of combining the high conductivity of IL electrolytes with the desirable mechanical properties solid polymer electrolytes, we investigated structure–property correlations in lithium-ion conducting PEGylated gels that were imbibed with ILs. In this work, the lithium ion conductivity and mechanical properties of membranes prepared by crosslinking poly(ethylene glycol) (PEG) diacrylate with a T_g-modifying comonomer, such as ethyl acrylate, were investigated using electrochemical impedance spectroscopy and dynamic mechanical analysis. Three different comonomers and six different ILs (all with the bistriflamide anion) were screened for incorporation in these gels. Consistent with our previous study [2], the extent of charge delocalization in the cation was found to greatly affect transport properties such as viscosity and conductivity of the IL. An optimal IL/comonomer combination was identified based on gel ionic conductivity measurements at different temperatures (in the range of 0 to 100 °C). The resulting composite gel electrolyte had good thermal, mechanical, and electrochemical properties. In this talk, the synthesis and characterization of these PEGylated gels will be discussed.

[1] Rock, S.; Lin, W.; Crain, D.; Krishnan, S.; Roy, D. Interfacial Characteristics of a PEGylated Imidazolium Bistriflamide Ionic Liquid Electrolyte at a Lithium Ion Battery Cathode of $LiMn_2O_4$, *ACS Appl. Mater. Interfaces* **2013**, *5*, 2075–2084.

[2] Lebga-Nebane, J. L.; Rock, S.; Franclemont, J.; Roy, D.; Krishnan, S. Thermophysical Properties and Proton Transport Mechanisms of Protic Ionic Liquids, *Ind. Eng. Chem. Res.* **2012**, *51*, 14084–14098.

388952 E. coli Pilot-Scale Fermentation and One-Step Purification of Amine Dehydrogenases Employing an Aqueous Two-Phase Extraction System

Abstract:

We have recently developed several amine dehydrogenases (AmDHs) that are of interest for production of enantiomerically pure amines as precursors for active pharmaceutical ingredients (Abrahamson et al. 2012 and 2013, Bommarius B et al, submitted). Their potential as synthetically interesting enzymes led us to develop a scalable, cost-effective pilot-scale fermentation and purification strategy.

Fermentations were carried out on 15 L scale with a mostly defined high-density medium and continuous glucose feed controlled by both the pO_2 value and the coupled growth rate of the *E. coli* host. This fermentation protocol led to 1 g/L pure amine dehydrogenase with an overall yield of 140 g/L wet cells.

To achieve simple purification on large scale, we have developed a one-step purification using an aqueous two-phase system (ATPS) with 9% PEG 6000 in the top phase and 16 % ammonium citrate in the lower phase. After homogenization the whole cell lysate was applied onto this ATPS and purification of AmDH was achieved through separation into the top layer, whereas cell debris was concentrated at the interface and contaminating proteins remained in the bottom phase. This ATPS is scalable and ammonium citrate is both cheap and environmentally friendly, even on large scale. We currently work on a method to recycle PEG 6000.

In summary, we present a straightforward and affordable protocol to produce and purify AmDH for industrial scale.

References:

M.J. Abrahamson, E. Vazquez-Figueroa, N.B. Woodall, J.C. Moore, A.S. Bommarius, "Development of a Novel Amine Dehydrogenase for Synthesis of Chiral Amines", *Angew. Chem. Intl. Ed.* **2012**, *51*, 3969-3972

M.J. Abrahamson, J.W. Wong, A.S. Bommarius, "The rapid evolution of an amine dehydrogenase biocatalyst for the asymmetric production of chiral amines" , *Adv. Synth. Catal.*, **2013**, *355*, 1780-1786

B. Bommarius, M. Schürmann, A.S. Bommarius, "A chimeric amine dehydrogenase offers surprising properties not found in either parent", submitted

Multi-Objective Optimization for Solid Waste Management Systems

José Ezequiel Santibañez-Aguilar,[1] Juan Martínez-Gómez,[1] José María Ponce-Ortega,[1] Fabricio Nápoles-Rivera,[1] Medardo Serna-González[1], and Mahmoud M. El-Hawagi[2, 3]

[1]*Chemical Engineering Department, Universidad Michocana de San Nicolás de Hidalgo, Morelia, Michoacán, México, 58060*

[2]*Chemical Engineering Department, Texas A&M University, College Station, TX, USA, 77843*

[3]*Adjunct Faculty at the Chemical and Materials Engineering Department, King Abdulaziz University, Jeddah, Saudi Arabia.*

Several options for reusing the municipal solid waste have been reported; however, integral approaches for optimizing the entire supply chains associated to the municipal solid waste have not been considered. Supply chains associated to waste management might help to reduce the waste generation in an environmentally and economically sustainable way. Particularly, multi-objective optimization approaches accounting for economic, environmental, social and safety issues in the entire life cycle have not been reported. Therefore, this work presents a mathematical programming model to the optimal planning for reusing municipal solid waste taking into account economic, environmental, social and safety issues simultaneously in a distributed framework. The optimization model is able to select the processing technologies, consumers, cities producing wastes, amount of recycled waste, products and location of processing facilities.

The optimization problem is formulated as a multi-objective problem that considers three different objectives: the net annual profit, the amount of reused wastes and the total fatalities generated with the considered risks. The addressed problem considers several cities with a given municipal solid waste production, these wastes can be utilized to obtain different useful products like energy, recycled material, olefins, fuels, etc. The waste is stored in several disposition centers or dumps; these dumps can affect the population's safety in different ways. The first one is the fatalities associated for consuming polluted water, since the pollutants in dumps can be transported through leaching from the surface to shallow well that can serve as water source to the population. In addition, the risk is associated the intoxication by the gas emissions caused by the burning of a given dump; this risk is taken into account by the implementation of a gas dispersion model. One of the main considerations is that each city is divided in several grids named sites, each one of these sites have different waste production, population and may have processing plants and dumps, this division is accomplished because each dump can affect in different magnitude each one of the sites, since each site is located at different distance from the dump considered.

The economic objective function is the net annual profit and considers all the costs of the entire supply chain; in this context, the net annual profit is the revenue from sales minus the costs for operation, capital, transportation, disposing and separating.

$$\begin{aligned} NETPROFIT = {} & REVENUE - OPCOST - CAPCOST^{total} \\ & - TRANSPCOST - DISPCOST - SEPCOST \end{aligned} \tag{1}$$

The safety objective function is based on the total fatalities generated by intoxication and leaching. For the fatalities for intoxication, a frequency factor ($f^{frequency}_{factor}$) to obtain the number of times that the landfill is burned yearly is considered. The total fatalities are equal to the sum of the total fatalities ($F_{k,s}^{leaching\ fatalities}$) for leaching plus the total fatalities for intoxication ($F_{g,d,a}^{intoxication\ fatalities}$).

$$F^{total\ fatalities} = f^{frequency\ factor} \sum_{g}\sum_{d}\sum_{a} F_{g,d,a}^{intoxication\ fatalities} + \sum_{k}\sum_{s} F_{k,s}^{leaching\ fatalities} \tag{2}$$

The total consumed waste is a way to take into account the environmental impact although there are many forms to assess the environmental impact; this is an adequate way for the case of the municipal solid waste because one of the main impacts is for the huge amount of waste that is disposed in landfills. In this way, the total consumed waste is equal to the total waste that is sent to the processing facilities from dumps divided by the total waste that is produced in the each site of all cities.

$$WASTECONS = \sum_{wn \in WASTES} \sum_{d \in DUMPS} W_{wn,d}^{out\ dump} \Big/ \sum_{wn \in WASTES} \sum_{s \in SITES} W_{wn,s}^{produced} \tag{3}$$

A case study for planning the waste management in the city of Morelia in Mexico was considered. Figure 1 shows the tradeoffs between the number of fatalities and the net annual profit; notice the number of fatalities can decrease until 21 % from the maximum number of fatalities while the change in the profit is about 19 %. It is important to note that the fatalities for leaching are the most important in the proposed case study. Furthermore, Figure 1 shows that the consumed waste is almost constant when the Pareto curve is carried out with the net annual profit and the number of fatalities, and this way the number of fatalities is not function of the consumed waste. Results show that it is possible to implement a distributed processing system to reuse municipal solid waste in an economically attractive way. In addition, results can be used for governments to make decisions about the waste disposal and define the amount of waste that must be reused to obtain several products. It should be noted that results include the supply chain configuration.

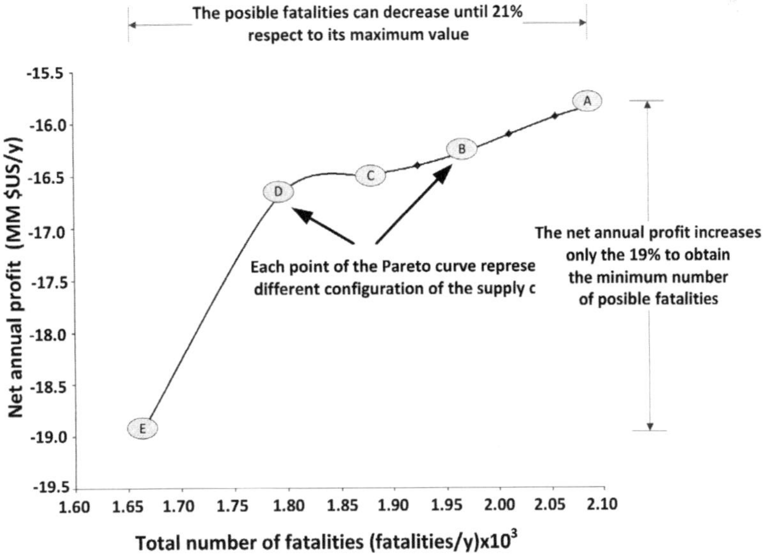

Figure 1. Pareto curve between the net annual profit and the number of fatalities.

384303 Global Optimization for Sustainable Design and Synthesis of Algae Processing Network for CO2 Mitigation and Biofuel Production Using Life Cycle Optimization

Abstract:

Renewable biofuel is considered as a promising fossil fuel substitute due to its potential for large-scale production and carbon mitigation. Among various biomass feedstocks with increasing attention in the recent decades, algal biomass offers significant advantages over conventional energy crops [1]. Algae have high area productivity and accumulate lipid materials up to 80% of its dry cell weight. They are able to utilize non-arable land and waste water for cultivation. Additionally, the hydrocarbon algal biofuels converted via hydroprocessing accommodate the specifications of drop-in fuels, which can be directly incorporated into the existing infrastructure [2]. In order to reduce environmental impacts and improve the overall economic performance, systematic modeling and optimization frameworks can be utilized to identify and assess the sustainable design and synthesis of a processing network [3-5]. Therefore, applying such a framework to the algae processing network provides valuable insights into the technology pathways in the context of sustainability. The goal of this research is to identify and assess the optimal sustainable design and synthesis of an algae processing network with both economic and environmental concerns.

In this work, we propose a superstructure of an algae processing network including 7,800 processing routes and 11 processing sections, namely cultivation, harvesting, primary dewatering, secondary dewatering, storage, cell disruption & hydrothermal liquefaction, lipid extraction, upgrading, remnant treatment, electricity generation, and steam generation. Feed gas, off-gas, makeup nutrients, and recycled materials are fed into the cultivation system and mature algae product from cultivation is concentrated by several technologies. First, 10% of the dilute slurry is harvested by auto flocculation and dissolved air flotation. The water content is further reduced by primary and secondary dewatering. Most of the reduced materials are recycled to the cultivation facility. In order to keep downstream facilities operating continuously, we employ a storage tank to temporarily store part of the enriched algae biomass for the production at night [6]. Next, the enriched microalgae paste undergoes cell disruption and lipid extraction or HTL so that most of the lipid materials are separated from algal cells. The lipid materials are then upgraded to biodiesel or renewable diesel depending on the specific conversion technologies. The rest of the algal body or remnant is fed to an anaerobic digester and decomposed into biogas, aqueous nutrients, and unreacted solids. Additionally, we couple electricity and steam generation into the network to satisfy the onsite energy consumption and reuse the resulting off-gas.

Based on this superstructure, a multi-objective mixed-integer nonlinear programming (MINLP) model for the sustainable design and synthesis is developed following the life cycle optimization frame work that integrates the multi-objective superstructure optimization scheme with life cycle assessment and techno-economic analysis [3, 4, 7-10]. The model consists of two objective functions and simultaneously minimize the unit annualized cost and the unit Global Warming Potential (GWP). As the final biofuel product can be biodiesel or renewable diesel depending on the specific upgrading technology, the unit objective functions are associated with the unit gasoline gallon equivalent (GGE) of the specific biofuel produced. To efficiently solve the problem, we propose a global optimization strategy that integrates a branch-and-refine algorithm based on successive piecewise linear approximations and an exact parametric algorithm based on Newton's method [11-15]. The multi-objective optimization problem results in two Pareto-optimal curves for biofuel production and biological carbon sequestration, respectively. Three optimal processing routes are identified. From the biofuel production perspective, the most environmentally sustainable processing route achieved a unit GWP of 16.520 kg CO2-eq/GGE and a unit annualized cost of $9.712/GGE, while the most economical processing route resulted in a unit annualized cost of $7.017/GGE and a unit GWP of 26.791 kg CO2-eq/GGE. If the algae processing network is utilized for biological carbon sequestration, the unit carbon sequestration and utilization cost ranges from $10.43 /t of CO2 to $1.64 /t of CO2, corresponding to unit GWP's from 352.41 kg CO2-eq/t of CO2 to 412.90 kg CO2-eq/t of CO2, respectively.

References

[1] J. Ferrell and V. Sarisky-Reed, "National algal biofuels technology roadmap," US Department of Energy College Park, MD2010.

[2] D. Yue, F. You, and S. W. Snyder, "Biomass-to-bioenergy and biofuel supply chain optimization: Overview, key issues and challenges," *Computers & Chemical Engineering,* 2013.

[3] D. J. Yue, M. A. Kim, and F. Q. You, "Design of Sustainable Product Systems and Supply Chains with Life Cycle Optimization Based on Functional Unit: General Modeling Framework, Mixed-Integer Nonlinear Programming Algorithms and Case Study on Hydrocarbon Biofuels," *Acs Sustainable Chemistry & Engineering,* vol. 1, pp. 1003-1014, Aug 2013.

[4] B. Wang, B. H. Gebreslassie, and F. Q. You, "Sustainable design and synthesis of hydrocarbon biorefinery via gasification pathway: Integrated life cycle assessment and technoeconomic analysis with multiobjective superstructure optimization," *Computers & Chemical Engineering,* vol. 52, pp. 55-76, May 10 2013.

[5] B. H. Gebreslassie, R. Waymire, and F. You, "Sustainable design and synthesis of algae-based biorefinery for simultaneous hydrocarbon biofuel production and carbon sequestration," *AIChE Journal,* vol. 59, pp. 1599-1621, 2013.

[6] J. Gong and F. You, "Optimal Design and Synthesis of Algal Biorefinery Processes for Biological Carbon Sequestration and Utilization with Zero Direct Greenhouse Gas Emissions: MINLP Model and Global Optimization Algorithm," *Industrial & Engineering Chemistry Research,* vol. 53, pp. 1563-1579, 2014/01/29 2014.

[7] F. Q. You, L. Tao, D. J. Graziano, and S. W. Snyder, "Optimal design of sustainable cellulosic biofuel supply chains: Multiobjective optimization coupled with life cycle assessment and input-output analysis," *Aiche Journal,* vol. 58, pp. 1157-1180, Apr 2012.

[8] F. Q. You and B. Wang, "Life Cycle Optimization of Biomass-to-Liquid Supply Chains with Distributed-Centralized Processing Networks," *Industrial & Engineering Chemistry Research,* vol. 50, pp. 10102-10127, Sep 7

2011.

[9] B. H. Gebreslassie, M. Slivinsky, B. L. Wang, and F. Q. You, "Life cycle optimization for sustainable design and operations of hydrocarbon biorefinery via fast pyrolysis, hydrotreating and hydrocracking," *Computers & Chemical Engineering,* vol. 50, pp. 71-91, Mar 5 2013.

[10] Q. Zhang, J. Gong, M. Skwarczek, D. Yue, and F. You, "Sustainable process design and synthesis of hydrocarbon biorefinery through fast pyrolysis and hydroprocessing," *AIChE Journal,* vol. 60, pp. 980-994, 2014.

[11] F. Q. You and I. E. Grossmann, "Stochastic Inventory Management for Tactical Process Planning Under Uncertainties: MINLP Models and Algorithms," *Aiche Journal,* vol. 57, pp. 1250-1277, May 2011.

[12] Z. X. Zhong and F. Q. You, "Globally convergent exact and inexact parametric algorithms for solving large-scale mixed-integer fractional programs and applications in process systems engineering," *Computers & Chemical Engineering,* vol. 61, pp. 90-101, Feb 11 2014.

[13] F. Q. You, J. M. Pinto, I. E. Grossmann, and L. Megan, "Optimal Distribution-Inventory Planning of Industrial Gases. II. MINLP Models and Algorithms for Stochastic Cases," *Industrial & Engineering Chemistry Research,* vol. 50, pp. 2928-2945, Mar 2 2011.

[14] D. J. Yue and F. Q. You, "Planning and Scheduling of Flexible Process Networks Under Uncertainty with Stochastic Inventory: MINLP Models and Algorithm," *Aiche Journal,* vol. 59, pp. 1511-1532, May 2013.

[15] F. Q. You, P. M. Castro, and I. E. Grossmann, "Dinkelbach's algorithm as an efficient method to solve a class of MINLP models for large-scale cyclic scheduling problems," *Computers & Chemical Engineering,* vol. 33, pp. 1879-1889, Nov 12 2009.

391695 Towards Integration of Life Cycle Assessment within a Multi Objective Supply Chain Optimization Modeling Framework

Abstract:

This paper presents advancements in integration of multiple environmental indicators (CO_2e, Human health, Ecosystem health) based on life cycle assessment methods within the framework of supply chain modeling and optimization. We present results related to the optimal design of supply chain networks with consideration of environmental and economic objectives. The environmental objective is measured through life cycle assessment indicators. IMPACT 2002+ methodology is used to evaluate the environmental scores. Global warming, human health damages and ecotoxicity impacts are considered for evaulating the environmental impacts. On the other hand, minimization of total production cost and maximization of net present value of the supply chain is another goal. A multi objective mixed integer linear programing (MILP) model is developed for optimal design of green supply chain. Multi objective optimization is accomplished through non dominated sorted genetic algorithm. The resulting Pareto-optimal curves demonstrate the tradeoff between economic and environmental pathways. This method considers life cycle emissions from cradle to gate for each process and suggests alternatives technologies for the reduction of these pollutants while considering financial implications. The method can be used to derive optimal solutions for scheduling and planning or could be used to design future green supply chains. The methodology is demonstrated via several case studies.

383858 Optimal Design of Microgrids with Innovative CHP Systems: Integration of Process Optimisation and Life Cycle Assessment

Abstract:

Optimal design of microgrids with innovative CHP systems: integration of process optimisation and life cycle assessment

Di Zhang, Sara Evangelisti, Paola Lettieri, Lazaros G. Papageorgiou

Department of Chemical Engineering, University College London, Torrington Place, London WC1E 7JE, UK

As an alternative to current centralised energy generation, microgrids are adopted to provide local energy with lower energy expenses and gas emissions by utilising distributed energy resources (DER). DER is defined as local resources related to the energy system, such as generators, boilers and energy storage devices. Micro Combined Heat and Power (micro CHP) units simultaneously produce heat and power for a single building. In order to meet renewable energy target in the UK, the installation of CHP units is considered because of their flexibility, reliability and safer operating conditions with high overall efficiencies. Several micro CHP technologies have been developed recently for applications at domestic scale, including fuel cell, micro turbine, internal combustion engine and sterling engine. The optimal design of DERs within microgrid plays an important role in promoting the penetration of microgrid. It attracts money research attentions from the economic view.

Different tools exist to evaluate the environmental impacts of a process or a service. Life Cycle Assessment (LCA) is recognised as one of the best tools available. Moreover, LCA is one of the most developed and widely used environmental assessment tools for comparing alternative technologies when the location of the activity is defined. Studies have been published on carbon footprint of micro – CHP units rather than a holistic environmental appraisal. A few studies have focused on the environmental impacts of different micro-CHP technologies analysing the functioning of single units rather than microgrids.

Limited studies in the literature have been focused on coupling environmental and economic sustainability in a multi-objective optimisation model: the majority of them focused on the operation of a single unit. To our knowledge, no studies have been published on the optimal design of microgrids considering environmental and economic metrics. In this work, the optimal design of microgrids with innovative CHP units is addressed while involving both concerns. A multi-objective modelling framework for the optimal design is presented. The trade-off between the two conflicting objectives is obtained by solving the proposed model using the ε-constraint method.

This work considers a general microgrid, which involves several participant sites of different building types. Macrogrid is available to provide electricity to the participant in the microgrid and extra electricity can also be sold back to the macrogrid when it benefits. The candidate technologies include CHP generators (with different capacities and heat-to-power ratios), boilers, thermal storage and a macrogrid power connection. Micro CHP technologies considered in this work are Proton exchange membrane (PEM) Fuel Cell, Solid Oxide Fuel Cell (SOFC), Internal Combustion Engine (ICE) and Stirling Engine (SE). Turn-key costs of CHP generators are based on the CHP types as well as the capacity range. The microgrid and the macrogrid are interacted and constrained through exporting or importing electricity.

The goal of the LCA in this work is to estimate the environmental impacts of four different CHP systems with additional thermal energy storage (TES) working in a distributed generation network. The life span considered is 15 years. The impact categories considered in this study are global warming potential (GWP) and acidification potential (AP). The output of LCA is used as the input for the optimisation model. A multi-objective mixed-integer linear programming (MILP) approach is developed for this problem. The key decision variables include unit capacities and resources utilised. They are determined by minimise the total equivalent annualised cost (EAC), GWP and AP of all participants, subject to equipment capacity constraints, CHP ramp limit constraints, energy demand constraints, CHP selection constraints, thermal storage constraints and transfer price level constraints.

The proposed model is implemented for different microgrid modes. For each mode, optimal design of the microgrid is obtained with (a) the candidate technologies selected and their capacities, (b) energy resources consumed, (c) energy production plan, (d) thermal energy storage plan. The trade-off between the economic and environmental objectives is analysed with a set of Pareto-optimal solutions.

Keywords: microgrids; life cycle assessment; multi-objective optimisation; ε-constraint method

Application of sustainability footprint and multicriteria optimization for the choice of sustainable chemical processes

AIChE Annual Meeting, Atlanta, GA, November 16-21, 2014
For submission to "Advances in Life Cycle Optimization for Process Development"
Rajib Mukherjee[1], Debalina Sengupta[2], Subhas K. Sikdar[3]

1. *Vishwamitra Research Institute, Chicago, IL*
2. *Texas A &M University, College Station, TX*
3. *NRMRL, US EPA, Cincinnati, OH,*

Abstract

A chemical product can be obtained by different process pathways involving varying amounts and types of resources, utilities, product, and byproduct formation. Evidently, the sustainability in terms of economic, environmental and social impacts of these processes will vary relatively. When such competing process options are presented, and compared to an existing process pathway, it is necessary to evaluate and identify the most sustainable process among these competing options. Sustainability of a chemical process is generally evaluated with indicators that need process and chemical properties data. These indicators individually reflect the impacts of the process on the respective areas of sustainability. In order to choose among several alternative processes an overall comparative analysis is essential.

To make a decision on the most sustainable process, the indicators need to be rationally aggregated into a single index. In this research, the suite of sustainability indicators includes traditional economic indicators to represent the economic dimension, environmental indicators, and human health indicators to represent the social dimension. The environmental and human health indicators were computed using the TRACI impact characterization method. Process flow and economic data were used to compute the indicators in these categories. These indicators were then aggregated to compute the sustainability footprint (D_e) of each process and evaluate the relative sustainability of the competing processes.

This method of process selection is applied to five competing processes for methanol production from carbon dioxide. These are novel chemical processes which involve carbon dioxide as raw material, and a uniform industrial scale capacity for production of methanol. The processes differ in the amount of raw material required, byproducts formed, and utilities used. Based on process data, these processes are compared using the sustainability footprint method. This alone should give us information to evaluate the best process as the one which has the minimum sustainability footprint, D_e.

However, we wanted to see if the minimization of the sustainability footprint, D_e, can be used as an objective function for a mixed integer optimization problem. The aim in solving the multi-criteria problem will be to select the most sustainable option among the five competing processes, and also determine the optimum values of process parameters which will enable the same.

Based on our results, this research will demonstrate how an aggregate sustainability footprint index can be utilized to compare between competing processes, as well as how it can be integrated into a mixed integer problem formulation to determine the optimum process conditions which help in determining the most sustainable process among competing options.

Sustainable Biorefinery Supply Network using a Concept of Eco- and Total Profit – Case Study of the European Union

Lidija Čuček,[a] Mariano Martín,[b] Ignacio E. Grossmann,[c] Zdravko Kravanja[a]

[a]*Faculty of Chemistry and Chemical Engineering, University of Maribor, Smetanova ul. 17, Maribor, Slovenia, lidija.cucek@um.si, zdravko.kravanja@um.si*

[b]*Department of Chemical Engineering, University of Salamanca, Salamanca, Spain, mariano.m3@usal.es*

[c]*Department of Chemical Engineering, Carnegie Mellon University, Pittsburgh, USA, grossmann@cmu.edu*

Abstract

Nowadays the shortages of energy and water resources, the increasing prices, the demand for and dependence on fossil fuels and a growing environmental consciousness, have become societal priorities. These concerns have supported the development of alternative transportation fuels and the use of renewable sources of energy. Thus, governments across the globe have set targets on fossil fuels' substitutions over the next decade. In order to be able to meet these targets, not only process designs for the productions of alternative fuels are needed, but also the development of the networks which account for all the stages from the energy source to the final product. Biomass, in particular, is a challenging resource because of its seasonality, its degradation over time, its low density, regional availability, growth yield, and the large area required to grow it. However, biomass can provide a reasonable alternative to crude-based gasoline and diesel in the short and medium terms, although the competitiveness of biomass as an energy source is highly dependent on its supply chain. Lately, several studies have extended the supply-chain studies to actual regions across the US, but only for a limited number of raw materials and fuels, focusing mainly on lignocellulosic resources and in most of the cases only for bioethanol production, although FT-products have also been accounted for [1,2].

This contribution deals with a sustainable large-scale biorefinery supply network by considering both economic and environmental objectives. A biorefinery's supply network accounts for different biomass sources (algae, lignocellulosic materials, corn, wheat, cooking oil), conversion technologies (biochemical, thermochemical, thermo-biochemical, oil extraction, transesterification) and includes first, second, and third generations of biofuels, such as bioethanol, biodiesel, hydrogen, Fischer-Tropsch diesel, and green gasoline [3] accounting for water consumption and availability and energy integration within the network. Multi-period synthesis of an optimally-integrated biorefinery supply network [4] was performed through maximization of the economic objective, and applied on the case study of the European Union [5].

In this contribution the synthesis has now been extended to simultaneously accounting for environmental impact. The concept of Eco- and Total profit [6] was used for this purpose, which offers two important advantages over conventional multi-objective approaches: i) the concept accounts for both, burdening and unburdening effects [6] on the environment, as Eco-Profit is defined as the difference between Eco-Benefit (unburdening) and Eco-Cost (burdening), expressed in monetary terms. ii) by applying the concept where multi-objective solutions are reduced into single-objective ones as Total profit is composed of both, economic and Eco-profits. This enables the obtaining of sustainable solutions directly, which is especially important when solving supply network problems at the continental level when the models are huge. The solutions obtained for the case study of the European Union are those alternatives that would unburden the environment the most while having significant profit.

Acknowledgments

The authors acknowledge the financial support from the Slovenian Research Agency (Programs P2-0032 and P2-0377) and Centre for Advanced Process Decision-making.

References

[1] Elia, J.A., Baliban, R.C., Floudas, C.A. Nationwide energy supply chain analysis for hybrid feedstock processes with significant CO_2 emissions reduction. AIChE Journal. 2012; 58(7), 2142-2154.

[2] Marvin, A. W., Schmidt, L. D., Benjaafar, S., Tiffany, D. G., & Daoutidis, P. Economic Optimization of a Lignocellulosic Biomass-to-Ethanol Supply Chain. Chemical Engineering Science. 2012;67, 68-79.

[3] Martín M, Grossmann IE. On the systematic synthesis of sustainable biorefineries. Industrial & Engineering Chemistry Research. 2013;52(9):3044-64.

[4] Čuček L, Martín M, Grossmann IE, Kravanja Z. Multi-period Synthesis of Optimally-Integrated Biomass and Bioenergy Supply Network. Computers and Chemical Engineering. 2014:doi: 10.1016/j.compchemeng.2014.02.020.

[5] Čuček L, Martín M, Grossmann IE, Kravanja Z. Large-Scale Biorefinery Supply Network – Case Study of the European Union. In: J. J. Klemeš, P. S. Varbanov, P. Y. Liew, editor. Computer Aided Chemical Engineering: Elsevier; 2014, accepted for publication.

[6] Kravanja Z, Čuček L. Multi-objective optimisation for generating sustainable solutions considering total effects on the environment. Applied Energy. 2013;101:67-80.

Planning and Scheduling Industrial Waste Management Using Knowledge Based Lagrangean Decomposition

Elisabet Capón-García[a], Edrisi Muñoz[b], José Miguel Laínez[c], Antonio Espuña[d], Konrad Hungerbühler[a], and Luis Puigjaner[d]

[a]*Chemistry and Applied Biosciences, Institute for Chemical- and Bioengineering, Swiss Federal Institute of Technology, Zurich (ETHZ)*
[b]*Information Technologies, Centro de Investigacion en Matematicas A.C., Guanaguato, Mexico,*
[c]*Department of Industrial & Systems Engineering, University at Buffalo, Buffalo, NY,*
[d]*Chemical Engineering Department, Universitat Politècnica de Catalunya - ETSEIB, Barcelona, Spain*

Waste liquid effluents originated from industrial activities contain contaminants and certain species, such as organic solvents, with material and energy recovery potential. Therefore, not only does industrial waste management stand as an end-of-pipe problem, but also as a potential origin of resources, which can result insignificant benefits if material and energy integration strategies are used. The simultaneous planning and scheduling of industrial waste management allows improved resource utilization and reduced environmental impact. However, rigorous mathematical models usually result in large scale problems. In order to tackle the problem complexity, decomposition techniques based on information flows between a master and a set of sub-problems are widely applied (Maravelias and Sung, 2009, Calfa et al., 2013).

In this work, we exploit this capability of ontologies to address the optimal integration of planning and scheduling using a Lagrangean decomposition approach. Thus, ontologies improve information sharing and communication in the enterprise and can even represent holistic mathematical models facilitating the use of analytic tools and providing higher flexibility for model building. Therefore, the scheduling/planning sub-problems are created for the waste management facility and their dual solution information is shared by means of the ontological framework. Furthermore, this work presents a multi-period waste management multi-objective optimization, considering economic and environmental issues. The behavior of waste treatment units is included in the optimization problem as black-box models based on industrial practice.

A case study of the planning and scheduling of industrial waste facilities is presented to show the advantages and limitations of the proposed approach in a multi-objective environment considering economic and environmental issues using the decomposition strategy. The proposed framework leads to reduced computation effort and better solutions in terms of solution quality, since waste stream scheduling is integrated in decision-making.

Acknowledgements
Financial support received from the Spanish "Ministerio de Economía y Competitividad" and the European Regional Development Fund, both funding the research Project

SIGERA (DPI2012-37154-C02-01), and the “Agència de Gestió d'Ajuts Universitaris i de Recerca (2014SGR1092)” is thankfully acknowledged.

References

Calfa, B. A., Agarwal, A., Grossmann, I. E., Wassick, J. M., 2013. Hybrid bilevel-lagrangean decomposition scheme for the integration of planning and scheduling of a network of batch plants. Industrial & Engineering Chemistry Research 52 (5), 2152-2167.

Maravelias, C. T., Sung, C., 2009. Integration of production planning and scheduling: Overview, challenges and opportunities. Computers & Chemical Engineering 33 (12), 1919-1930.

391026 Identifying the Preferred Subset of Alternatives for Environmental Improvements Via an MILP Approach Based on the Analytic Hierarchy Process

Abstract:

The combined use of multi-objective optimization (MOO) and LCA has recently gained wider interest in process systems engineering. This approach provides as output a set of Pareto solutions that represent the optimal trade-off between the economic and environmental concerns considered in the analysis. From this set of optimal alternatives, decision-makers should identify the ones that better fulfill their preferences. Generating a large and representative enough subset of Pareto points to aid decision-making is challenging. This task is particularly difficult in problems with a large number of (environmental) objectives, as is the case when incorporating LCA principles in MOO.

In this work we present a mixed-integer linear programming method that simplifies MOO problems by concentrating on determining only a reduced number of Pareto points that are particularly appealing. Our approach, which relies on the analytic hierarchy process (AHP), identifies a set of weights to be assigned to the environmental objectives so as to translate them into a single aggregated indicator that reflects to the maximum extent possible the decision-makers' preferences. For every such combination of weights, we solve a single objective problem that optimizes the corresponding weighted sum of objectives, thereby generating only a subset of alternatives (Pareto solutions) reflecting preferences with a large degree of consistency. We illustrate the capabilities of our approach through its application to the design of supply chains for bioethanol production.

Time dependent dielectric failure has become a pivotal aspect of interconnect design as industry pursues integration of sub-22nm process-technology nodes. Literature has provided key information about the role played by individual species such as electrons, holes, ions and neutral impurity atoms. However, no mechanism has been shown to describe how such species interact and influence failure. Current leakage relaxation in low-*k* dielectrics was studied using bipolar field experiments to gain insight into how charge carrier flow becomes impeded by defects within the dielectric matrix. Leakage current decay was correlated to injection and trapping of electrons. We show that current relaxation upon inversion of the applied field can be described by the stretched exponential function. The kinetics of charge trapping events are consistent with a time-dependent reaction rate constant, where $0<\beta<1$. Such dynamics have previously been observed in studies of charge trapping reactions in amorphous solids by Hamill *et al.*[1].We explain the relaxation process in charge trapping events by introducing a nonlinear charge trapping model. This model provides a description on the manner in which the transport of mobile defects affects the long-tail current relaxation processes in low-*k* films.

1. Hamill, W. H. and Funabashi, K. *Phys. Rev. B*,vol. 16 (1977).

387149 Development of Acid Doped PBI Electrolyte Membrane

Abstract:

Current proton-exchange membrane fuel cell (PEMFC) technology employs Nafion membranes, which suffer from costly humidity/water management requirement and low CO tolerance. High temperature PEMFCs are challenging and offer many benefits of fast reaction rate, low ohmic resistance, high tolerance to fuel impurities and high efficiency in combined heat and power (CHP) applications. We prepared PBI membrane doped with phosphoric acid (PBI/PA) for the application in high temperature PEMFCs. Detailed fuel cell performance tests were conducted using membrane electrode assembly (MEA) prepared using PBI/PA and Pt/C at different operating conditions, such as pressures, temperatures, and H_2/O_2 gas compositions. Preliminary results show that the phosphoric acid doped PBI membrane exhibited high ion conductivity (low ohmic resistance) of 0.11 and 0.24 S/cm at 120 and 160 °C, respectively, compared to 0.09 S/cm at 80 °C for Nafion. The power density output of PBI/PA MEA equipped fuel cell increased from 0.92 W/cm^2 to 1.15 W/cm^2 when the pressure was increased from 1 to 2 atm at 160 °C. Furthermore, only a slight drop in voltage of about 31 mV at 0.3 A/cm^2 current density was observed across the PBI/PA membranes. Preliminary results show that PBI/PA membrane is capable of operating fuel cells at working temperature greater than 120 °C without the need of water management. Some of these results will be highlighted in this presentation.

Uncertainty Quantification of VLE models for an MEA System

Josh Morgan[a], Benjamin Omell[a], Debangsu Bhattacharyya[a], Charles Tong[b], David C. Miller[c]

[a] *Department of Chemical Engineering, West Virginia University, Morgantown, WV 26506, USA*

[b] *Lawrence Livermore National Laboratory, Livermore, CA 94550, USA*

[c] *National Energy Technology Laboratory, 626 Cochrans Mill Rd, Pittsburgh, PA 15236, USA*

Uncertainties in the vapor-liquid equilibrium (VLE) model of an MEA-H_2O-CO_2 system can significantly reduce the accuracy of prediction from the process modes that use these VLE models. This works focuses on uncertainty quantification (UQ) of the VLE model of an MEA-H_2O-CO_2 system by using a systematic, generalized methodology that can be deployed to other properties models as well.

The deterministic VLE model is a rigorous electrolyte-NRTL model. The electrolyte speciation model has been developed based on the experimental data. The system is represented as a single solvent (H_2O) while CO_2 and MEA are represented as Henry's components. In the deterministic model, approximately forty parameters have been regressed to fit the experimental data. The objective of this work is to quantify parametric and model-form uncertainties of this VLE model. Starting with prior beliefs about parametric uncertainties, a Bayesian method is used to derive informed posteriors by using the experimental data of VLE, heat capacity, heat of absorption. To reduce the computational cost, response surface models are developed as surrogate models for the Bayesian inference method that uses MCMC method with Gibbs sampling. For down-selecting the parameter space, a sensitivity matrix approach based on computing analytical/numerical derivatives of the VLE model with respect to variables and parameters is developed. Finally the posterior distributions of the properties parameters are propagated through the process model to obtain the uncertainty distribution of the key process variables in the absorber and stripper. The UQ analysis provides important information about uncertainties in the parameter space. The VLE and other associated properties models are developed in Aspen Properties with user models developed in FORTRAN while the UQ analysis is performed using PSUADE. Our study shows that the uncertainties in the VLE model can significantly affect the prediction accuracy of key variables such as extent of CO_2 capture and reboiler heat duty for MEA-based CO_2 capture systems.

388552 Process Technology Development for Isocyanatoorganosilanes Production

Abstract:

Isocyanatoorganosilanes are of great economic importance. They are industrially used as adhesion promoters between organic polymers and inorganic materials. The compounds are sometimes termed coupling agents or crosslinkers.

Commercial isocyanatoorganosilanes are primarily made via thermal cracking of corresponding carbamatoorganosilanes with alcohol as the by-product. There are numerous side reactions occur along with thermal cracking reaction causing low yield and low efficiency. Therefore, there has been great interest in an economic method for preparing isocyanatoorganosilanes in high yield and purities. Extensive research work has been done on Process Technology Development for Isocyanatoorganosilanes Production which led to many patent disclosures. A variety of aspects, such as, gas/liquid phase cracking, the type of reactors, catalytic or non-catalytic cracking reaction, and separation technologies, have been described in these patents in terms of process improvement. This present work will summarize and review the patent trend for industrial isocyanatoorganosilanes production.

Computer aided solvent design for extraction of a pharmaceutical intermediate

Madakashira Harini†, Jhumpa Adhikari†, Santosh B. Noronha† and K. Yamuna Rani*

†*Department of Chemical Engineering, Indian Institute of Technology Bombay, Mumbai-400076*
**Chemical Engineering Division, Indian Institute of Chemical Technology, Hyderabad-500007*

A majority of common solvents that are extensively used in chemical industry for reaction, formulation and separation have potential health hazards. The use of volatile organic solvents in industry can be assessed by Sheldon E-factor, (kg of waste per kg of product) and ranges from 25 to 100 for pharmaceutical industry with a production of 10^3 tonnes/year.[1] To reduce the E-factor, the need to minimize the usage of solvent in a process arises with regard to health and environmental concerns. In addition, the usage of a green solvent that can replace these hazardous conventionally used solvents will be able to reduce the impact on environment to a larger extent. Computer aided molecular design[2] (CAMD) approach is generally adopted to design a solvent reducing the need for carrying out experiments performed on a trial and error basis. The CAMD approach employs structure property correlations and an optimization technique to find a potential solvent for a physical separation process. In this study, CAMD approach has been applied to a case study where an aliphatic/ aromatic solvent has been designed for efficient extraction of R-phenylacetylcarbinol (PAC), a pharmaceutical intermediate from an aqueous phase. Current industrial practice involves the use of the solvent toluene.

The properties which have been considered in our CAMD approach for the design of a solvent to replace toluene include distribution coefficient, solvent loss, solvent power, selectivity, octanol water partition coefficient, viscosity, boiling point and toxicity. These properties have been predicted using group contribution models and the constraints on these properties are built by choosing toluene as a reference solvent. A variant of the genetic algorithm, namely non-dominated sorting genetic algorithm[3] (NSGA-II) along with the UNIFAC[4] (universal functional activity coefficients) model has been employed for optimization. A total of 64 UNIFAC groups (including repetitions) have been considered for the current study and the combinations of these groups lead to the design of different types of aliphatic solvent molecules such as aldehydes, ketones, ethers, carboxylates and esters. In the case of the design of an aromatic solvent, only the feasible combinations of functional groups have been attached to a fixed aromatic ring. UNIFAC based structural constraints are further employed to ensure the design of a structurally feasible molecule. A multi objective optimization, minimizing solvent loss while maximizing distribution coefficient, solvent power and selectivity, has been performed to arrive at globally best task specific solvents. The CAMD approach yielded a pareto set of optimal aliphatic solvents including, amyl acetate, 2-methyl penta 1,3-diene, methyl hexene and ethyl pentene; whereas the optimal aromatic solvents obtained are propyl benzene, 1-phenyl propanone, ethoxy benzene and propenyl benzene.

Gibbs ensemble Monte Carlo simulations at constant temperature of 298.15 K and a constant pressure of 101.325 k Pa, have been employed to study the extraction of PAC from the aqueous phase by the solvents identified above, using the MCCCS *towhee* package[5]. Simulation results confirm the property estimation predictions and hence, the short-list of solvents chosen for further investigations. Finally, the suitability of the solvents designed using CAMD for PAC recovery are confirmed using experiments. Accounting for economic considerations, the CAMD derived aromatic solvents are found cheaper when compared to the aliphatic solvents.

References:

1. Grodowska, K. & Parczewski, A. Organic solvents in the pharmaceutical industry. *Acta poloniae pharmaceutica* **67**, 3–12 (2010).
2. Harper, P. M. , Gani, R., Kolar, P., Ishikawa, T. Computer-aided molecular design with combined molecular modeling and group contribution. *Fluid Phase Equilibria* **158-160**, 337–347 (1999).
3. Deb, K., Member, A., Pratap, A., Agarwal, S. & Meyarivan, T. A Fast and Elitist Multiobjective Genetic Algorithm. *IEEE transactions on evolutionary computation* **6**, 182–197 (2002).
4. Magnussen, T. UNIFAC Parameter Table for Prediction of Liquid-Liquid Equilibria. *Industrial Engineering Chemistry Process Design and Development* **20**, 331–339 (1981).
5. MCCCS Towhee. Available at: http://towhee.sourceforge.net.

Assessing the Environmental Impact of a Flare Minimization Approach using CO2 Emissions and Wobbe Index Calculator – An Ethylene Process Case Study

Fahd M. Mohammed, Kazi Monzure Khoda, Fadwa T. Eljack*
Department of Chemical Engineering, College of Engineering,
Qatar University, P.O. Box-2713, Qatar
Email: Fadwa.Eljack@qu.edu.qa

Flaring is a common practice in industrial process for various reasons such as safety, maintenance and during emergency process situations. It is a high-temperature oxidation process that is used to burn combustible components, mostly hydrocarbons of waste gases from industrial operations[1]. Flare emissions contribute to more than 50% of the annual Certified Emissions Reductions (624 million tons) according to Kyoto's Clean Development Mechanisms[2]. Emissions due to flaring activities led to high impact on the climate and environment that cannot be ignored by researchers, environmentalists and policy makers. Hence a number of environmental policies and international standards such as Kyoto protocol and The United Nations Environment Programme (UNEP) have been and are currently being implemented in developed and developing countries around the globe. GHG calculators are created and used by different industries to report CO2 and other GHGs emissions to comply with governing authorities. Therefore, a standard and acceptable GHG calculator is important to quantify those emissions, which led us to develop one. The calculator is developed as part of an ongoing project that proposes mitigation methods for flare reduction during abnormal process operations, also known as process upsets. The aim of this work is to propose an effective mass balance based methodology for tracking CO2 emissions from fuel combustion and process flare during both normal and abnormal operation. It is based on the EU commission guideline for the monitoring and reporting of greenhouse gas emissions pursuant to Directive 2003/87/EC of the European Parliament and of the Council[3]. An Ethylene process base case is presented in this paper to highlight the usefulness of this tool. The case assesses the impacts of using a stand by cogeneration system for utilizing flare streams during upsets and minimizing the overall process CO2 emissions. The merits of the GHG calculator are that it has the ability to estimate the emissions at different upset scenarios while each scenario has multiple flare sources; it has the flexibility of estimating the wasted energy of process flare streams using international accredited standards from the EU commission such as Deutsches Institut für Normung standards . In addition, the calculator is able to estimate the carbon tax and credit using the latest carbon tax rates at various focal countries around the globe. Finally, in this work we will show the potentiality of COGEN system as an alternate solution to reduce GHG emissions and to recover the energy content from the flare streams which was emitted previously. Initial results showed that for our case study, both the cost of fuel and the cost of CO_2 tax can be reduced through the utilization of flare streams from abnormal situations. Therefore, COGEN system is economically and environmentally viable option for abnormal situations management.

[1] Industrial Flares - US Environmental Protection Agency
http://www.epa.gov/ttnchie1/ap42/ch13/final/c13s05.pdf

[2] The world bank - Russia, Kazakhstan Lead Way to Reduce Gas Flaring and Lower Emissions
http://web.worldbank.org/WBSITE/EXTERNAL/TOPICS/EXTSDNET/0,,contentMDK:22949562~menuPK:64885113~pagePK:7278667~piPK:64911824~theSitePK:5929282,00.html

[3] Official Journal of the European Union L 229/1
http://eur-lex.europa.eu/legal-content/EN/TXT/?uri=OJ:L:2007:229:TOC

378703 Optimal Sizing of Cogeneration (COGEN) System to Manage Flares from Uncertain Sources during Abnormal Process Operations

Abstract:

Optimal Sizing of Cogeneration (COGEN) System to Manage Flares from Uncertain Sources during Abnormal Process Operations

Monzure-Khoda Kazi, Fahd Mohammed, Fadwa Eljack*

Department of Chemical Engineering, College of Engineering,

Qatar University, P.O. Box-2713, Qatar

Email: Fadwa.Eljack@qu.edu.qa

Flaring is a very common scenario in different industrial plants. Industries usually do flaring to reduce the risk during abnormal situations, to maintain the product quality or to operate safely during process start up and shut down. Negative effects of those flares are not hidden to modern society. Nowadays, everybody is concern about the environmental and social effects of those industrial flaring. Especially, the emissions of green house gas and other toxic materials during flaring have a significant impact on the environment. Therefore, various protocol and steps i.e., Kyoto protocol, The United Nations Environment Programme (UNEP), have been taken to mitigate the impact of flaring due to process upsets. Although negative impacts of flaring are known, these flare streams have significant amount of heating value. These streams can be considered as waste stream from the industrial plants and can be used as supplement of fresh fuel feed to generate require heat and power using COGEN system. Cogeneration (COGEN) system is one of the effective and efficient approaches where waste streams can be used as fuel to generate require heat and power within a process itself. In our earlier works, we showed the possible reduction of fuel demand and CO_2 emissions by using cogeneration system where flaring streams were recycled as fuel feed (Kamrava et al.). In this paper, our objective is to develop an optimization framework to determine the optimal size of a cogeneration system to manage flares from uncertain sources such as upsets streams. The ultimate results of using cogeneration system are usually reduction in overall cost of the process and GHG emissions. An optimization formulation is developed for indentifying the optimum equipment size and operating parameters of the system, the optimum energy produce for the process from cogen system by minimizing the overall cost and emissions of green house gases. The optimization formulation also helps to find out the mixing ratios of different flare streams available to cogen system, which essentially reduce the utility cost of the process. Multi-objective tradeoffs between the economic, environmental, and energetic aspects are presented through Pareto fronts. A stochastic optimization technique based on genetic algorithm is used to solve this discrete multi-objective optimization problem. This Multi-objective optimization problem was also solved using global optimization toolbox in LINGO. A base case study of ethylene plant is used to demonstrate the applicability of the proposed procedure of cogeneration sizing.

References:

Kamrava, S., Gabriel, K., El-Halwagi, M., & Eljack, F. Managing abnormal operation through process integration and cogeneration systems. Clean Technologies and Environmental Policy, 1-10.

381111 Dynamic Simulation Targeting Emission and GHG Reduction Under Abnormal Operations: Start-up of an Ethylene Plant By Total Recycle and Intermediate Storage

Abstract:

Flaring is the safety way for any chemical processing facilities to protect personnel, equipments and their surrounding environment especially during process upsets. Flared gases contain natural gas, highly reactive volatile organic compounds (HRVOCs) and in some cases, carbon monoxide (CO), react with atmospheric oxygen in combustion reactions, and theoretically create carbon dioxide (CO_2) and water. CO_2 generated from flaring is considered a major industrial source in greenhouse gases (GHG). Besides, flare emissions release volatile organic compounds (VOCs) and pollutants into the atmosphere, which react with nitrous oxide (NOx) in sunlight to generate ground-level ozone and smog. This phenomenon often raises concern within the chemical process industry (CPI) such as chemical plants or oil and gas processes. Chemical processes, especially refineries and petrochemical plants, are usually designed to utilize their on-site produced gas in a balance. Hence, more gas recovery from flaring also brings better thermal efficiency and more environmentally friendly operations. Flaring is very essential during turnaround operations as a large quantity of inert and off-specification gases needs to be ejected until the whole system reaches normal conditions. Flare minimization (FM) has been practiced not only during regular operations but also when processes are going through start-ups, shutdowns and malfunctions (SSM).

A typical startup of an ethylene plant sends approximately 400,000 - 1,000,000 lbs of materials to flare and generates 200,000 lbs of CO and VOCs. Hence, managing flare emission during startup, shutdown and malfunction of an ethylene plant are very important. This research continues the previous work on introducing a generic approach of using dynamic simulation for emission and GHG reduction under abnormal operations, demonstrated by a case study of an front-end de-ethanizer ethylene plant start-up. The base case employs recycling off-specification streams from distillation columns to charge gas compressor (CGC) inlet as substitute partially for furnace feed in order to shorten the total start-up time and reduce the amount of hydrocarbons sent to flare. However, there is significant flaring occurring in this scenario during the transient period when the sub-sections are connected and brought from after-commissioning status to normal operation. Therefore, the optimal case presented in this work investigates different approaches of using temporary storage units to accumulate and recycle this flaring quantity to either furnace feed or CGC inlet, ultimately targeting flareless start-up for the whole ethylene plant. Large-scale dynamic simulation, emission characterization, planning/scheduling, and result evaluation are carried out simultaneously to achieve the desired outcome. It is proven that the outcome contributes a great deal in evaluating new strategies and predicting process dynamic behaviors. It also helps achieving more precise timing to control the recycle loop; hence, flaring is potentially minimized while the start-up time is shortened.

383055 Electrochemical Approach for Removal of H2S and Inorganic Nutrients in Anaerobic Digestion

Abstract:

Anaerobic digestion (AD) is a commercial waste treatment process that can also generates biogas for energy and heating use. Sulfate (SO_4^{2-}) in organic waste feedstock is reduced by sulfate-reducing bacteria (SBR), e.g., *Desulfovibrio desulfuricans*, to either stay in liquid effluent or emit into biogas as hydrogen sulfide (H_2S). Hydrogen sulfide decreases the value of the biogas and poses environmental challenges because it is odorous, toxic, and corrosive. Another issue regarding AD is that the effluent still contains high levels of total nitrogen and phosphorus.

Considering that electrochemical reactors are generally easy and cost-effective to incorporate and operate and can be precisely controlled, this study coupled AD systems with electrochemical reactors to reduce concentrations of sulfide, total nitrogen, and phosphorus. Different anode materials, including carbon cloth, stainless steel mesh, nickel foam, and mixed metal oxide (MMO) coated titanium mesh, were evaluated for their electro-activity for sulfide removal via cyclic voltammetry in various types of media, including phosphate buffer solution, Na_2S solution, and AD influent. Independent variables, such as position of electrode (immersed, half-immersed, and in headspace), applied voltage (0, 0.5 V, 1 V, 2 V, 3 V), and temperature were optimized for hydrogen sulfide removal. Under the similar condition, electro-activity for phosphorus accumulation on cathode was optimized. Batch reactors of anaerobic digestion were performed with the selected anode materials, and biogas production rate, oxidation-reduction potential of medium, and hydrogen sulfide level in biogas were periodically monitored. Total nitrogen and total phosphorus levels were analyzed at the end of the batch reaction.

Two mechanisms of anodic sulfide removal were elucidated and quantified: first, at an anode potential less than the required level for oxygen generation, the sulfide generated by SRB can be reduced to elemental sulfur which deposits on the surface. Second, higher anode potential created micro-aerobic conditions due to oxygen gas formation, and it reduced SBR activity and therefore decreased the rate of sulfide buildup. This oxygen generation also transformed sulfide to more oxidized forms. Meanwhile, the micro-aeration condition boosted the activity of ammonia- and nitrite-oxidizing bacteria, generated more nitrate as feedstock for denitrification and resulted in a reduction in total nitrogen level in AD effluent. Microbial and archael analysis for regular and electrochemical reactors were carried out by 16SrRNA gene amplification, cloning, and sequencing. The sequences were compared with the GenBank databases to determine the approximate phylogenetic affiliations, and the phylogenic tree was constructed using the neighbor-joining method. The shift in microbial community between regular and electrochemical reactors, especially the sulfate-reducing bacteria, ammonia- and nitrite-oxidizing bacteria, as well as denitrifying bacteria was elucidated. This study demonstrated that electrochemically assisted AD reactors successfully reduced hydrogen sulfide, total nitrogen, and total phosphorus concentrations in biogas and AD effluent.

384397 Phosphorus Recovery from Manure By Thermochemical-Biological Process

Abstract:

Phosphorus is an essential nutrient and fertilizer component to support the plant growth, but excess concentration of phosphorus in the animal manure can cause phosphorus accumulation in the soil and water eutrophication. This study investigated the phosphorus recovery from manure via several processes with different combinations of multiple conversion steps, including a thermochemical pretreatment, an anaerobic digestion and a fungal cultivation for phosphorus accumulation. In the first step, animal manure was thermochemical pretreated or partial anaerobic digested to release orthophosphate and sugars. The effects of pretreatment temperature, time, catalyst type and loading on orthophosphate and sugars release were studied. Then, a fungal strain (*Mucor circinelloides*) with high phosphorus accumulation capability was cultured in the pretreated manure or digested manure to form biological phosphorus fertilizer and the factors that affect phosphorus removal and storage by fungi were studied.

392009 Improving Process Sustainability and Profitablity for a Large U.S. Gray Iron Foundry

Abstract:

Energy savings and sustainability have been an important topic in many industrial processes. Limited energy resources and increasing electricity prices make foundry processes less competitive in today's market. Generally in foundry processes, the energy costs represent 12% of the total revenue. Most of the energy use is directly related to melting and heat treatment operations. Integrating heat and power systems in a foundry can lead to energy conservation and improve process sustainability and profitability. The work presented here is an in-depth energy management analysis using comprehensive mathematical models to identify potential energy savings through process optimization and improved control strategies.

In this work, energy usage data has been collected and analyzed from the Caterpillar (CAT) gray iron foundry in Peoria, IL. A detailed mass and energy balance of the foundry process has been completed in Aspen Plus^{TM}. This model has helped identify strategic energy intensive areas for optimization. Based on Aspen results, a dynamic process simulation using MATLAB Simulink^{TM} has been developed and used to evaluate various energy-saving control strategies. Results show that process optimization and efficient process control, up to 25% energy savings is possible in the foundry.

Based on this work, recommended strategies have been made to reduce energy use by up to 25%. These recommendations are expected to improve process sustainable and reduce its impact on the environment. These changes have the potential to save CAT more than $4 million annually.

377570 Process Retrofit of LNG Regasification Based on Exergy Analysis

Abstract:

Retrofit designs of existing LNG regasification process are suggested. Conventional LNG regasification process is analyzed by exergy decomposition method, followed by the same analyses on three alternative processes. That is, four independent processes are designed, analyzed, and evaluated. Three of them, i.e., conventional LNG regasification process, dual Brayton cycle, and Organic Rankine Cycle, are based on the process presented in literature and the other one is newly integrated, retrofitted scheme for LNG regasification, with a gas-to-liquid (GTL) process. The overall rational exergetic efficiencies are, in order, 0.847, 0.890, 0.849, and 0.868, respectively.

The importance of distillation columns continues to increase both in the traditional petro-chemical industry and in the sustainable sector with renewable resources and energy. The key role they play in the chemical and petrochemical industries and the quest to make them more energy efficient has made distillation process a high priority for all stake holders in the industries. For instance there are about 40,000 distillation columns in the US alone. These consume about 40-60% of the total energy usage in the chemical and petrochemical industries and 6% of the total US energy. Also, major costs of operations second only to the cost of crude in the refinery is energy and 35% of these is consumed in the crude distillation unit. Usually the order of economic importance in the control of distillation column is product quality, process throughput and utility reductions and often traded off between them has to be made. This paper presents a new methodology for optimising the efficiency of the crude distillation unit without trading off the product quality and process throughput. The method uses a neural network based strategy for the modelling and optimisation of distillation columns incorporating the second law of thermodynamics. Second law analysis indicates how well a system is performing compared to the optimum possible performance and hence gives a good indication of the actual energy use of a process. Real time optimisation of distillation columns is made feasible by using neural network models which can be quickly developed from process operation data. The computation time in neural network model evaluation is very short making them ideal for real-time optimisation. Bootstrap aggregated neural networks are used in this study for enhanced model accuracy and reliability. Aspen HYSYS was used for the simulation of the distillation systems. Neural network models for exergy efficiency and product quality are developed from simulated process operation data and are used to maximise exergy efficiency while satisfying product quality constraints. There is great potential for energy recovery in the analysed processes. The recoverable energy is a sum of internal and external exergy losses. The economic analysis of the recoverable energy reveals the energy saving potential of the method. The green house gas (GNG) analysis also affirms it's sustainability nature. It is expected that the application of the methods will aid the design and operation of energy efficient crude distillation column.

Effect of NaCl and $CaCO_3$ on the density, conductivity, and pH of MEG solution: Implication to the design of MEG regeneration process

Yooil Jeon, Suyoul Park, Jiheon Ryu, Yutaek Seo, KAIST

Abstract

Mono-ethylene glycol (MEG) is widely used as a hydrate inhibitor to prevent formation of hydrates in offshore gas fields and used to be regenerated in MEG regeneration process on-board platform. Typical distillation of water from MEG solution can be applied to purify the MEG, however the presence of various salts in produced water may induces difficulties in operating the process. Therefore salts such as NaCl and $CaCO_3$ are required to be removed efficiently to avoid the fouling of the regeneration process and to improve the energy efficiency. Through understanding on the physical properties of salt ions is required to better design of the MEG regeneration process. In this work, density, conductivity, pH and solubility of salt were studied in the MEG solutions containing both NaCl and $CaCO_3$. The concentration of MEG can be easily estimated from the concentration of NaCl, conductivity and density measurement. However the presence of $CaCO_3$ induces slight change in those properties, thus pH measurement is required to supplement the density and conductivity data for estimating the concentration of MEG. And it is essential to optimize distillation column considering both salt precipitation and wt% of MEG. To optimize the distillation column, thermodynamic properties such as salt solubility, vapor pressure of the fluid containing salts are required. Therefore, the salt solubility, vapor pressure of mixture (MEG+Water) with NaCl is experimentally measured. From the obtained physical properties of MEG solution, we develop the novel process to distillate the MEG solution. In the design of distillation column, the temperature has to be controlled carefully to increase the concentration of MEG in bottom product and to avoid the precipitation of NaCl especially inside the re-boiler. We propose the novel design of MEG regeneration while avoiding the precipitation of NaCl by investigating experimentally and by simulating the process using process simulation software, HYSYS V7.3 and Aspen Plus V7.3.

Introduction

MEG (Monoethylene Glycol) is usually used as a gas hydrate inhibitor, and it is essential to regenerate and re-inject this into well because great amount of MEG is consumed to control the gas hydrate problems. MEG is recovered as rich MEG from the topside. Rich MEG is a water solution, which is about 50 wt% of MEG with rich water. This is regenerated as lean MEG through MEG regeneration process. Lean MEG is high concentration of MEG.

In the production of subsea oil and gas, formation water is also produced. Formation water contains salts[1]. This causes problem in MEG regeneration process. If problem arises, the whole process is stopped[2]. And it can result in tremendous loss. It is important to know the salts equilibrium under MEG regeneration conditions in order to prevent problems related to deposition and scale precipitation.

That is, MEG regeneration process is to separate water and salts from the mixed solution, which is composed of MEG, water, and salts. Using the difference in vapor pressure to separate water is called re-concentration and to separate salts is called reclamation process.

As it shows from figure 1 and figure 2, density of MEG solution increases along with increase of NaCl and MEG concentration. NaCl concentration was calculated from the conductivity. If MEG effect and salts effect are separated, assumption of MEG concentration is possible inversely. Sandengen introduced this method in his paper[3].

But there is a limit to monitor the MEG system with only these results. Ions which are included within the formation water are Sodium (Na^{+}), Potassium (K^{+}), Calcium (Ca^{2+}), Magnesium (Mg^{2+}), Barium (Ba^{2+}), Strontium (Sr^{2+}), Iron (Fe^{2+}), Chloride (Cl^{-}), Sulfate (SO_4^{2-}), Alkalinity as HCO_3^{-}, and Bromide (Br^{-}) etc.[1]. Salts can be separated into low soluble salts and high soluble salts. High soluble salts are dissolved in water well like NaCl and KCl. Low soluble salts cannot be dissolved in water well like $CaCO_3$ and $MgSO_4$ [4].

There are two methods in MEG regeneration. One is the full stream concept, and the other is slip stream concept[5]. Many studies were conducted to explain this process. In the full stream concept, high soluble salts and low soluble salts are not sorted. The reclamation process flashes the feed solution. In the slip stream concept, water is separated in the re-concentration process first, and then from part of it, high soluble salts are removed using flash. Core of this concept is to tolerate some of high soluble salts. On the other hand, since low soluble salts are at high risk of scale problems, it is removed in the pre-treatment process beforehand. Enough studies on low soluble salts need to be conducted. Only measurements of density and conductivity are not enough, and they have limits.

Table 1 shows the density and conductivity values by NaCl depending on existence of $CaCO_3$. $CaCO_3$ is a typical low soluble salt. For $CaCO_3$, 0.05 mol per kg of solvent was put. Solubility of $CaCO_3$ is very low. The value of solubility is 000066 g per 100 g H_2O[6]. So an experiment was conducted with enough amount of $CaCO_3$ concentration to saturate the solution. $CaCO_3$ was added to pure water, difference in density between the two is either 0.0001 or the same. So it is hard to measure the deposition of $CaCO_3$ in MEG system from the density. When NaCl concentration increases, conductivity also increases. The amount of NaCl, dissolved in the solution can be estimated. But as it shows from the measured values, in the system that has $CaCO_3$, NaCl, and MEG, effect of $CaCO_3$ on the whole system are slight. As it was previously explained, since enough amount of $CaCO_3$ exists in the solution, $CaCO_3$ is saturated in every condition. The molecular weight of $CaCO_3$ is much heavier than that of NaCl, but it has slight effect because the amount is so small.

That is to say, movement of low soluble salts in MEG system cannot be explained with conductivity and density. Therefore, pH was applied in this paper. The pH is a critical variable in MEG system because the pH is affected by CO_2, $CaCO_3$, NaOH, and $NaHCO_3$ etc. which are the important factors of MEG process[7].

In this study, it is about MEG, water, $CaCO_3$, NaCl, CO_2 system. Solution, which includes these, has chemical reactions as follows

$CO_2 + H_2O \leftrightarrow H_2CO_3$ (1)

$H_2CO_3 \leftrightarrow H^{+} + HCO_3^{-}$ (2)

$HCO_3^{-} \leftrightarrow H^{+} + CO_3^{2-}$ (3)

$CaCO_{3(s)} \leftrightarrow Ca^{2+} + CO_3^{2-}$ (4)

$H_2O \leftrightarrow H^{+} + OH^{-}$ (5)

CO_2 dissolves in water, and it effects on pH. Only part of CO_2 dissolves in water, and it produces carbonic acid (1). Equilibrium constant of this reaction has very small value. Less than 1% of dissolved CO_2 in water produces carbonic acid. This carbonic acid has the reactions (2, 3). Due to the produced H^+ ion, it becomes acidic. Water also degraded into H^+ and OH^- (5).

pH is the function of hydrogen ion concentration. Hydrogen ion produced from the reaction above are represented in equations (2, 3, 5). $CaCO_3$ is dissolved in water and separated into Ca^{2+}, CO_3^{2-} (4). CO_3^{2-} has influence on the equation (3). In other words, hydrogen ion concentration of the solution above is same as follow (6).

$$2[Ca^{2+}] + [H^+] = [HCO_3^-] + 2[CO_3^{2-}] + [OH^-]$$

$[HCO_3^-]$, $[CO_3^{2-}]$ and $[OH^-]$ can be expressed as hydrogen ion concentration through equilibrium. As a result, concentration of $[Ca^{2+}]$ can be expressed as hydrogen ion concentration. In other words, hydrogen ion concentration within the solution can be measured from pH measurement, and $CaCO_3$ solubility can be easily and promptly predicted from this. In this study, Properties of $CaCO_3$ were analyzed from pH measurement. pH was measured as it changes the condition; the temperature from room temperature to 80°C, MEG concentration from 0 wt% to 90 wt%, NaCl concentration from 0 to 0.9 mol/kg solvent.

Also, solubility of high soluble salts is significant factor in MEG regeneration process. Solubility of high soluble salts is increased as temperature increase. But, water is vaporized as temperature increase. As a result, there is a risk of precipitation of high soluble salts by decreasing an amount of solvent in distillation column. Solubility of NaCl is measured in operating condition of distillation column of 80 wt% MEG and 90wt% MEG at 140°C. From the obtained physical properties of solubility of high soluble salts in MEG solution, novel process to avoid the precipitation of high soluble salts is proposed.

List of figures

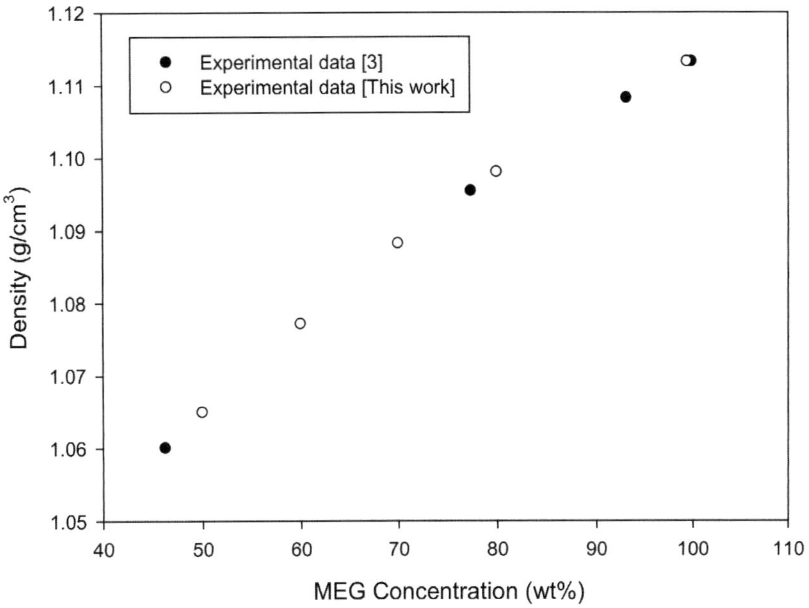

Figure 1. Density of solution as function of MEG concentration (wt%) at 20℃

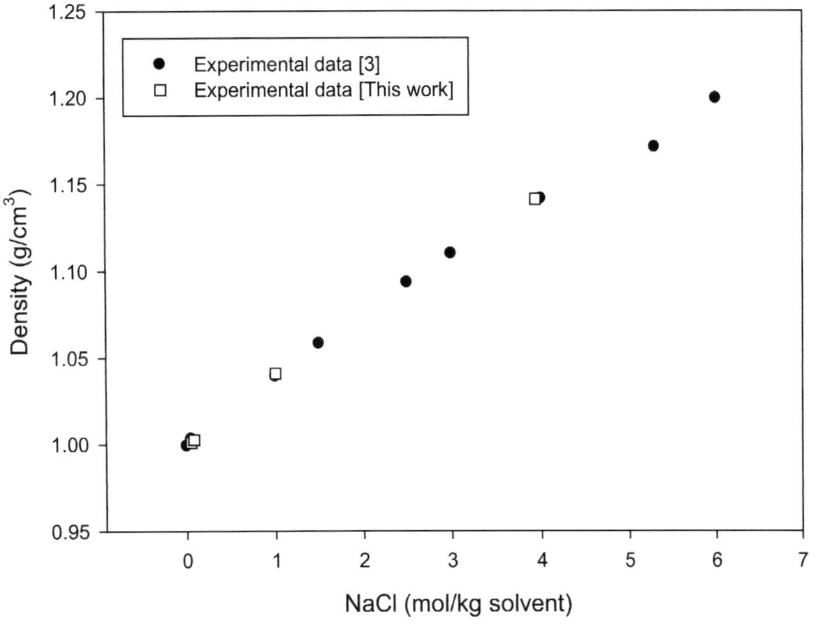

Figure 2 Density of solution as function of NaCl (mol/kg solvent) at 20℃

List of tables

Table 1. Density and conductivity of MEG solution with NaCl and $CaCO_3$

MEG (wt%)	NaCl (mol/kg)	$CaCO_3$ (mol/kg)	ρ (g/cm^3)				k at 25℃ (μs/cm)
			15℃	20℃	25℃	50℃	
0.0	0	0			0.9970		0.9710
0.0	0.003	0			0.9971		497
0.0	0.009	0			0.9972		1015
0.0	0.024	0			0.9979		2840
0.0	0.05	0	1.0011	1.0002	0.999	0.9898	5470
0.0	0.08	0	1.0027	1.0017	1.0005	0.9914	9220
0.0	1	0	1.041	1.0385	1.0368	1.0267	83700
0.0	4	0	1.1412	1.1388	1.1364	1.1236	214000
0.0	0	0.05			0.9970		43.3
0.0	0.003	0.05			0.9971		539
0.0	0.009	0.05			0.9973		1057
0.0	0.024	0.05			0.9980		2890
16.7	0.8	0	1.0541	1.0524	1.0504		50100
37.5	0.63	0	1.0749	1.0725	1.07		25800
37.5	0.83	0	1.0826	1.0803	1.0777		32500
50.0	0	0		1.0649	1.0620	1.0483	4.05
50.0	0.003	0			1.0622		178.3
50.0	0.009	0			1.0624		375

50.0	0.024	0			1.0629	904
50.0	0.5	0	1.0863	1.0835	1.0807	16030
50.0	0	0.05			1.061	11.94
50.0	0.003	0.05			1.0642	184.6
50.0	0.009	0.05			1.0646	383
50.0	0.024	0.05			1.0654	926
60.0	0	0		1.0771		
70.0	0	0		1.0882		
80.0	0	0		1.098		
100.0	0	0		1.1132		

References

(1) Satman, A.; Ugur, Z.; Onur, M.; The effect of calcite deposition on geothermal well inflow performance. *Geothermics* **1999**, *28*, 425–444.

(2) Kaasa, B.; Marheim, N.; Jacobsson, J.; Scale Removal from the Condensate Stripper at Kollsnes. *Oil Field Chemistry Symposium* **2007**, 25–28.

(3) Sandengen, K.; Kaasa, B. Estimation of Monoethylene Glycol (MEG) Content in Water + MEG + NaCl + $NaHCO_3$ Solutions. *J. Chem. & Eng. Data* **2006**, *51*, 443–447.

(4) Flaten, E. M.; Seiersten, M.; Andreassen, J. P.; Induction time studies of calcium carbonate in ethylene glycol and water. *J. Chem. & Eng. Data* **2010**, *88*, 1659–1668.

(5) Addicks, L; Method for regeneration and reclamation of mono ethylene glycol using a vacuum slip stream. PCT: Patent Cooperation Treaty, PCT/NP2010/000006, **2010**.

(6) Lide, D. R., Ed. *CRC Handbook of Chemistry and Physics*, 71st ed.; Chemical Rubber Co.: Boca Raton, FL, **1990**.

(7) Sandengen, K.; Kaasa, B.; Østvold, T.; pH Measurements in Monoethylene Glycol (MEG) + Water Solutions. *Ind. & Eng. Chem. Res.* **2007**, *46*, 4734–4739.

(8) Flaten, E. M.; Seiersten, M.; Andreassen, J.-P.; Growth of the calcium carbonate polymorph vaterite in mixtures of water and ethylene glycol at conditions of gas processing. *J. Cryst. Growth* **2010**, *312*, 953–960.

(9) Segnit, E. R.; Holland, H. D.; Biscardi, C. J.; The solubility of calcite in aqueous solutions - I The solubility of calcite in water between 75℃ and 200℃ at CO_2 pressures up to 60 atm. *Geochimica et Cosmochimica Acta* **1962**, *26*, 1301-1331.

(10) Coto, B.; Martos, C.; Pena, J. L.; Rodríguez, R.; Pastor, G.; Effects in the solubility of CaCO3: Experimental study and model description. *Fluid Phase Equilibria* **2012**, *324*, 1-7.

(11) Wolf, M.; Breitkopf, O.; Puk, R.; Solubility of calcite in different electrolytes at temperatures between 10℃ and 60℃ and at CO_2 partial pressures of about 1kPa. *Chemical Geology* **1989**, *76*, 291-301.

(12) Robbins, C. W.; The $CaCO_3 - CO_2 - H_2O$ system in soils. *J. Agronomic Education* **1985**, *14*, 3-7.

(13) Millero, F. J.; Milne, P. J.; Thurmond, V. L.; The solubility of calcite, strontianite and witherite in NaCl solutions at 25℃. *Geochimica et Cosmochimica Acta* **1984**, *48*, 1141-1143.

(14) Smith, F. L.; Harvey, A. H.; Avoid Common Pitfalls When Using Henry's Law. *CEP: Chemical Engineering Progress* **2007**.

(15) Serpa, F.; Vidal, R.; Filho, J.; Solubility and Thermodynamic Properties of Carbon Dioxide in MEG/Water Mixtures. *III Iberoamerican Conference on Supercritical Fluids Cartagena de Indias (Colombia)*, **2013**.

(16) Osvaldo Chiavone-Filho, Peter Rasmussen.; Solubilities of Salts in Mixed Solvents. *J. Chem. Eng. Data* **1993**.

(17) H. M. Trimble.; Solubilities of Salts in Ethylene Glycol and in Its Mixtures with Water. Industrial and Engineering Chemistry **1931**.

381303 Development of DME Production Process Using Self-Heat Recuperation

Abstract:

The demand for dimethyl ether (DME) will continue increasing, since DME is categorized as one of the green energy sources. In fact, DME contains 34.8% oxygenates and has higher heat value than natural gas. Moreover, it is less toxic to human beings and can be stored as a liquid fuel with pressurising or cooling. DME is commonly produced by two methods; indirect and direct DME synthesis methods. In the indirect method called double stage method, methanol is produced and then converted to DME. On the other hand, in direct method called single stage method, DME is synthesised from syngas. The main advantage of direct method is reported that the CO transformation rate is higher than the indirect method. Thus, many investigators try to find or create new catalysts for direct method. However, this CO high transformation phenomenon occurs due to series reactions taking place in the reactor, leading to shifting the reaction equilibrium to the product side.

So, as well as other chemical processes, many DME production processes in the indirect method have reactant recycle system with product separation in order to increasing the overall conversion ratio in the reactor. Distillation processes have often been used for this product separation from reactant recycle stream in industry due to the large throughput. However, the distillation processes are well known as energy consuming processes since the large latent heat has to be provided to reach at the gas-liquid equilibrium in the column. Thus, many chemical engineers and investigators have been managing to produce a catalyst which achieves high conversion from reactant to production in the reactor for whole process optimization.

Recently, authors have developed a self-heat recuperation technology based on exergy destruction minimization to reduce the energy consumption of chemical processes. Applying the self-heat recuperation technology to chemical processes, not only the latent heat but also the sensible heat of the process stream can be circulated into the processes without any heat addition. As a result, the energy consumption and exergy destruction of a process can be greatly reduced in the steady state. In fact, this technology has been applied to several chemical processes as case studies and shows large energy saving potentials in these processes. From these previous studies, it found that thermal and separation processes are suitable processes for self-heat recuperation technology.

In this research, we investigated the feasibility of applying self-heat recuperation technology to DME production process and developed an innovative process for DME production process from the energy saving point of view. By installing the self-heat recuperation technology to the DME production process, the energy consumption of the thermal and separation processes can be greatly reduced. This process has a great potential to produce green fuel, DME with low energy consumption. In addition, if we will integrate the proposed DME production process based on self-heat recuperation with the methanol production process based on self-heat recuperation as the series of processes from syngas to DME, this series of processes has great potential instead of current direct DME production method from syngas.

Energy Integration and Cost Analysis of Ethyl Benzene Production Plant using Pinch Technology

Saket Kumar[1], Subhajit Majumder[2*]

[1]*Under-Graduate Student, Department of Chemical Engineering*
Birla Institute of Technology and Science (BITS), Pilani- 333031, Rajasthan, India
Email: sktkumar8@gmail.com

[2]*Lecturer, Department of Chemical Engineering*
Birla Institute of Technology and Science (BITS), Pilani- 333031, Rajasthan, India
Email: subhajit@pilani.bits-pilani.ac.in; subh.jit@gmail.com

EXTENDED ABSTRACT

The petro-chemical industries are facing great challenges in terms of energy conservation and cost reduction. While fuel prices continue to climb, energy conservation remains a prime concern for these industries. It is always a common approach to install a feed-effluent heat exchangers around a reactors and distillation columns to integrate the energy in the overall process. For last few decades, pinch technology has been gaining significant importance to improve the efficient energy usage in process industries. Pinch technology follows a systematic procedure for energy saving in processes. At a starting point, thermodynamic concepts are invoked while calculating the minimum heating and cooling requirements in the process. Minimum number of heat exchangers required for the process is calculated to obtain the minimum energy requirements. These calculations can be performed without finalizing any heat exchanger network (HEN). The minimum energy requirements and minimum number of heat exchangers provide targets for the subsequent design of complete heat exchanger network which helps to determine the total annual cost of heat exchangers involved in the process.

Present study deals with the minimizing the total annual cost required for an ethyl benzene (EB) process plant using Pinch Technology. In production process, EB is produced by alkylation of ethylene and benzene, and the product is used to produce styrene monomer. Aspen Plus (V8.4) is used to simulate the entire process. HEN for the process is specified using Aspen Energy Analyzer (V8.4). Analysis of the HEN reveals that there is a possibility to improve energy usage in the process and to reduce the load on utilities, thereby reducing the operating costs. A modified HEN is proposed by adding three new heat exchangers. It reduces the annual energy cost by 25.5%. It decreases the overall utility requirement, thereby saving energy and capital requirements. In order to implement the modified network, cost analysis of the process is done using both *Guthrie's correlation* and *Aspen correlation*. A trade-off is considered between new capital investment and operating cost savings. The base case has a return on investment of 17.76% and payout period of the initial investment 3.33 yrs. Considering the modification, the new return on investment comes out to be 24% and a payout period of 2.7 yrs. The modification shows a huge rise in profits and most importantly energy savings by reducing utility load. This retrofitted model, if implemented can be a benchmark for industries towards an efficient energy usage.

Keywords: *Heat exchanger network, Pinch analysis, Annual energy cost, Guthrie's correlation, Aspen correlation, Payout period, Return on investment.*

* Corresponding author. Tel.: +91-1596-515872; Fax: +91-01596-244183
E-mail address: *subhajit@pilani.bits-pilani.ac.in; subh.jit@gmail.*

Sustainable Urea Manufacture Using Novel Techniques Utilizing Alternative Raw Material Sources

SUST-E Urea Process

Taibah Jaffery
Supervisor: Dr. Arshad Hussain

Fatima
Fertilizer Company Limited

Sponsor

The speaker's visit to the 2014 AIChE Annual Meeting has been Sponsored by:

Fatima Fertilizer Company Limited

* This work has been conducted entirely by the National University of Sciences and Technology, Pakistan. Fatima Group does not own a share in this technology.

2

Fatima Group

- One of the largest business groups in Pakistan with diversified businesses in fertilizers, textiles, and sugar industries.
- Primary business includes manufacture of compound fertilizers- annual capacity 2.1 MTonnes
- Fatima Group has experience in undertaking large-scale green-field projects in Pakistan
- FFL was conceived at a cost of M USD 750 in December 2003

Fatima Group

Fertilizers

Sugar

Trading

Textiles

Pakarab Fertilizers | Fatima Fertilizer | Fatima Sugar Mills | Reliance Commodities | Reliance Weaving | Fazal Cloth Mills

Energy

Fatima Energy

3

Fatima Fertilizers Company
Vision, Mission & Core Values

Vision

To be a world class manufacturer of fertilizers and ancillary products, with a focus on safety, quality and contribution to national economic growth and development. We will care for the environment and the communities we work in while continuing to create shareholders' value.

Mission

- To be the preferred fertilizer company for farmers, business associates and suppliers through quality and service.
- To provide employees an exciting, enabling and supportive environment to excel in, be innovative, entrepreneurial in an ethical and safe working place based on meritocracy and equal opportunity.
- To be a responsible corporate citizen with a concern for the environment and the communities we deal with.

Core Values

Integrity: Our actions are driven by honesty, ethics, fairness and transparency
Innovation: We encourage creativity and recognize new ideas
Teamwork: We work collectively towards a common goal
Safety, Health and Environment: We care for people and communities around us
Customer Focus: We believe in listening to our customer
Excellence: We strive to excel in everything we do
Valuing People: We value our people as our greatest resource

4

Production

- **Final Products**
  - Urea 1,500 TPD
  - Calcium Ammonium Nitrate (CAN) 1,400 TPD
  - Nitro Phosphate (NP) 1,200 TPD
  - Nitro Phosphorous Potash (NPK) 1,000 TPD
- **Intermediary Products**
  - Ammonia 1,500 TPD
  - Nitric Acid (NA) 1,500 TPD
- **Off Sites & Utilities**
  - Power plant - 52 MW ISO Rating
  - Product Handling & Storage
  - 47 KM Natural Gas Pipeline
  - 12 KM Raw Water line & Pumping Station with own power
  - Railway Siding
- Debottlenecking
  - Ammonia 1800 TPD

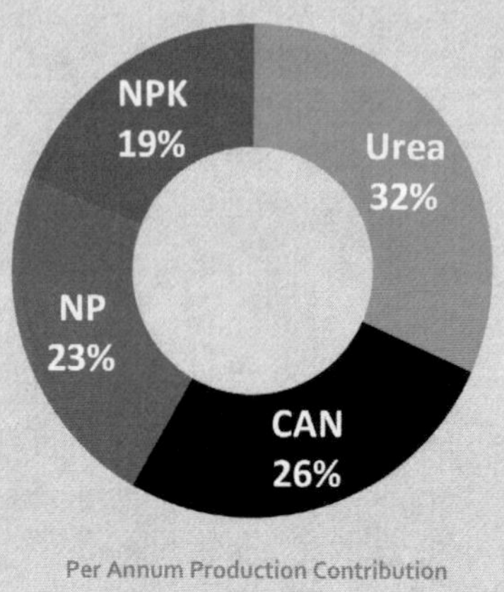

Per Annum Production Contribution

5

Major Milestones

- Apr. 2011 — NP Plant Production
- Apr. 2010 — NA Plant Production
- Mar. 2010 — Urea Plant Production
- Feb. 2010 — Ammonia Plant Production
- Nov. 2009 — CAN Plant Production
- Oct. 2009 — Ammonia Furnace 1st Fire
- Financial Closure achieved and LC's established — 26th Nov. 2006
- Construction & Commissioning Contract with CNCEC — Sep. 2006
- Contract with Sojitz signed — 27th Jun. 2006
- Ground Breaking Ceremony — Apr. 2006
- Gas Supply / Purchase Agreement Signed — 12th Jul. 2005
- Gas Allocation — 4th Sept. 2004
- Date of incorporation — 23rd Dec. 2003

6

Fatima Fertilizer Company Limited — FFL Site Safety

PSM

Lagging Indicators

Target 0.27

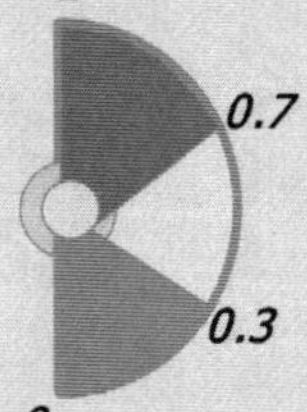

Total Recordable Injury Rate (Year to date)

FFL 0.00
Cont. 0.087
Comb. TRIR = 0.061

Leading Safety Indicators Sco

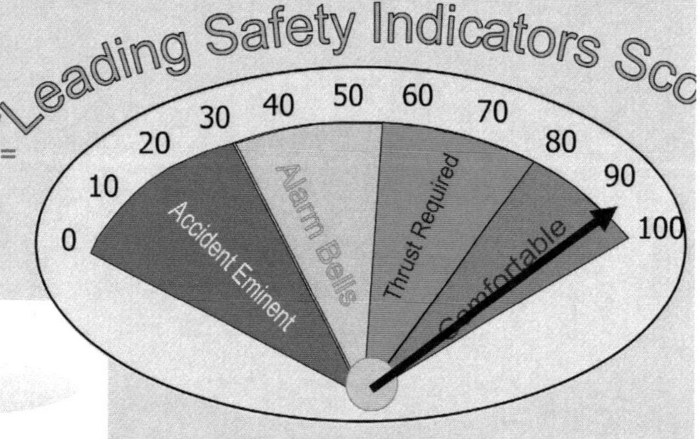

Current PSM Level at 2013 start

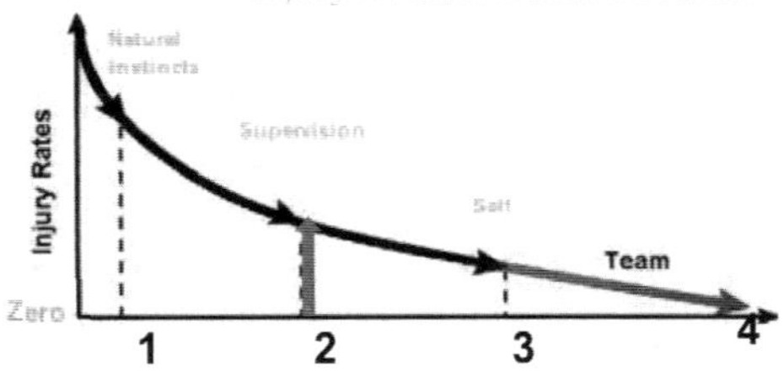

- SAFE WORKING HRS FFL = 6.80 Million
- SAFE WORKING HRS CONT. = 13.89 Million
- SAFE WORKING HRS COMBINE = 20.7 Million

Current 93

Target 80

7

Fatima Fertilizer Company Limited — Future Projects

Debottlenecking

- Ammonia DBN
- Phase 1 completion in Dec 2015 - 1500-->1650 TPD
- Phase 2 completion in 2017 - 1650-->1800 TPD

Midwest Fertilizers Company Project

- Ammonia 1500 TPD
- NA 1500 TPD
- UAN1500 TPD

8

Objective

- To make the Urea manufacturing process inherently independent of non-renewable feed sources (Natural Gas, Coal, etc)
- Sustainable and environment friendly production with minimum net energy requirement
- Minimizing carbon emissions in the atmosphere - Goal of regulating authorities: achieving 90% recovery of CO_2 with maximum 35% increase in electricity cost [1]
- Carbon Sequestration

9

Challenges

- Clean Hydrogen Source
- Clean Energy Source
- Flue Gas Separation into Carbon dioxide and Nitrogen

10

Proposed Solution

- SUST-E Urea Manufacturing Process
- Al-steam reaction for combined production of Hydrogen and Energy
- Membrane Separation of Flue Gases from Thermal Power Plant, Cement Industry, etc.
- Replace Front-end (Steam reforming)
- Renewable energy e.g. CSP for additional energy requirements

$$2Al + 3H_2O \rightarrow Al_2O_3 + 3H_2 + \text{Heat}$$

$$\text{Flue Gas} \rightarrow CO_2 + N_2$$

$$N_2 + 3H_2 \rightarrow 2NH_3 + \text{Heat}$$

$$2NH_3 + CO_2 \rightarrow NH_2COONH_4 \leftrightarrow CO(NH_2)_2 + H_2O$$

Scope

- Providing a workable design for the new synthesis route
- Investigation of technical feasibility of 7000 TPD Urea and 4000 TPD Ammonia production
- Investigation of economic feasibility of 7000 TPD and 4000 TPD production
- Providing direction for further research
- Tools – MATLAB, ASPEN PLUS

Process Development

Process Research & Innovation	Pilot Studies	Technology Transfer and Manufacturing	Process Intensification & Microprocess Engineering	Product Design

12

The Sust-E Urea Process

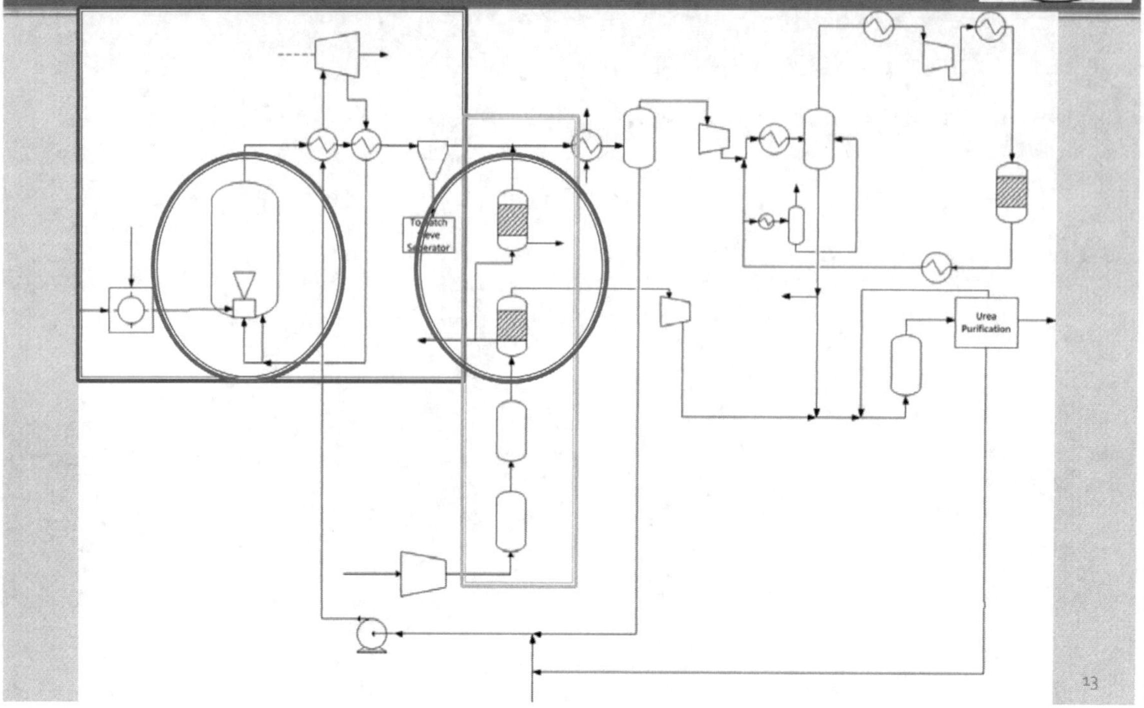

13

Raw Materials Chosen

- Aluminum ingots (grinding tool based on Watanabe's work) [2]
- Flue Gas from coal based plants
- Water
- Energy source – Al/H_2O reaction

14

Aluminium- Steam Reactor

- $2Al + 3H_2O \rightarrow Al_2O_3 + 3H_2$ above 480 °C [3-6]
- Reactants are **fed** at only 600 °F (316 °C) and steam is fed at 150 psig via nozzle
- The Reactor is a combustor (in plug flow)
- **15442** tubes of 10" dia and height of 4ft, arranged in 16 floors, 34 rows of 29 tubes each
- Temperature of 2900 K achieved(> Al melting temperature) [4-5]
- The residence time is **60 ms** and conversion is 95%.[4-5]
- Heat of reaction **1100-1250 MW**, Heat Loss = 405-680 MW
- ΔH > **540 MW**

15

Membrane Separation

- Coal based flue gas
- Two stage, co-current process using PVAm membrane [7-8]
- Nitrogen purification using Carbon Molecular Sieve Membrane (Celo550) [9]
- Multi-component model used to for designing using MATLAB
- Membrane Area (m^2): PVAm Stage1= 2594122, Stage2= 627118
- Celo550 Area ~ 27,922,908.05 ft^2
- 94.6 % pure CO_2 & 94.5 % pure N_2 recovered after membrane separation.
- Celo 550 counter-current membrane separator gives 99.99% pure nitrogen.

16

PFD for Membrane Separation

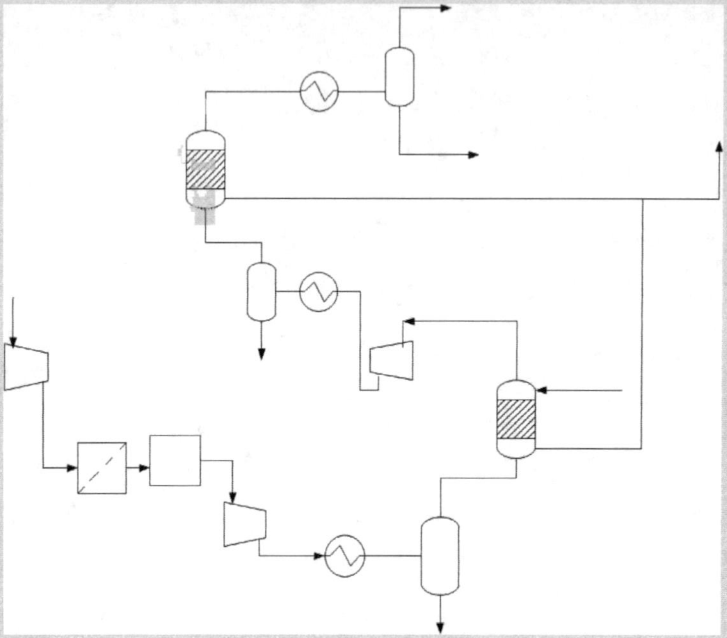

17

Process Integration

- Process Integration for reduction of energy and capital consumption.
- Reaction heat used to heat up feed as well as produce excess electricity
- Water vapor from exhaust steam turbine condensate used as sweep gas

18

Consumption and Production

Urea (99.8%)	7000 TPD
Ammonia (including inventory)	4000
Aluminium	6670
High Purity Alumina	11973
H_2O	4265
H_2	710
N_2	3290
CO_2	5119
Excess N_2	30008
Flue gas	39852 (1206331.27 Nm^3/hr)

19

Energy Efficiency

- Released by exothermic reaction: 1100-1200 MW
- Electricity production: 315 MW including 129 MW required for membrane separation
- Requirement for Al-Steam Feed: 260-280 MW
- Total major plant electricity requirements (grinding, vacuum systems, etc): 174.3 MW
- Surplus: 140.7 MW

*Turbine and generator efficiencies have already been accounted for

20

Cost Feasibility

- Total Plant Investment: $ 138 billion (at current cost of Alumina) – will be reduced by:
  - Increase in supply of alumina refractory after establishment of Sust-E Urea Technology
  - More efficient reactor design with better heat recovery and lower surface area
- Highly expensive bi-product (High Purity Alumina)
- Payback time: 2 yrs
- ROR = $ 69 billion/yr (can be achieved at $32/kg of alumina (34% of current cost) and $0.20/kg of Urea)

21

Sustainability

- Inherently independent of non-renewable resources - capable of running on solar power
- 0.73kg CO_2/kg Urea consumed, hence CO_2 emissions reduced.
- Feed Aluminum can be produced using renewable energy sources
- Eliminates requirement for catalysts (and hence emissions related to their production)

22

Conclusion

- The process is technically capable of achieving process requirements (purity, temperature, pressure) for ammonia and urea manufacture
- Economically feasible
- Energy Efficient

23

Areas for Further Research

- Pilot Studies
- Optimization of reactor design for reduced heat loss and surface area (and hence cost)
- Cost Models for Aluminum and High Purity Alumina nano-particles after establishment of Sust-E Urea plants
- Scale-up of electrolysis cells using non-consumable cathodes for Aluminum production
- Higher efficiency membranes

24

References

- [1] Favre E, "Carbon dioxide recovery from post-combustion processes: Can gas permeation membranes compete with absorption?"
- [2] Watanabe M, Ximeng J, Ryuichi S, "Method for generating hydrogen utilizing activated aluminum fine particles." US patent. US 2006/0034756 A1; 2006
- [3] Petrovic J, Thomas G, "Reaction of Aluminum with Water to Produce Hydrogen – A Study of Issues Related to the Use of Aluminum for On-Board Vehicular Hydrogen Storage", U. S. Department of Energy
- [4] Foote, J. P, Thompson, B. R, Lineberry, J. T, "Combustion of Aluminum with Steam for Underwater Propulsion"
- [5] Franzoni F, et al., Combined hydrogen production and power generation from aluminum combustion with water: Analysis of the concept, International Journal of Hydrogen Energy (2010),
- [6] Golovitchev, V. I, Al/H2O Combustion, Report No. 2, Combustion Engine Research Centre, Chalmers University of Technology
- [7] Hussain, A., A single stage membrane process for CO2 capture from flue gas by a facilitated transport membrane, Journal of Separation Science & Technology 47 (2012) 1857-1865.
- [8] Hussain, A., Hägg, M.-B., A feasibility study of CO2 capture from flue gas by a facilitated transport membrane, Journal of Membrane Science 359 (2010) 140-148.
- [9] Campo M. C, Magalhães F. D, Mendes A, Separation of nitrogen from air by carbon molecular sieve membranes

25

Questions?

26

Energy Efficient Light Olefin Recovery: Absorption Versus Cryogenic Distillation

Pieter A. Reyniers, João F. dos Santos, Stephanie Saerens, Pieter Cnudde, Laurien A. Vandewalle, Thomas M. Deconinck, Steven J. Govaert, Philip de Smedt, Kevin M. Van Geem, Guy B. Marin

Laboratory for Chemical Technology, Ghent University, Ghent, Belgium

Abstract:

Current propene supply from conventional sources, i.e. steam cracking and fluid catalytic cracking is insufficient to fulfill the market demand. Therefore, on-purpose propene production routes have been developed to bridge the gap, one of which is methanol-to-olefins. This provides the industry with the opportunity to implement novel separation sections specifically tuned to the alternative production routes. In the current work, the absorption based separation section operated by Wison Engineering is compared with a traditional cryogenic demethanizer first separation section in terms of heat integration potential, operational expenditure and capital expenditure. No clear winner could be identified since the performance of the two separation schemes is very comparable. The minimum required hot and cold utility is quasi-identical. For operational expenditure, a difference of 7.9 % is calculated (1.6 M$/year on a total of 21.0 M$/year) in favor of the demethanizer first configuration. For capital expenditure, a difference of 5.2 % is calculated (3.1 M$ on a total of 59.5 M$) in favor of the absorption configuration. Other factors including sensitivity to variations in the feed, operational experience within a company, licensing costs could induce a preference of one scheme above the other.

1. Introduction

Demand for light olefins has increased significantly over the past decades and is expected to continue on this path in the upcoming years. Figure 1 illustrates the time-evolution of the ethene demand of the top 20 economies and the installed ethane production capacity. It is expected that in the coming years, demand increases more quickly than production capacity.

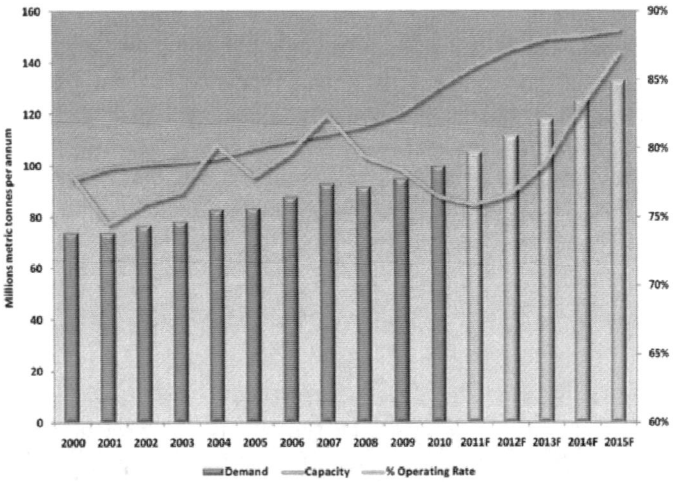

Figure 1: World ethene demand, capacity and operating rate (* Demand values limited to global top 20 economies) (Source: Global Data Petrochemicals eTrack) [1].

A similar trend is observed for the supply and demand of propene, see Figure 2. Demand has increased since 1995 and this trend will continue in the upcoming years. However, propene supply from traditional sources, i.e. steam cracking and fluid catalytic cracking (FCC) is insufficient to fulfill the market demand. Therefore on-purpose propene production routes have been developed to bridge the gap [2]. Of these on-purpose production routes, propane dehydrogenation and metathesis are expected to have the most significant contribution.

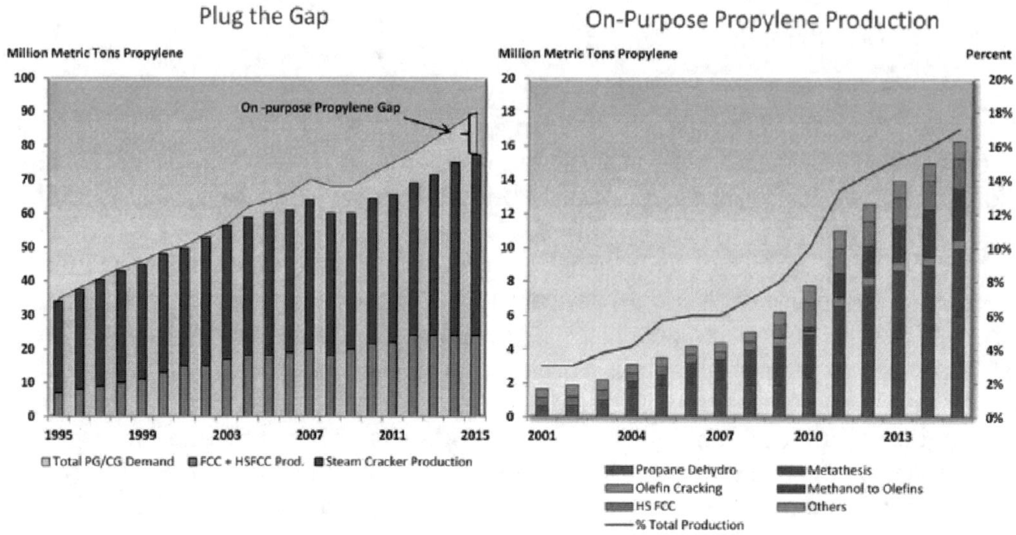

Figure 2: Gap between propene demand and supply via traditional routes, on-purpose propene production [1].

One of the other on-purpose production routes for light olefins is methanol-to-olefins (MTO). Methanol molecules are combined and dehydrated in a catalytic reaction yielding light olefins which can further react to higher olefins. The process is run in a setup similar to fluid catalytic cracking to allow continuous catalyst regeneration [3, 4]. The main advantage is a high carbon efficiency for propene/ethene (wt.) product ratios in the range of 1.2 – 1.8. The light ends content in an MTO effluent, i.e. H_2, CO, N_2, CH_4 is typically lower than in the effluent of a steam cracker. This reduces the load on the demethanizer column and opens the path towards alternative separation configurations. Implementation of these novel production routes for propene has triggered the development of alternative back-end units for olefin separation.

In traditional steam cracker setups, recovery of valuable light olefins from the cracker effluent is realized via cryogenic separation [5]. A commonly used cryogenic separation section consists of a demethanizer, deethanizer, depropanizer, C2-splitter and C3-splitter. Several lay-outs are possible which differ in the order of the distillation columns and the location of the hydrogenation unit. Low temperatures down to -140 °C are required for sufficient light olefins recovery and for hydrogen separation. These low temperatures are generated in a separate refrigeration section consisting of closed loop ethene and propene cycles. Mixed cooling cycles in which the cooling medium consists of a mixture of components is able to generate refrigeration over a wider temperature range using only one closed cycle, thus only requiring a single compression train [6, 7].

One of the leading firms in the development of alternative separation trains for the MTO process is Chinese based Wison Engineering [8]. Separation of light olefins is performed without refrigeration below -50 °C, i.e. no ethene refrigeration cycle is required while the traditional demethanizer column is replaced by a combination of a precutting column and an absorption column. Recovery of the remaining light olefins in the impure C1-stream is done via absorption in an open loop C3 cycle. Similar schemes have been patented by other firms including but not limited to Lummus Technology [9] and UOP LLC [10]. The absorption separation scheme has been commercialized by Wison Engineering in Nanjing, China [11]. The full scale plant has been operational since September 2013.

A comparison is made between a traditional cryogenic demethanizer first scheme and the novel absorption scheme as implemented by Wison Engineering. Only the equipment downstream of the main cracked gas compressor and the equipment in the refrigeration section is considered.

2. Simulation setup

A model feed composition representing an average MTO reactor effluent is used. It is assumed that all C6+ components have been liquefied during compression and are hence not present anymore in the feed. The feed composition is determined based on the Wison Engineering patent [8] and the work of Senetar and Romers [4] and Tallman and Eng [12].

Table 1: Model feed composition for an MTO reactor effluent stream after caustic treatment and dehydration.

Component	Element formula	Mass flow [kg/hr]	Mass fraction [wt%]
N_2	N_2	593	0.40
CO	CO	739	0.49
H_2	H_2	739	0.49
CH_4	CH_4	887	0.59
ETHYLENE	C_2H_4	50832	33.90
ACETYLENE	C_2H_2	15	0.01
ETHANE	C_2H_6	591	0.39
PROPYLENE	C_3H_6	67677	45.12
ME-ACETYLENE	C_3H_4	74	0.05
PROPADIENE	C_3H_4	74	0.05
PROPANE	C_3H_8	739	0.49
BUTANE	C_4H_{10}	1513	1.01
BUTENE	C_4H_8	12801	8.53
ISOBUTANE	C_4H_{10}	378	0.25
ISOBUTENE	C_4H_8	3405	2.27
BUTADIENE	C_4H_6	817	0.54
PENTENE	C_5H_{10}	5639	3.76
PENTANE	C_5H_{12}	627	0.42
2ME-BUTENE	C_5H_{10}	1410	0.94
2ME-BUTANE	C_5H_{12}	157	0.10
DME	C_2H_6O	296	0.20
Total		150003	100.00

Product boundary conditions, recoveries and purities are indicated in Table 2.

Table 2: Feed and product boundary conditions, recoveries and purities

Product stream	Temperature [°C]	Pressure [barg]	Recovery [%]	Purity [wt.%]
FUEL GAS	30	5	-	-
ETHYLENE	30	20	99.50	99.95
ETHANE	30	5	-	-
PROPYLENE	40	15	99.50	99.60
PROPANE	40	15	-	-
C4+	40	10	-	-

The flow schemes for the cryogenic demethanizer first configuration and the absorption configuration are given in Figure 3.

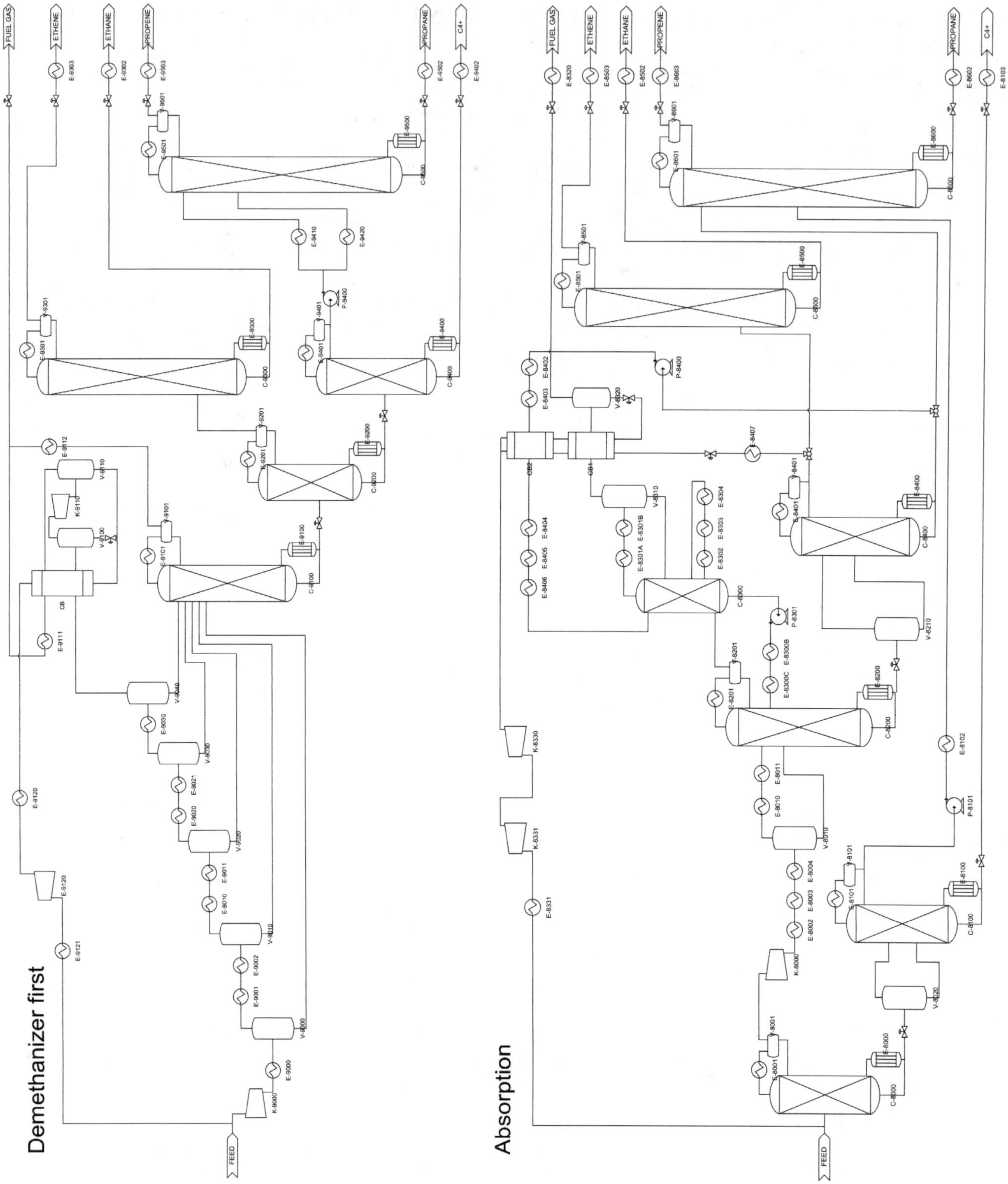

Figure 3: Cryogenic demethanizer first flowsheet (left) and absortion flowsheet (right).

The process simulations have been carried out in Aspen Plus V7.3 using the Soave-Redlich-Kwong (SRK) equation of state model. This property method is suited to simulate hydrocarbon system at pressures higher than 1 bar. This has been validated by comparing key pure component properties like evaporation enthalpy, density and liquid heat capacity for important components with experimental data provided by the National Institute of Science and Technology (NIST) [13].

The refrigeration section displayed in Figure 4 consisting of a closed propene cycle and if required, a closed ethene cycle, is also modeled explicitly. The required cooling duty at the different levels (P1 to P4 and E1 to E3) for each configuration is acquired from the process simulations. The results of the refrigeration section calculations are the mechanical duties of the ethene compressor and propene compressor on one hand and the duty of the propene condenser heat exchanger on the other hand.

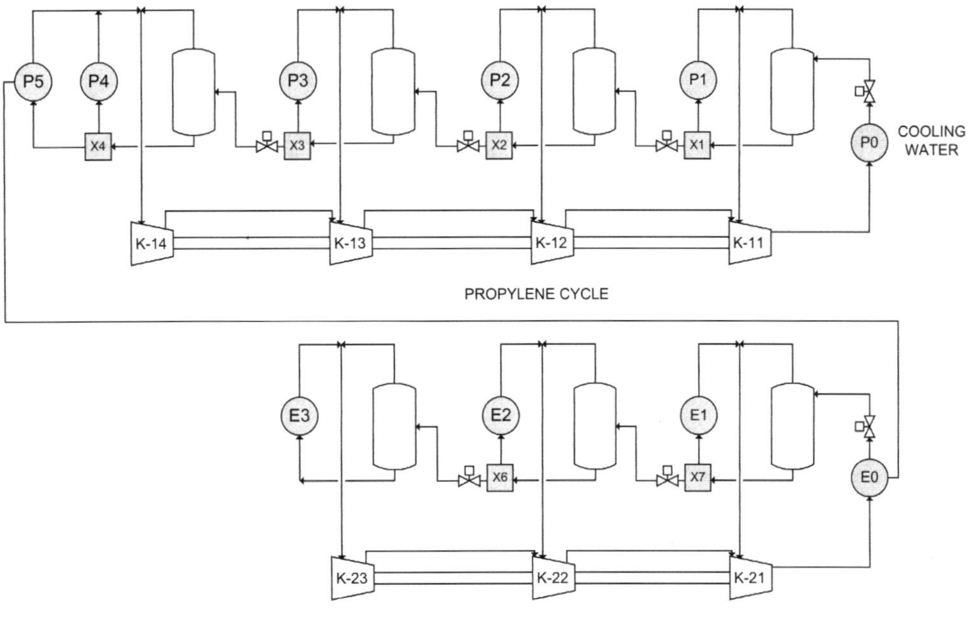

Figure 4: Refrigeration section flowsheet.

The potential for heat integration in both configurations is evaluated based on the spreadsheet program for pinch analysis developed by Kemp [14].

Operational expenditure (OPEX) is determined based on the calculated steam consumption and mechanical energy use. It is assumed that all mechanical energy is delivered via electro-motors with an efficiency of 80%. Low pressure steam at 5 barg is generated from natural gas with an efficiency of 85% [15]. The electricity and natural gas prices for industrial users in the region West South Central of the United States are taken in June 2014. This results in a cost of 5.85 $/GJ for steam [16] and 63.1 $/MWh for electricity [17]. For cooling water, only the electricity cost of the cooling water pump is considered.

Capital expenditure (CAPEX) is determined based on the Aspen Cost Estimator utility. All equipment is sized using this software via the build-in correlations. In the costing step, manual corrections are performed to include the correct construction material, i.e. for low temperature applications (-30 °C and lower), SS304 stainless steel is selected over plain carbon steel.

3. Simulation results and discussion

The results of the pinch analysis for both configurations is given in Table 3. The minimum temperature approach is assumed to be 10 K. Also grand composite curves are shown in Figure 5 for both configurations. The overall shape of the grand composite curves is similar since it is mostly determined by the heat duties of the distillation columns. Consequently, the pinch temperature is quasi-identical for both configurations. The demethanizer first scheme and absorption scheme differ in the setup of the C1-/C2+ separation and in the depropanizer setup. The heat demand without integration is comparable for the demethanizer first configuration and for the absorption configuration. This conclusion is also valid for the heat demand at full heat integration. Based on pinch analysis, it is not possible to identify a clearly advantageous configuration.

Table 3: Pinch analysis result for both configurations.

	Cold Pinch temperature [°C]	Hot utility [MW]	Cold utility [MW]
Demethanizer first	41.7		
No heat integration		88.60	83.64
Full heat integration		59.29	64.25
Absorption	40.4		
No heat integration		88.33	83.76
Full heat integration		60.07	64.64

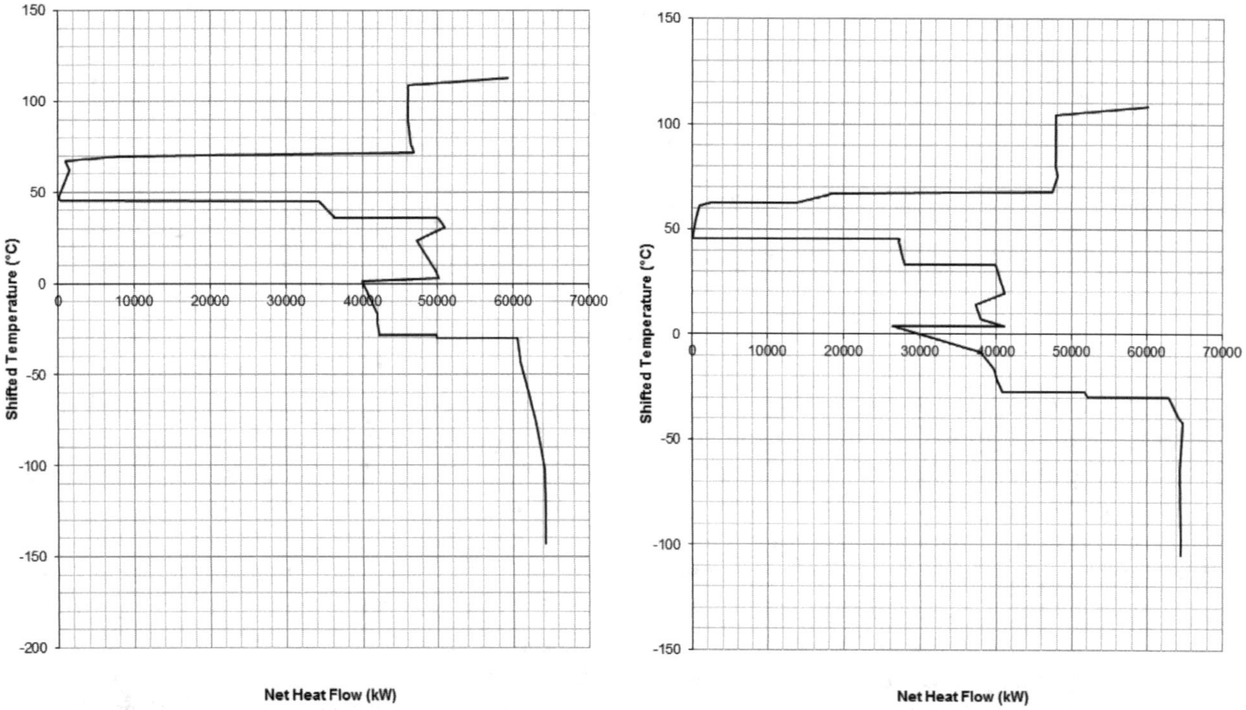

Figure 5: Grand composite curves for the demethanizer first configuration (left) and the absorption configuration (right).

The results of the operational expenditure (OPEX) calculations are given in Table 4. The required steam duty is nearly identical for both configurations resulting in a small difference in associated cost for steam. The total required mechanical duty in the absorption configuration is larger compared to the demethanizer first configuration. This is primarily caused by the larger compressor duty in the refrigeration section for the absorption configuration, even though only a closed propene cycle is installed with a single multistage compressor. The compressor duty of the separation section is higher for the demethanizer first configuration since the total feed is compressed to the highest pressure while in the absorption configuration, only the C2+ stream and part of the C3 stream are compressed. Based on operational expenditure considerations, the demethanizer first configuration performs slightly better than the absorption configuration.

Table 4: Operational expenditure for the demethanizer first configuration and the absorption configuration.

		Demethanizer first	**Absorption**
Mechanical			
Compressor duty separation	MW	2.15	1.67
Compressor duty refrigeration	MW	14.68	17.70
Pump duty	MW	0.98	1.04
Total mechanical duty	MW	17.81	20.42
Associated cost	M$/y	8.99	10.31
Steam			
Required duty	MW	71.32	73.37
Associated cost	M$/y	12.02	12.36
Total OPEX	M$/y	21.01	22.67

The results of the capital expenditure estimates are given in Table 5. The demethanizer first configuration is overall more expensive than the absorption configuration by about 3.1 M$, which is well below the uncertainty of the calculated capital expenditure.

In terms of the separation section, the major cause for the higher total cost for the demethanizer first configuration is the higher cost for the distillation columns. Although the number of columns is smaller the total cost is higher. The demethanizer column in itself is more expensive that the two-column alternative of precutting column and absorber column in the absorption configuration. The costs for the remainder of the equipment in the separation section of the demethanizer first configuration is comparable to the equivalent cost in the absorption section.

In terms of the refrigeration section, the cost of two multistage compressors in the refrigeration section causes the higher total cost for the demethanizer first configuration compared to the absorption configuration. Although the total cost of the compressors in the refrigeration section of the demethanizer first configuration is larger than the equivalent cost in the absorption configuration, the total duty of those compressors in the demethanizer first configuration is lower than the duty in the absorption configuration.

Table 5: Capital expenditure for the demethaniser first configuration and the absorption configuration [M$].

	Demethanizer first		**Absorption**	
	Separation	**Refrigeration**	**Separation**	**Refrigeration**
Distillation columns	33.69	0.00	31.33	0.00
Vertical vessels	1.06	0.54	0.62	0.39
Horizontal vessels	0.58	0.00	0.42	0.00
Centrifugal compressors	1.93	10.44	3.05	8.99
Turbo-expander	0.11	0.00	0.00	0.00
Centrifugal pumps	0.01	0.00	0.02	0.00
Shell & tube exchangers	10.13	2.80	10.33	3.02
Reboilers	1.37	0.00	1.36	0.00
Total	48.88	13.78	47.13	12.40
Grand Total	62.67		59.53	

4. Conclusions

Two configurations for the back-end of a methanol-to-olefins plant have been compared: the traditional demethanizer first configuration based on cryogenic separation and the absorption configuration. In the absorption configuration, having to provide cooling duty at temperatures below -40 °C is avoided by replacing the demethanizer column by a combination of a precutting column and an absorber column. This has the distinct advantage that only a closed loop propene refrigeration cycle is to be installed in contrast to the additional closed loop ethene cycle which is required for cryogenic separation.

For both configurations, the actual separation section and the refrigeration section have been modeled using Aspen Plus V7.3. The results are compared at three levels: heat integration potential, operational expenditure (OPEX) and capital expenditure (CAPEX). Upon comparison of both configuration with respect to heat integration, no distinctly favorable configuration could be identified. For both setups, the required hot and cold utility duties are quasi-identical in the case without heat integration and in the case with full heat integration. With respect to operational expenditure (OPEX), the absorption configuration is slightly more expensive than the demethanizer first configuration due to the higher compressor duty required in the refrigeration section. In terms of capital expenditure (CAPEX), the demethanizer first configuration is slightly more expensive than the absorption configuration due to the higher cost for the distillation columns.

Overall, the differences in heat integration potential, operational expenditure and capital expenditure are small. From an operating cost standpoint no clear winner can be identified. Other factors including sensitivity to variations in the feed, operational experience within a company, licensing costs could induce a preference of one scheme above the other.

5. References

1. CB&I Investor / Analyst Day. http://sec.edgar-online.com/chicago-bridge-iron-co-n-v/8-k-current-report-filing/2011/11/09/section6.aspx
2. Zimmermann, H., Propene. In *Ullmann's Encyclopedia of Industrial Chemistry*, Wiley-VCH Verlag GmbH & Co. KGaA: 2000.
3. Kempf, R. In *Advanced MTO: Breakthrough Technology for the Profitable Production of Light Olefins*, Middle East Downstream Week, Abu Dhabi, United Arab Emirates, 2013.
4. Senetar, J. J.; Romers, E., Scale-Up Of Advanced MTO Technology And Integrated OCP Technology. In *AIChE Spring National Meeting*, Chicago, IL, USA, 2011.
5. Zimmermann, H.; Walzl, R., Ethylene. In *Ullmann's Encyclopedia of Industrial Chemistry*, Wiley-VCH Verlag GmbH & Co. KGaA: 2000.
6. Mafi, M.; Amidpour, M.; Mousavi Naeynian, S. M., Comparison of low temperature mixed refrigerant cycles for separation systems. *International Journal of Energy Research* **2009,** 33, (4), 358-377.
7. Nogal, F. D.; Kim, J.-K.; Perry, S.; Smith, R., Optimal Design of Mixed Refrigerant Cycles. *Industrial & Engineering Chemistry Research* **2008,** 47, (22), 8724-8740.
8. Li, L.; Ni, J. A non-cryogenic separation method for lower hydrocarbon containing light gas. WO 2009012623 A1, 2009.
9. De Haan, S.; Kuzma, P. D. Absorber demethanizer for methanol to olefins process. US 20100105973 A1, 2013.
10. Corradi, J. T. Adsorbent Regeneration in Light Olefin Recovery Process. US 20120083634 A1, 2012.
11. Wison commercializes new methanol-to-olefins technology in China. *Hydrocarbon Processing* 2013.
12. Tallman, M. J.; Eng, C. N. In *Catalytic Routes to Olefins*, AIChE Spring National Meeting, New Orleans, LA, USA, 2008.
13. *NIST Standard Reference Database 103b - NIST ThermoData Engine Version 8.0 - Pure Compounds, Binary Mixtures, Reactions*, National Institute of Science and Technology: 2014.
14. Kemp, I. C., *Pinch Analysis and Process Integration: A User Guide on Process Integration for the Efficient Use of Energy*. Elsevier Science: 2011.
15. *Boiler efficiency guide - facts about firetube boilers and boiler efficiency*; Cleaver Brooks: 2010.
16. Natural Gas Prices. http://www.eia.gov/dnav/ng/ng_pri_sum_dcu_nus_m.htm (09/23/2014),
17. *Electric Power Monthly with Data for June 2014*; U.S. Energy Information Administration: 2014.

386128 Incorporating Measurement Lab Design Projects in an Undergraduate Chemical Engineering Laboratory Course

Abstract:

An initiative toward integrating process and product design concepts throughout the undergraduate chemical engineering curriculum has recently been undertaken at Vanderbilt University. In response to this challenge, an introduction to chemical process and product design concepts as part of the undergraduate laboratory experience has recently been implemented in the chemical engineering curriculum at Vanderbilt University. Specifically, half-semester "measurement lab" design projects have been included over the past two years in the junior chemical engineering laboratory course. The students work in groups on these projects, select their own project topic, design their experiments, collect data, perform analysis and present their results in a poster session at the end of the semester. Measurement lab sensors, analytical equipment, and software used for this project will be discussed. In addition, past example projects and their results will be highlighted. Finally, the impact of this effort on the senior chemical engineering capstone design projects will be discussed.

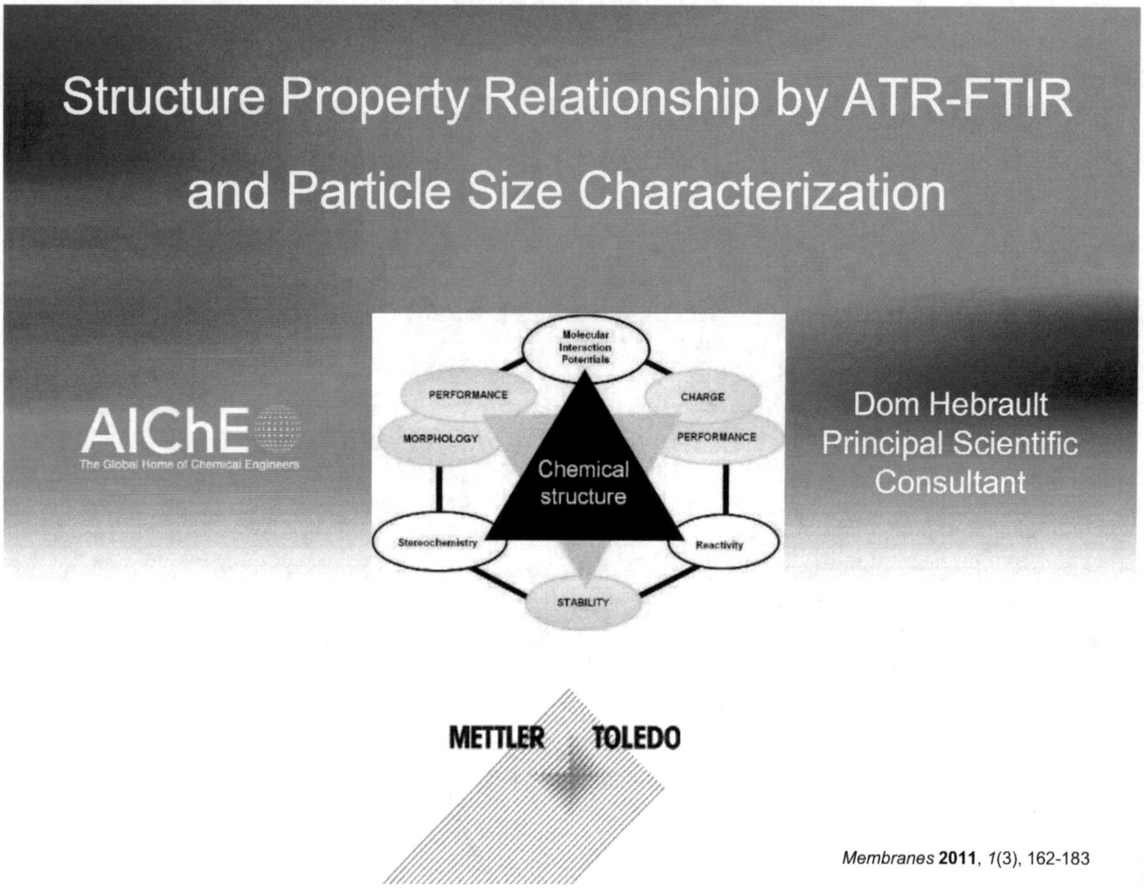
Structure Property Relationship by ATR-FTIR
and Particle Size Characterization
AIChE
The Global Home of Chemical Engineers
Molecular Interaction Potentials
PERFORMANCE
CHARGE
MORPHOLOGY
PERFORMANCE
Chemical structure
Stereochemistry
Reactivity
STABILITY
Dom Hebrault
Principal Scientific Consultant
METTLER TOLEDO
Membranes 2011, 1(3), 162-183

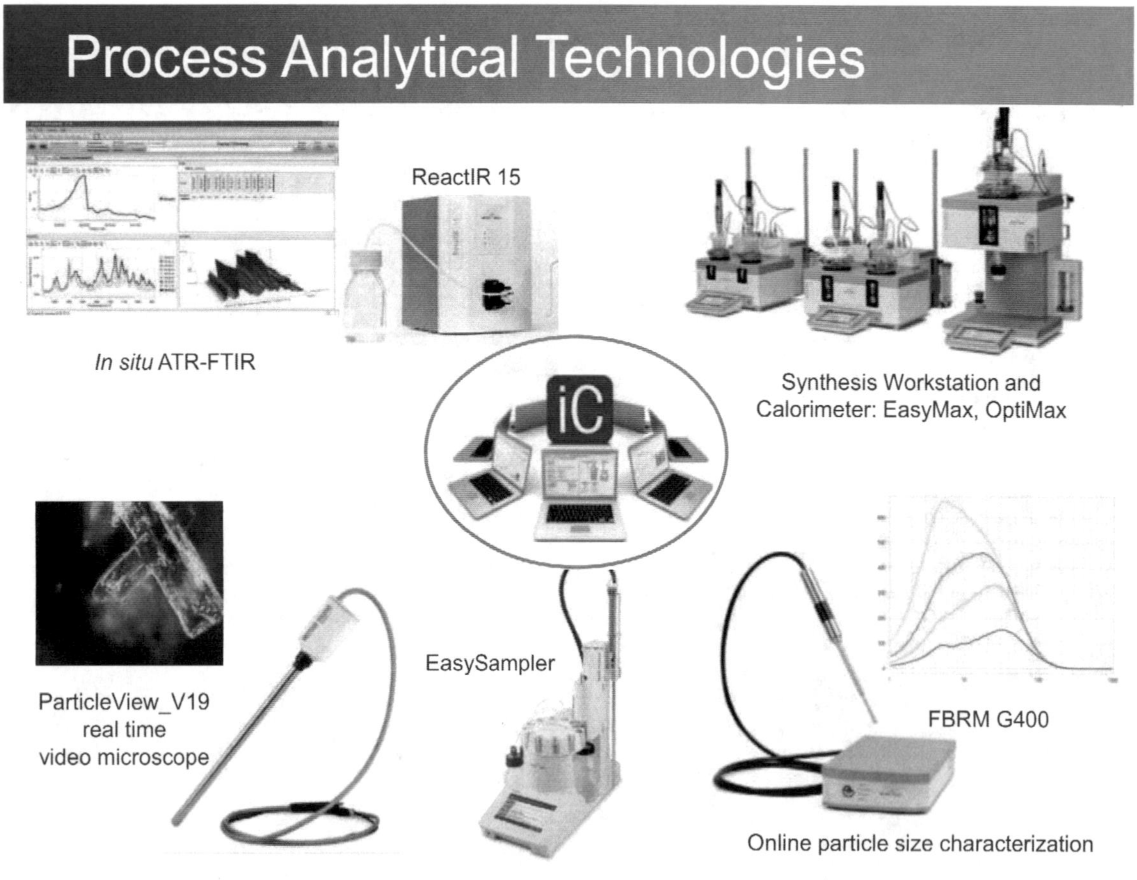
Process Analytical Technologies
ReactIR 15
In situ ATR-FTIR
iC
Synthesis Workstation and Calorimeter: EasyMax, OptiMax
EasySampler
ParticleView_V19
real time
video microscope
FBRM G400
Online particle size characterization

Inline Mid-IR Monitoring

- Monitor chemistry *in situ*, under all reaction conditions
- "Movie" of the reaction
- Determine reaction evolution, outcome and pathway

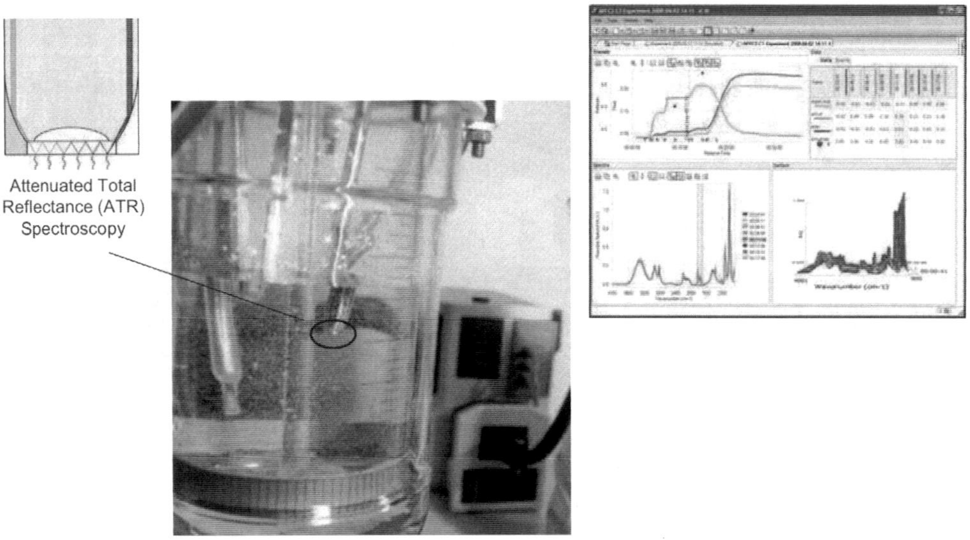

ReactIR Solutions from METTLER TOLEDO

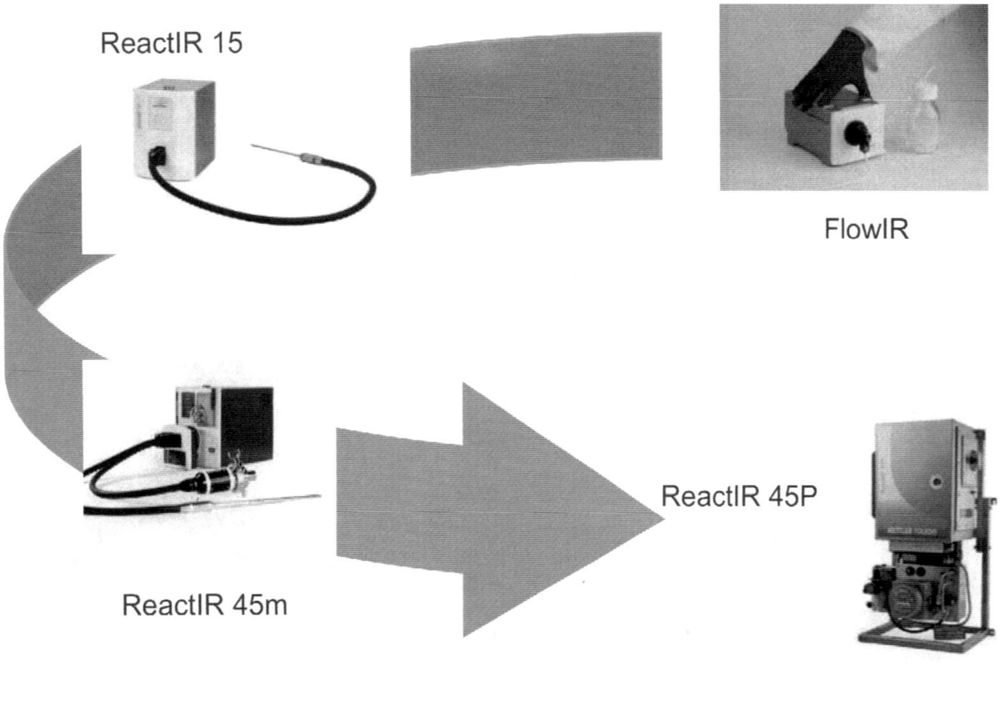

In Situ Particle Characterization Tools

FBRM® Technology
Focused **B**eam **R**eflectance **M**easurement

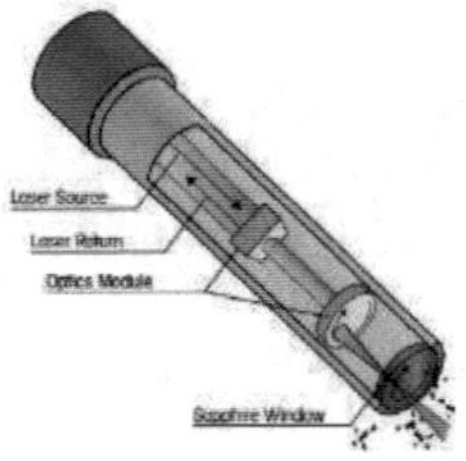

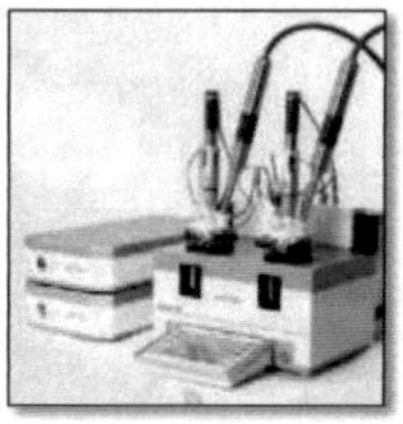

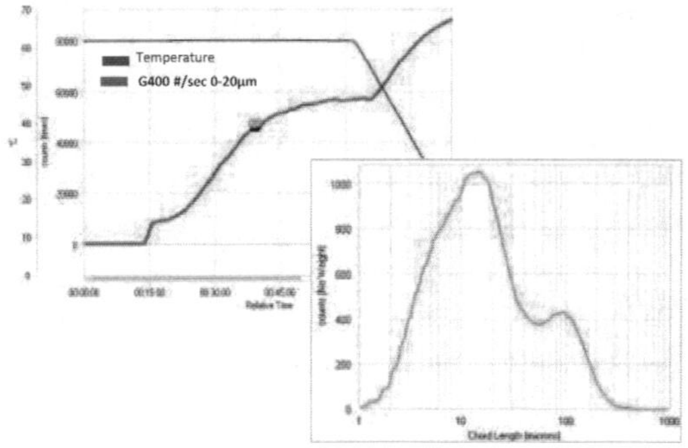

ParticleView™
Particle **V**ision and **M**easurement

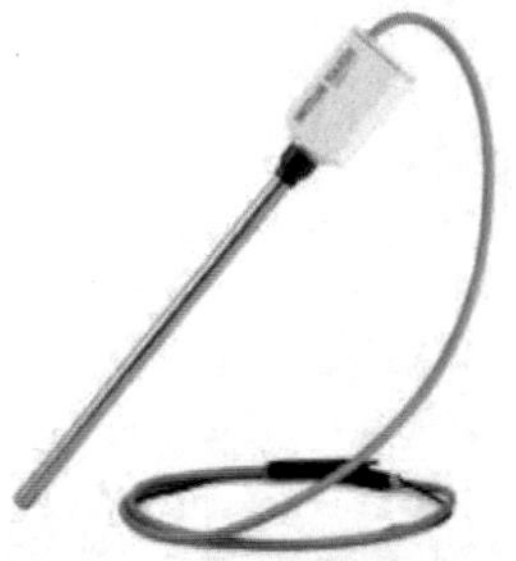

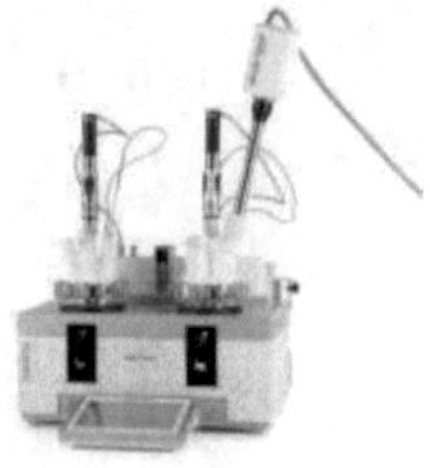

7

D2i or Data to information

iC Kinetics

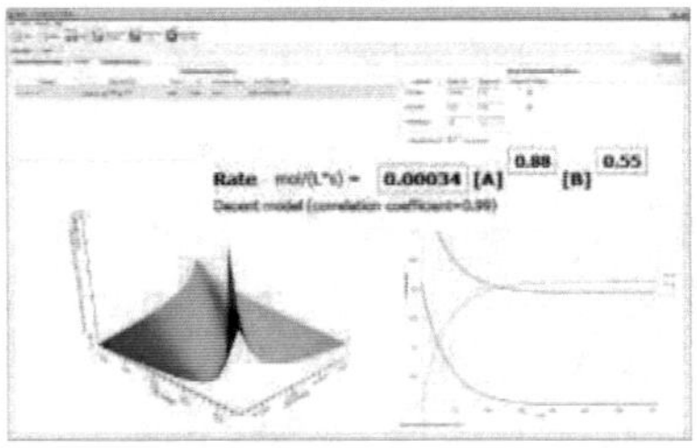

- A fast way to optimize chemistry
- Greater insight with fewer experiments
- Understand process robustness

ConcIRT Pro

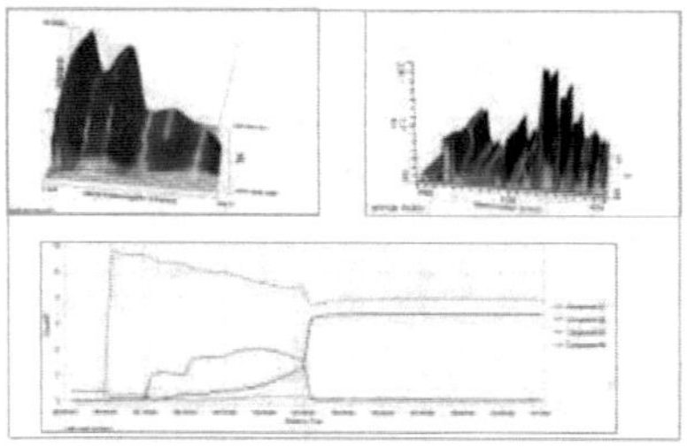

- Accurate trends for kinetics analysis by combining spectroscopy data from two techniques
- Improve process understanding by analyzing different batches simultaneously

iC PVM

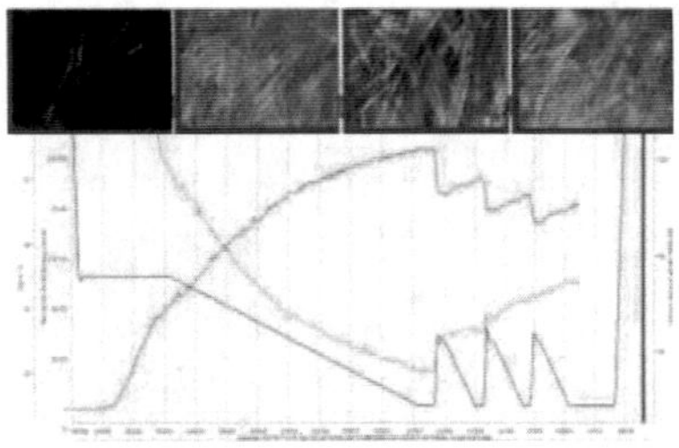

- Image-Based Process Analytics
- Enable simultaneous visual understanding and trending
- Trending based on particle size and concentration

iC Quant

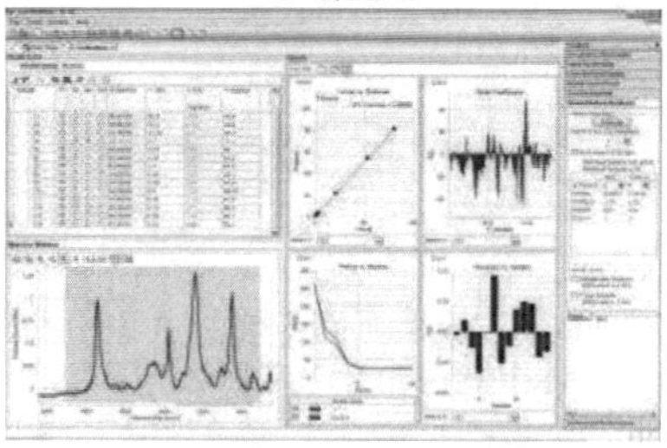

- Develop calibration models
- Predict concentration of key reaction species
- Understand reaction performance

Polymer Characterization Using SEC-FTIR

- Improve the quality of polyethylene (PE)
- Flow-FTIR coupled to GPC to deliver concentration for MW calculations, composition

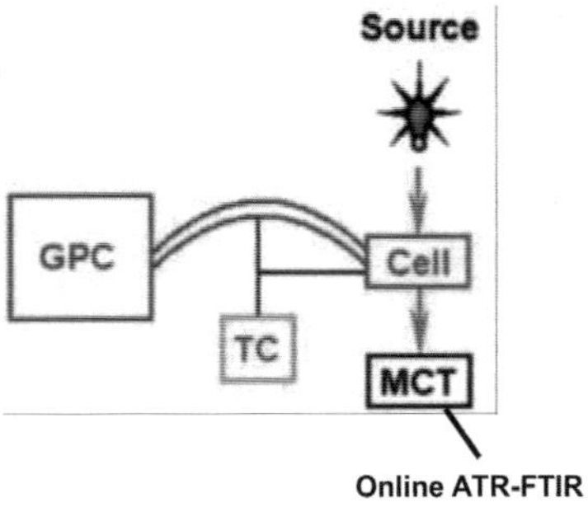

GPC/FTIR interface with heated FTIR flow cell

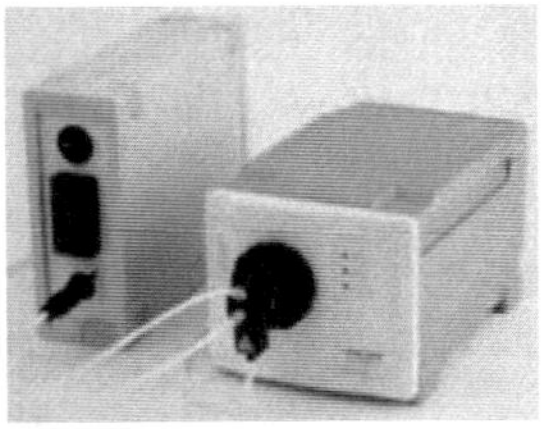

FlowIR Technology

- Measurement of–CH_3 *versus* >CH_2 groups
- Chemometrics-based approach to short-chain branching, SBC (distribution, assay)

Chevron Phillips Chemical Company LLC

Paul J. DesLauriers, D. C. Rohlfing, Chevron Phillips Chemical Company, *Macromolecular Symposia*, 282, 1, 136–149, 2009

Polymer Characterization Using SEC-FTIR

- Level of SCB → PE crystallinity, density, stress crack resistance
- Estimated using chemometrics for level of SBC as a function of MW

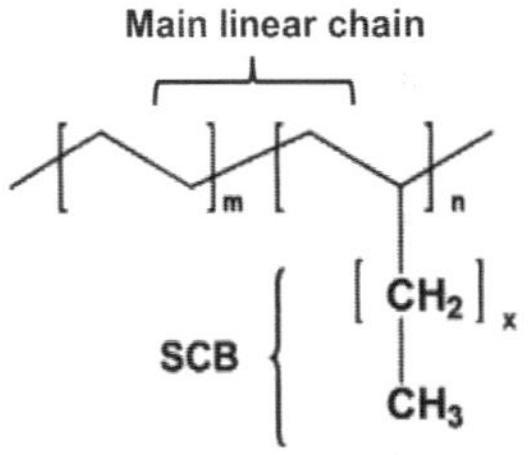

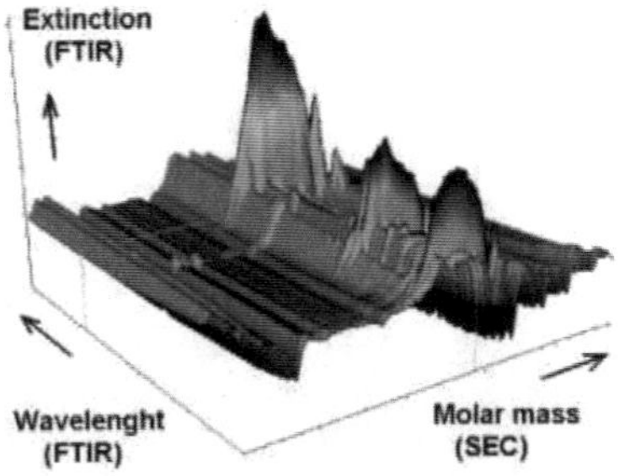

Plot of time-resolved FTIR spectra for PE

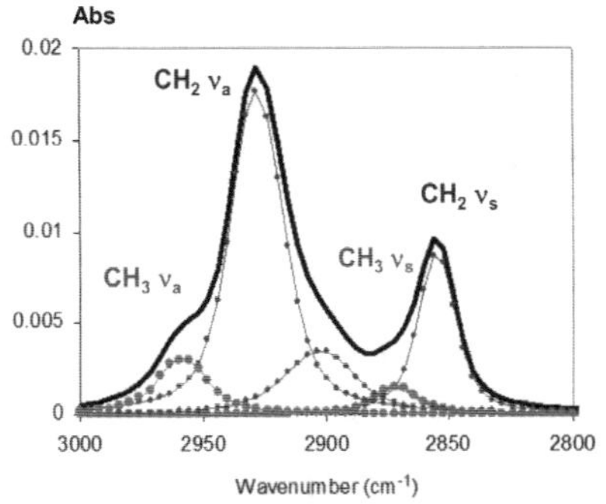

FTIR-based short chain branching (SBC) determination in PE

Chevron Phillips Chemical Company LLC

Paul J. DesLauriers, D. C. Rohlfing, Chevron Phillips Chemical Company, *Macromolecular Symposia*, 282, 1, 136–149, 2009

Catalytic Ethylene Polymerization

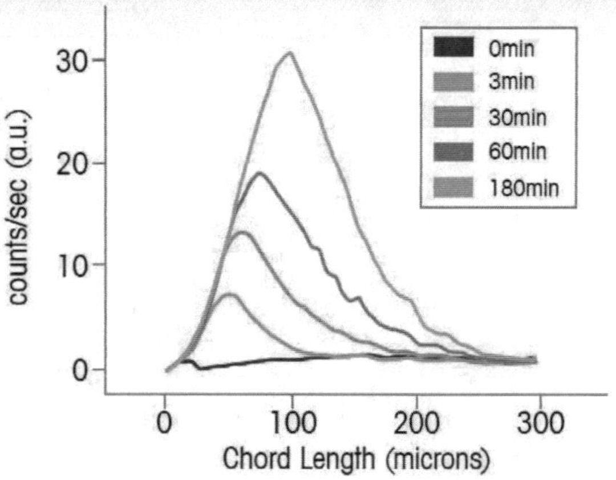

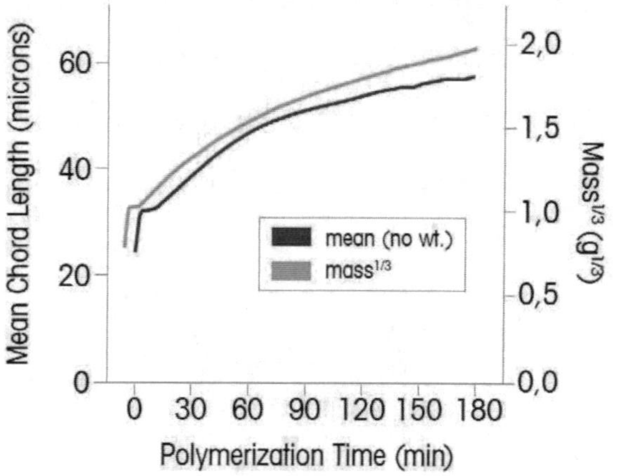

- FBRM, PVM to monitor polymer particle growth in catalytic ethylene slurry polymerizations
- FBRM mean size and mass uptake track polymerization in real-time

Rainer Xalter, Rolf Mülhaupt, FMF Institute, University Freiburg, Germany, *Macromolecular Reaction Engineering*, *4*, 1, 25–39, **2010**

Catalytic Ethylene Polymerization

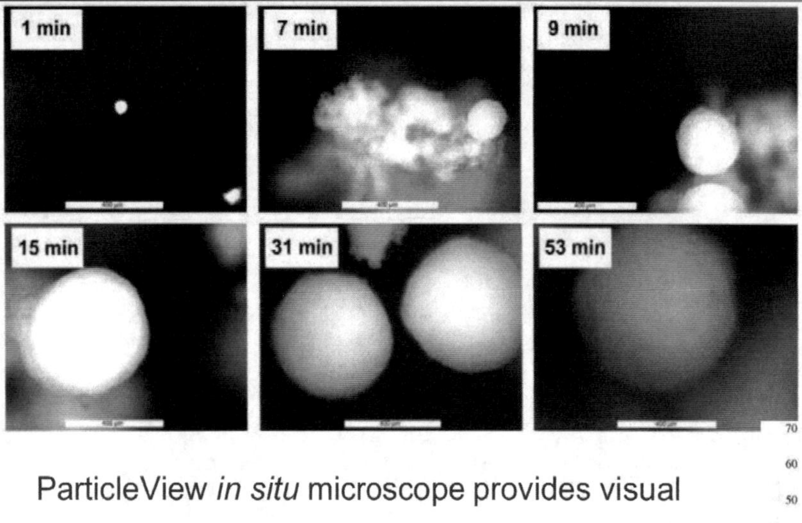

Particle size influences PE physical properties

ParticleView *in situ* microscope provides visual confirmation of particle formation, growth, and little aggregation

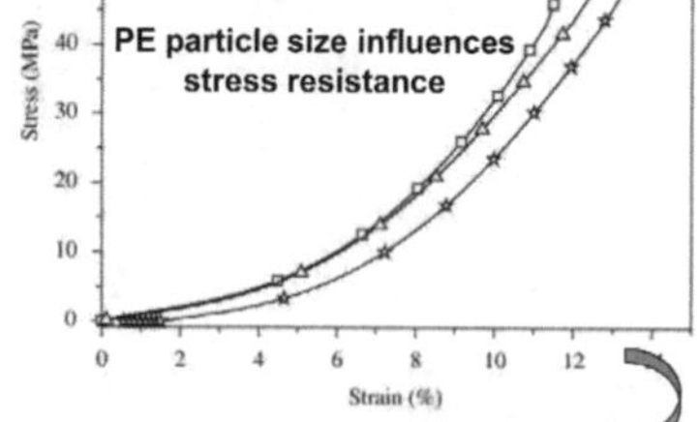

Gean Vitor Salmorial et.al; *Mat. Res.* 10, *2*, São Carlos Apr./June **2007**

Polyurethane Production

Polyurethane production is complex

1. Exothermic reaction of polyol and isocyanate → polymer matrix
2. Blowing agent, in the matrix, converts liquid droplets into gas bubbles

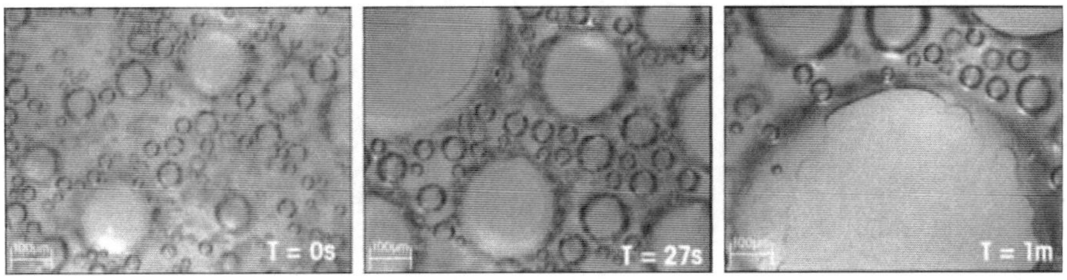

ParticleView *in situ* images

Coalescence < 1'; Unstable emulsion → Variation in foam size → Final product quality

<u>ParticleView</u>

- Monitor droplet distribution
- Screen surfactants
- Ensure more consistent and stable emulsion

Polyurethane Production

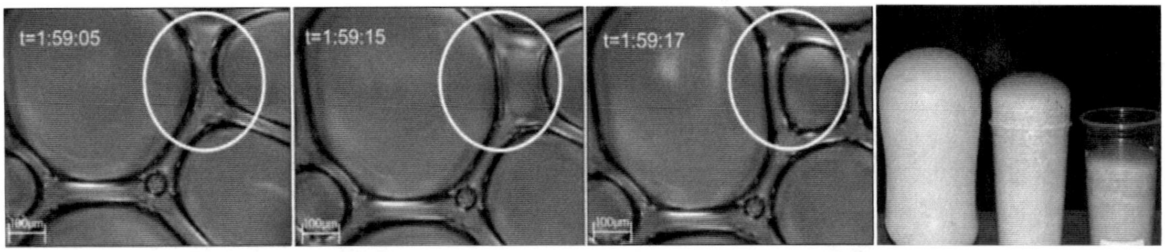

Spherical droplets → Non-spherical polyhedra (polymer struts), and trapped gas bubbles

ParticleView: Real-time monitoring of fast (seconds) bubble formation within polymer strut

Off-line analysis cannot capture fast dynamics and causes partial structure destruction

ParticleView provides a better understanding through *in situ* visual observation of the process with no sampling involved.

Acknowledgements

- James Rawlins and team, Mississippi Polymer Institute, University of Southern Mississippi, Hattiesburg, MS
- Paul J. DesLauriers, Chevron Phillips Chemical Company, Bartlesville, OK
- Rainer Xalter, Rolf Mülhaupt, FMF Institute, University Freiburg, Germany
- My colleagues in the reaction engineering team at Metter Toledo, Columbia, MD
- And YOU!

387597 Designing Products and Processes for Sustainability Using Greenscope: Application and Education

Abstract:

The U.S. Environmental Protection Agency has developed a beta software tool called GREENSCOPE (Gauging Reaction Effectiveness for the Environmental Sustainability of Chemistries with a multi-Objective Process Evaluator) to support sustainable processes and products. Evaluations of existing and new processes can be accomplished using GREENSCOPE with 140 indicators for objectives in four basis areas: environment, economics, energy, and efficiency. GREENSCOPE provides information on sustainability for processes, and through product/process decisions, about product sustainability.

Increasing sustainability can be approached through designing more sustainable products and educating those who design products and related facilities. As chemical engineers learn elements of product design and process systems engineering the opportunity exists to introduce sustainability concepts. The EPA's GREENSCOPE methodology and tool can be used to teach these concepts. This work describes example calculations of GREENSCOPE indicators for the oxidation of toluene and puts the indicators into context with best- and worst-case limits. The indicator calculations use data available from the process for the generation of oxidation products, thus transforming data into understandable information which describes sustainability. Targets for more sustainable process designs are understood, and enhancements can be considered to improve designs, either for real-world processes or for educational purposes.

390047 Development of QSPR Model Using 3D Molecular Descriptors Relating Solvent Structure to Crystal Morphology: An Efficiency Study of PCA, GA and Ann Algorithms

Abstract:

The paper outlines the development of a quantitative structure-property relationship (QSPR) that relates solvent structure to the morphology of ibuprofen crystals grown within that solvent. Morphology can be quantified by aspect ratio, and ibuprofen aspect ratio data was obtained for crystals grown in 16 different organic solvents. To provide a quantitative representation of the geometry optimized solvent molecules, 3D molecular descriptors were generated. Three different empirical force fields were used to estimate the three-dimensional structure of the solvent molecules while their effect on the developed models is analyzed. The descriptor data matrix, containing a multitude of descriptor types, is reduced in size, using Principal Component Analysis (PCA), Genetic algorithm (GA), Artificial Neural Network (ANN), and also combination of such algorithms for regression into linear models. The Singular Value Decomposition (SVD) procedure was used for PCA due to its increased stability in primary component prediction. The predictive capabilities of these models were also analyzed through means of internal and external validation methods.

391439 Advancements in Computer Aided Design of Task Specific Ionic Liquids

Abstract:

This presentation provides details on recent advancements related to computer-aided ionic liquid design method. In particular, consideration of environmental, ecological and health impacts along with thermophysical properties within the design framework is discussed. The proposed methodology is able to screen a variety of cations, anions and alkyl groups to arrive at optimal ionic liquid structures for a given application. Group contribution methods are used for ionic liquid property prediction while optimization methods are used to screen all possible structure. The designed ionic liquids are not only optimal for the given application but also have reduced environmental burden. The utility of the method is demonstrated through interesting case studies.

390590 Simulation and Optimization of Ethyl Benzene Production Using Aspen Plus Platform

Abstract:

Ethyl benzene is an aromatic hydrocarbon and is also important in the petrochemical industry and as an intermediate in the production of styrene, which is utilized for making polystyrene, it is a common plastic material. The study deals with optimization of the process for the production of ethyl benzene by liquid-phase benzene alkylation. This process involves the reaction of benzene with ethylene to form ethyl benzene. Ethylene reacts with ethyl benzene to form undesired product di-ethyl benzene, if the temperatures of a reactor or concentrations of ethylene are high. Di-ethyl benzene reacts with benzene to form ethyl benzene. Di-ethyl benzene is the highest-boiling component in the system; it comes out the bottom of two distillation columns. The recycling benzene is more expensive. The economic optimum steady-state design is developed that minimizes total annual cost. Therefore, it offers a classic case of an engineering invention and optimization of a procedure. The purpose of this study is to develop an optimum design for the ethyl benzene process considering the reactor size and the amount of benzene recycled. Optimization of the process led to an increase in the yield of ethyl benzene by 0.8832%.

356991 Effects of Synthetic Polymers on Mechanical Strength and Water Absorption of Hardened Gypsum Pieces Made from Calcium Sulfate Hemihydrates

Abstract:

α- or β-Calcium sulfate hemihydrate (α-H or β-H) is commonly used as gypsum-modeling material for ceramics, which require high mechanical strength and water absorption. Polymers may affect the gypsum performance and ratio of water to hemihydrate (W/H). In this experiment, the effect of W/H, the addition and structure of some synthetic polymers on the properties of hardened gypsum piece were investigated. It was found that α-H and β-H are hardly workable in common W/H region; their performances are strictly relevant to W/H. At a W/H as high as 0.7, the more hydrophilic polymers provided an increase in the mechanical strength and water absorption; four sulphur polycarboxylates synthesized were found to maintain or enhance water absorption of β-H, but only the one without carboxyl group led to an increase of the mechanical strength. The results suggest that keeping sulphur and PVA residues in the molecular design will be necessary for polymer structure applied in gypsum modeling material.

Catacarb Carry over incident causing gigantic production loss and Fire Incident

Muhammad Hashim, Fatima Fertilizer Company Ltd. Sadiqabad Pakistan

Abstract

From 31st August to 15th September 2012 Fatima Fertilizer Ammonia Plant faced massive Catacarb foaming in Towers subsequently solution carry over to Methanator 4 times during this period causing gigantic Production Loss and downstream Expander System Fire Incident. This paper describes how the problem was analyzed, the sequence of events, which tests were performed after consulting with Vendor and how the problem was eventually solved.

Introduction

Fatima Fertilizer Ammonia Plant is designed by CF Braun & Co. Alhambra California built in 1967. This plant owned by Exxon and Kemira in 1985. Fatima took over this plant from Netherlands to Pakistan in 2007 that started successful production in 15th March 2010 with design capacity 1500 MTPD. This paper encourages you to answer following questions;

1. Importance of activated carbon filtration how badly it can hit the Catacarb solution if not in service during normal operation.
2. Importance of Color index measurement besides of other tests in Catacarb Analysis.
3. Methantor trip Logic Modification in order to avoid Methantor run away reaction
4. Expander trip logic modification to avoid oil console over pressurization / Fire Incident

Incident Description

Ammonia Plant was in restart after 10 days planned shutdown, while restarting the ammonia plant from 29th August 2012, various problems were encountered including foaming in Catacarb causing excessive carryover of solution repeatedly to Methantor, Methantor run away reaction and Cold box Expander process gas breakthrough from HP drain to oil console causing huge fire incident. Issues were resolved accordingly and Production resumed on 15th Sept. 2012.

Sequence of Event:

31st August at 1730 hrs; Catacarb solution carried over with process gas to downstream exchangers and Methantor R-104 due to foaming in CO2 absorber. R-104 was immediately isolated, depressurized and drained.

1st September at 1430 hrs; Catacarb solution carryover was again observed due to foaming causing high level of CO2 absorber downstream knock out vessel D-133.

2nd September at 1030 hrs; Due to poor performance of Methantor feed effluent exchanger E-120A/B. Flushing of E-120A/B was carried out with turbine condensate.

3rd September at 1540 hrs; Even with Excessive shock dozing of antifoam in catacarb foaming could not be controlled and Methantor bed temperature ran away to maximum value on first bed TI-025B 516 °C due to high CO_2 slip from absorber. Methantor trip security I-6 was manually actuated due to high temperatures of Methantor.
4th September at 0915 hrs; Cold box expander located at downstream of Methantor; Huge Fire was occurred at C-103 oil console due to oil splashing on hot steam lines caused by over-pressurization of oil console due to process gas at -140°C entrance into sour oil HP drain line because C-103 was not isolated timely from field during emergency trip of Methantor because of catacarb carry over again Consequently Expander seal, some instrument cables and instruments around C-103 were damaged which were replaced.
12th September at 1850 hrs; Catacarb foaming still not came in control even taking all control measures air injection, reducing circulation rates, increasing anti foam doze etc. Vendor E&A was also consulted but in vain so it was decided to replace 40% of the total catacarb charge.
15th September at 2330 hrs; Ammonia production was resumed under low load operation with controlled foaming and bearable flooding.

Root Cause analysis of Catacarb Foaming and Expander FIRE

1. **Solution foaming tendency:**

   Fatima CATACARB system was equipped with a carbon filter to treat around 30 m3/hr of solution continuously. The best defense against foaming was extensive carbon and mechanical filtration. Fatima had kept the activated carbon filter idle since start up based on their Lab Results. Without carbon filtration, the solution foaming tendency had been rising; Organic contaminants gradually accumulated in the solution, for example: derivatives of heavier hydrocarbons, amine degradation products, glycols from antifoam, alcohols from shift reactors etc. They reduced solution surface tension and increase foaming tendency.

2. **Passivation**

   Fatima did passivate the system during startup by circulating the solution for 41 hours. Due to lack of accurate available inhibitor information from lab, the effectiveness of passivation was unknown. Fortunately, foaming from high iron and suspended solids alerted the operation to increase air injection and later performed another passivation. Consequently, the system regained corrosion protection with iron dropped to 60-70 ppm. Otherwise, if corrosion proceeds, it not only creates heavy foam but also may damage the semi-lean tower.

3. **Local Corrosion:**

   Review of operating procedures and solution chemistry before shutdown did not show any abnormality however local corrosion might be resulted from bed churning in semi-lean regenerator during regeneration due to excessive liquid level. Since no solution was returned to the storage, excessive liquid level in the semi-lean regenerator submerged the reboiler vapor return nozzles and caused water hammering on packed beds and packing. Through violent scratching and abrasion against carbon steel vessel walls and among carbon steel rings, passivation films breached and minor corrosion starts. Lab results clearly indicated high iron and suspended solids. However, the analytical method for available inhibitor was interfered by dark solution color and did not detect any difference.

4. **Suspended solids:**

Once corrosion happens, Fe+2 will fast buildup in the solution. Through oxidation, Fe+2 is converted to Fe+3 and precipitated from the solution as iron oxides.

4.1 Laboratory sample results were misguided the process team when sample were sent to nearby industries then Suspended solids were reported 1560 ppm against the recommended value <100 ppm while Fatima Lab was reporting 119 ppm during foaming incident due to difference in analytical method.

4.2 Color index method was not being in used since commissioning as it was not mentioned in Vendor analysis guidelines and during foaming incident when color index analysis were carried out it reached to 12800 Hz

5. **Foaming tendency by Shake Test:**

Foam shake test help operators to develop close understanding of frothiness of solution but this technique was not being used. Only foaming tendency was being checked by sparger test. Shake test usually offers advance warning against pending problems.

Laboratory Results:

Test	Catacarb Results Comparison (Engro / Fatima)								
Date: 5th Sep		Catacarb Lean T-101			Catacarb Semi-Lean T-102			Engro Semi-Lean Tower	
	Date	by Fatima	By Engro		by Fatima	By Engro		Engro Results	
Time (Hrs)		0500	Value	unit	0500	Value	UOM	03-09-2012	05-09-2012
K_2CO_3	%	19.5			14.3				
$KHCO_3$	%	8.4			18.0				
Equivalent K_2CO_3	%	25.3	24.27	CC	26.7	26.0	%	26.8	26.2
V^{+5}	%	0.39	0.29	%	0.30	0.29	%	0.4	
V^{+4}	%	0.01	0.09	%	0.02	0.08	%	0.04	
Iron Total	ppm	*68	64	ppm	*91	62.01	ppm	53.2	
Amine	%	0.83	0.98	%	0.90	1.05	%		1.1
Chlorides	ppm		22	ppm		24.0	ppm		
Formate	N		1.39	%		1.49	%	2.5	2.5
Foaming(Height)	ml	Nil	4	cm	Nil	4.0	cm		2.5
Foaming(Break Time)	Sec	Nil	9.1	sec	Nil	9.6	sec		5.0
pH								10.1	10.0
Color	Hazen	22820	1.192	Abs	23360	1.313	Abs	0.365	0.366
S.Solids	ppm	119	1240	ppm	166	1560.0	ppm	42.0	

6. **Methantor Trip Logic:**

Due to Catacarb carry over during flooding from CO2 Absorber CO/CO2 slippages increased temperature and Methantor trip logic was built in only on high temperatures but no tripping upon high level of upstream KO drum as per latest design practices.

7. **Expander Trip Logic:**

Upon Methantor tripping there was no built in trip logic of Expander C-103 tripping. Once Methantor tripped seal gas to Expander was cut, It was operator actions to trip C-103 from DCS and close expander inlet outlet valves from field in order to avoid inadvertent mixing of oil and cold process gas. As its mixing could cause freezing and cold process gas ingress into oil console system could cause console over pressurization and Fire Incident.

Root Cause:

1. Different methods for analyzing Total Vanadium and Suspended solids misguided the whole TEAM when samples compared with the nearby industries then came to know the BIG difference between both results.
2. Putting Carbon filtration system out of service since commissioning thinking no need of it till the time laboratory results are normal whereas mechanical filtration system remained in service.
3. No built in Methantor Trip Logic upon high level of Absorber KO drum caused Runaway reaction
4. No built in Expander Trip Logic upon Methantor trip caused oil freezing in Expander and delay in operators action to isolate C-103 caused FIRE incident.

Corrective Actions:

1. Installation of one more carbon filter and Put both carbon filters into continuous service with an additional upstream mechanical filter of 5-10 microns retention for suspended solids and a downstream filter of 1-2 microns retention for carbon fines. Normal Results of Fe and SS were achieved in late October 2012.
2. Replacement of activated carbons at least once every 6 months.
3. Analyzing available inhibitor according to E&A's newly revised redox titration method in future.
4. Performing shake test on semi-lean sample to get foam time in each shift to get advance warning against foaming issue
5. Monitor iron and keep it below the suggested 100 ppm against possible corrosion by monitoring total Vanadium contents and injecting air as per requirement.
6. Monitor suspended solids and keep it below the recommended 100 ppm against possible corrosion through stringent monitoring and control on Filtration Mechanism
7. Drain solution back to storage if plant outage is more than a week.
8. Regenerate solution after shutdown by keeping reboiler in operation for 2-4 hours so that bicarbonate conversion drops below 20%.
9. Clean the open sump from time to time and provide a SS liner to avoid in-leaks of contaminants in next Turnaround.
10. Modification of Methantor trip logic upon high level of Absorber downstream KO drum high level.
11. Expander C-103 logic modification to automatically trip C-103 on

11.1 DP across seal gas and HP seal oil (PDI-080) < 0.2kg/cm2

11.2 DP across seal gas and expander casing (PDI-081) drops below 0.1kg/cm2.

12. Installation of degasifying tank at sour oil HP drain line to avoid Oil console over pressurization and consequent fire due to higher seal gas venting.

Conclusion

Although the foaming and flooding instance was very costly and inconvenient we had to suffer production loss from 31st August to 12th October partially or completely but it served as a good reminder that CATACARB needs adequate cares to sustain safe operation. Fortunately, there was no physical damage to equipment.

Cold box Operational Problems causing significant Production Loss

Muhammad Hashim, Fatima Fertilizer Company Ltd. Sadiqabad Pakistan

Abstract

From 23rd March to April 2014 Fatima Fertilizer Cold box operation became mystery for us, We faced cold box frequent transition from Good mode to Bad mode operation with unknown reasons causing significant Production Loss. Investigation on the basis of Why Tree Analysis was carried out and Root cause was identified. This paper describes how the problem was analyzed, the sequence of events, which tests were performed to rule out one by one all issues and how the key factor was eventually solved.

Introduction

The feed gas, containing 3% nitrogen, is cooled to a temperature around -140° C in the first heat exchanger section by heat exchange with the two effluent streams. It then passes through the turbo expander to provide the net refrigeration. The feed is subsequently cooled to around -175°C in second section of the heat exchanger and flows then into the rectifier column where the liquid portion (mainly nitrogen) drops into the bottom and the Vapor flows upward. Liquid from the bottom is flashed at reduced pressure via the Joule-Thompson valve; the resulting drop in temperature provides the driving force in the condenser. Reflux from this condenser washes methane and argon out of the up flowing vapors. As a consequence, the rectifier produces a hydrogen-nitrogen mixture in the ratio three to one, with only little residual argon and methane. This stream and the waste gas stream from the condenser flow upward through both exchangers and cool the feed.

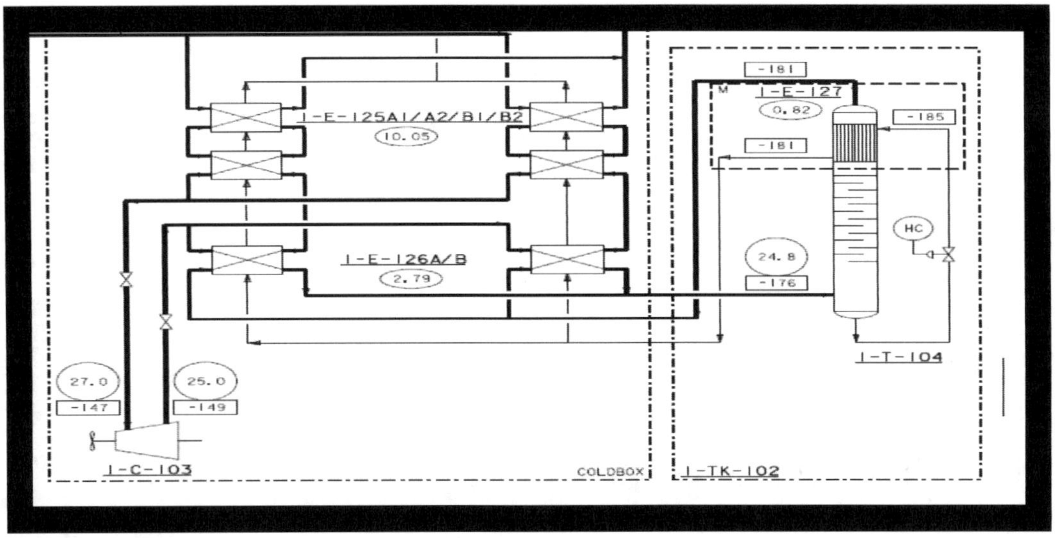

Problem Description

Cold box feed gas inlet path frequent frosting in E-125A2/B2 and bad mode operation in E-126 A/B used to observe each after 1~2 months due to un-known reason resulting in gradual decrease in performance of middle exchangers. Ultimately heat transfer duty of middle exchangers E-125A2/B2 drop below a certain value consequently level in tower (T-104) start depleting rapidly. Due to drop in level it becomes difficult to maintain H/N ratio for Ammonia Synthesis leading to back end shutdown followed by cold box warm up. Production to Production loss is 0.17 Billion after every 1~2 months period.

Sequence of Event

On 23rd March 2013 Delta T across E-125A2/B2 exchangers dropped badly and duty shifted on upper exchangers. Consequently delta P across expander increased and level in tower start depleting sharply, H/N ratio changed from 3:1 to 2.5:1 and ammonia production stopped. Once again Bad Mode situation happened after Turnaround on 27th May 2013 this time temperature profile across both cores varied surprisingly shown in graphs.

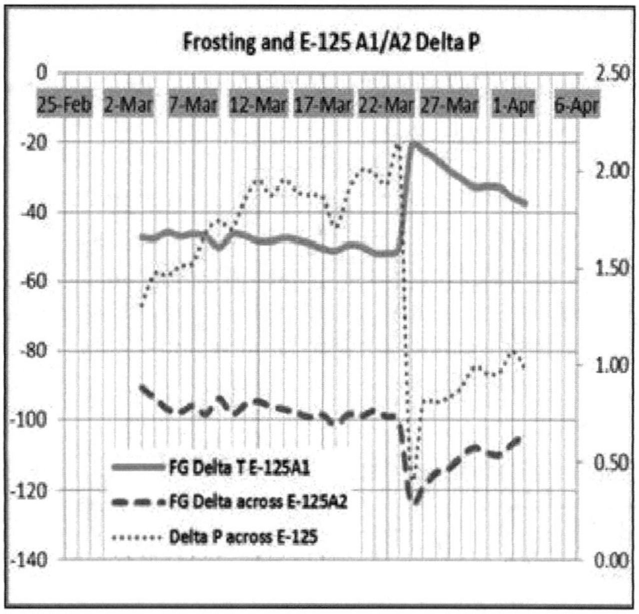

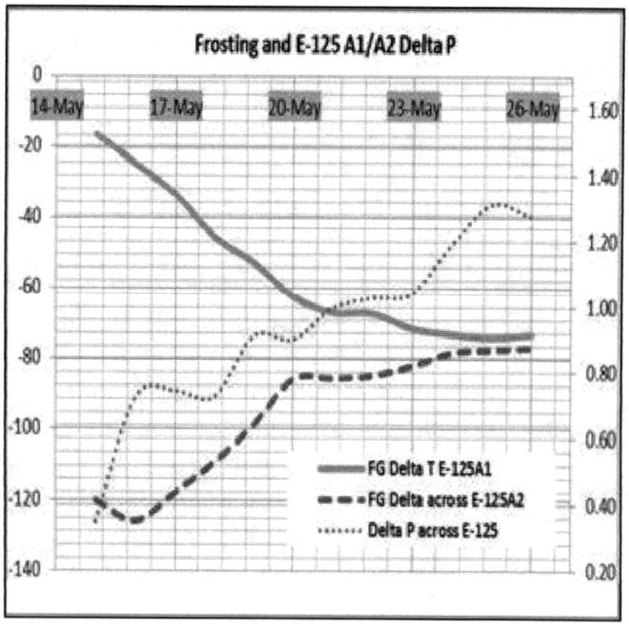

Possible Causes

Possible causes for Exchangers bad mode operations were think through to pin point for addressing the actual root causes. Following causes are noted;

1. Moisture slip from upstream driers section causing frosting in Cold box
2. Ammonia slip from upstream driers section causing frosting in Cold box
3. Carbon dioxide slip from upstream driers section causing frosting in Cold box

4. Driers desiccant dust from filters FIL-104A/B outlet
5. C-103 Expander oil film in Cold box exchangers E-126A/B
6. Waste gas Mal distribution from E-127 to E-126 A/B
7. Mechanical Failure of Cold Box Exchangers

Why Tree Analysis

Based upon above possible causes Why Tree Analysis was carried out and root cause was find out.

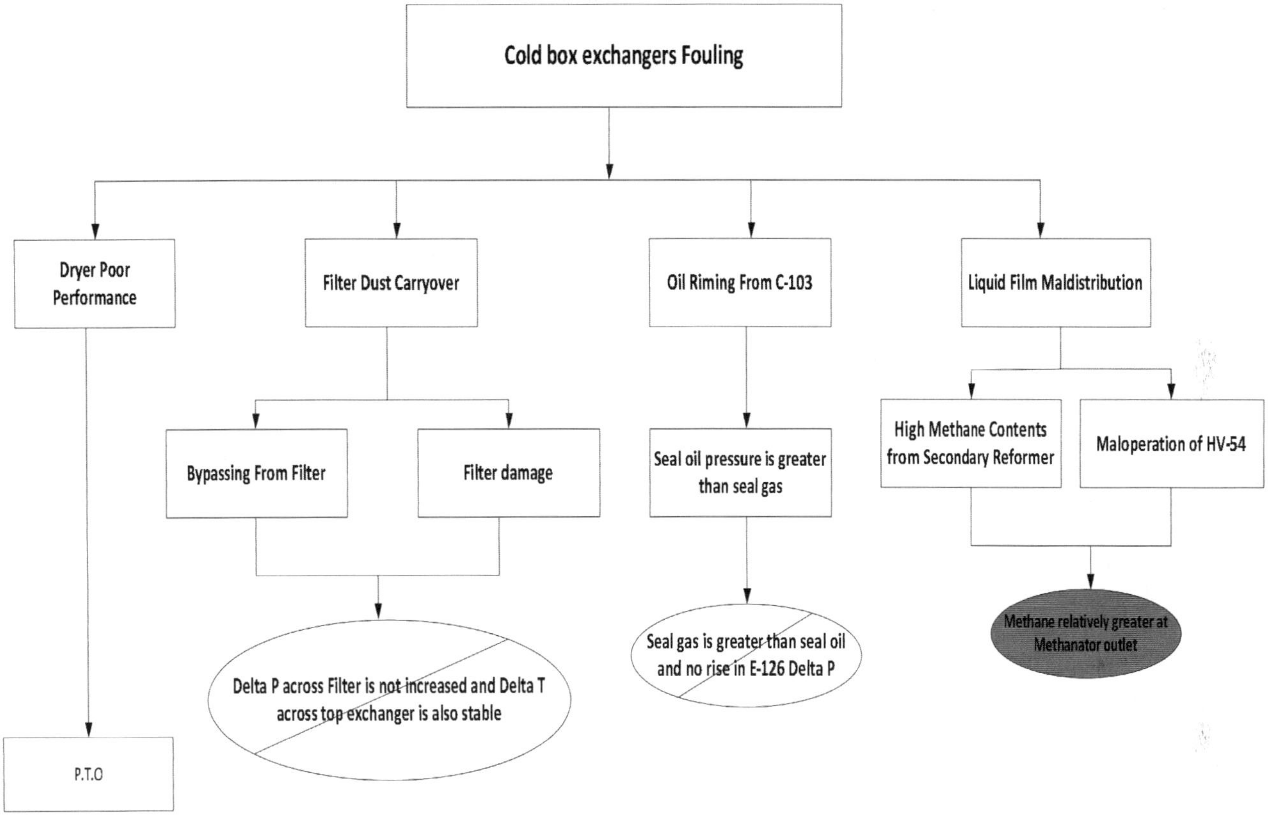

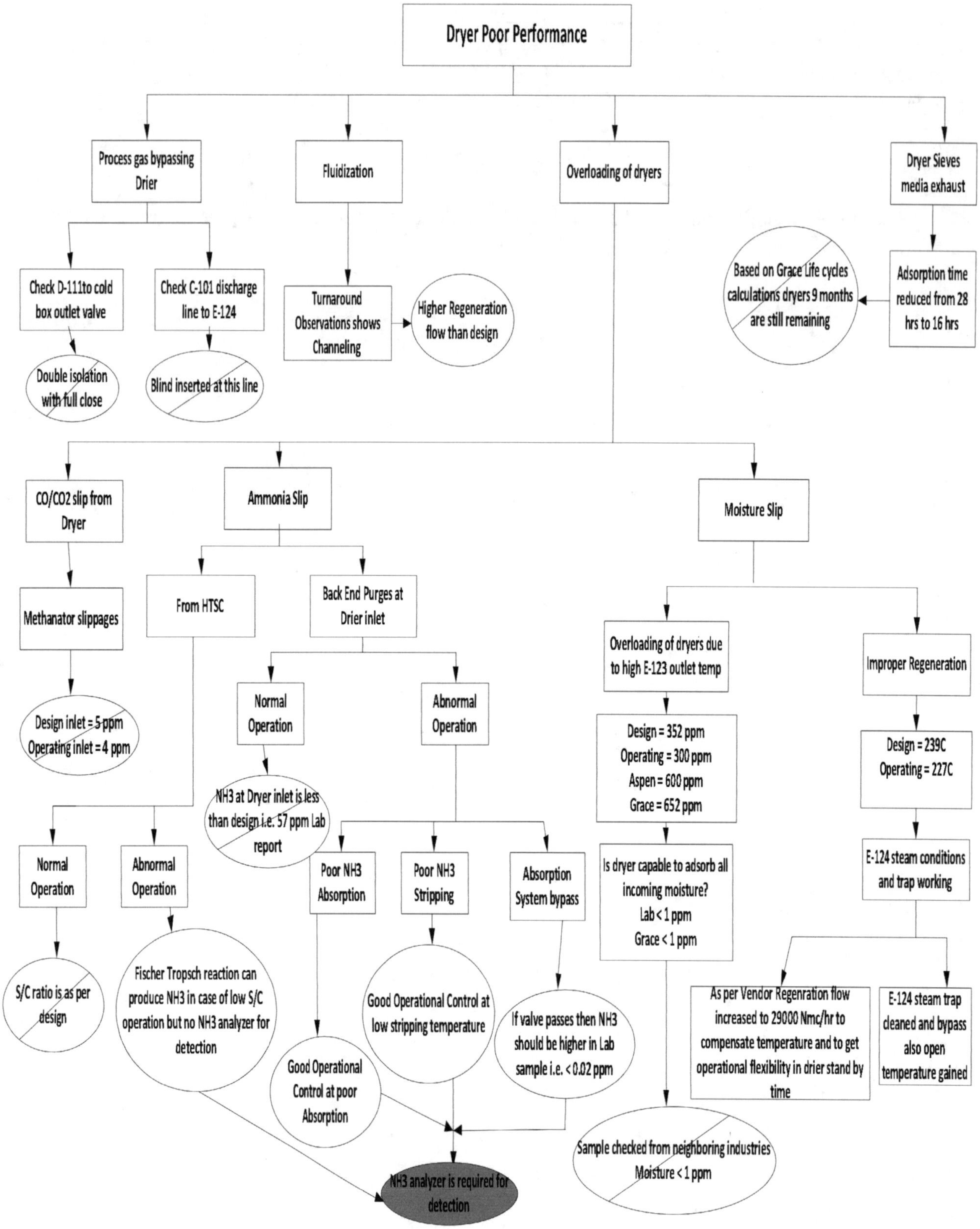
Dryer Poor Performance
Process gas bypassing Drier
Fluidization
Overloading of dryers
Dryer Sieves media exhaust
Check D-111to cold box outlet valve
Check C-101 discharge line to E-124
Turnaround Observations shows Channeling
Higher Regeneration flow than design
Based on Grace Life cycles calculations dryers 9 months are still remaining
Adsorption time reduced from 28 hrs to 16 hrs
Double isolation with full close
Blind inserted at this line
CO/CO2 slip from Dryer
Ammonia Slip
Moisture Slip
Methanator slippages
From HTSC
Back End Purges at Drier inlet
Overloading of dryers due to high E-123 outlet temp
Improper Regeneration
Design inlet = 5 ppm
Operating inlet = 4 ppm
Normal Operation
Abnormal Operation
Design = 352 ppm
Operating = 300 ppm
Aspen = 600 ppm
Grace = 652 ppm
Design = 239C
Operating = 227C
NH3 at Dryer inlet is less than design i.e. 57 ppm Lab report
Normal Operation
Abnormal Operation
Poor NH3 Absorption
Poor NH3 Stripping
Absorption System bypass
Is dryer capable to adsorb all incoming moisture?
Lab < 1 ppm
Grace < 1 ppm
E-124 steam conditions and trap working
S/C ratio is as per design
Fischer Tropsch reaction can produce NH3 in case of low S/C operation but no NH3 analyzer for detection
Good Operational Control at low stripping temperature
If valve passes then NH3 should be higher in Lab sample i.e. < 0.02 ppm
As per Vendor Regenration flow increased to 29000 Nmc/hr to compensate temperature and to get operational flexibility in drier stand by time
E-124 steam trap cleaned and bypass also open temperature gained
Good Operational Control at poor Absorption
Sample checked from neighboring industries
Moisture < 1 ppm
NH3 analyzer is required for detection

Recommendations related to Operational Control

Based upon above RCA following recommendations were extracted that derived the whole plan.

1. Heat of vaporization of methane is 125Cal/gr @ -80C vs. 37 Cal/gr @ -80C for N2. so Keep Secondary Reformer Gas to Air ratio in such a way that Methane slip remain within design limit. Lower methane slip is beneficial for cold box.
2. Operate Cold box with waste gas pressure as low as possible to achieve maximum vaporization from Cold box Exchangers.
3. Drier regeneration gas flow should be kept as per design 20000 NMc/hr rather to operate at 29000 NMc/hr during cooling cycle to avoid channeling in Drier.
4. Drier End of Run sampling frequency regime for NH_3 analysis and cross verification on Moisture Analyzer.
5. ΔP across Driers outlet Filters monitoring to avoid dust fusion in cold box
6. Operating seal gas pressure greater than seal oil.

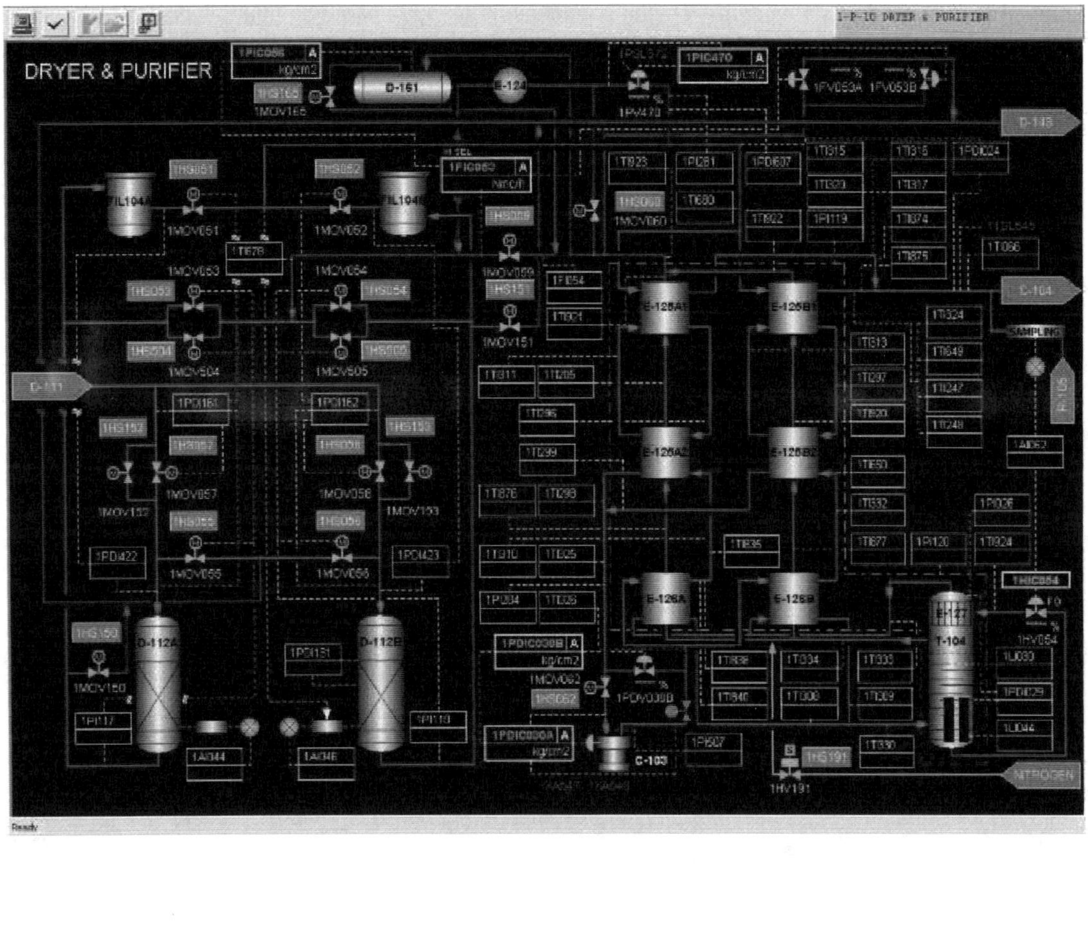

Recommendations related to Engineering Control

Following suggestions are proposed for Drier and Adsorbent Vendors for sustainable Cold box plant operations.

1. Moisture, CO_2 and NH_3 analyzers should be installed at outlet of drier for good operational control at cold box inlet instead of Lab sampling regime.
2. Drier depressurization should always be kept in downward flow to avoid bed lifting.
3. Drier regeneration heating path should be design with upward flow and cooling path with downward flow so that during downward flow flow rate can be increased and time of regeneration may reduce.
4. For Drier pressurization and Depressurization control valves should be proposed instead of gate valves for controlled pressurization and depressurization otherwise channeling may occur.
5. Drier upstream moisture separator should be design with dual density demister pads to avoid chances moisture slip from drier.
6. Drier inlet must contain distribution nozzle for gas equal distribution on bed.
7. Drier adsorbents should be designed in such a way that stand by drier timing should be ~ 10 hours for creating margin in operational flexibility.

All designers to close the above Engineering Controls points in design phase for better services to customer.

Conclusion

Although cold box frosting and bad mode operational instance was very costly and inconvenient we had to suffer production loss from 23rd March to April 2014 partially or completely but it served as a good reminder that Drier needs adequate cares to sustain safe operation of Cold box. Fortunately, there was no physical damage to Cold box equipment.

Urea Manufacture by Novel Techniques Utilizing Alternative Raw Material Sources

SUST-E Urea Process

Taibah Jaffery
Supervisor: Dr. Arshad Hussain

Fatima
Fertilizer Company Limited

Sponsor

The speaker's visit to the 2014 AIChE Annual Meeting has been Sponsored by:

Fatima Fertilizer Company Limited

*This work has been conducted entirely by the National University of Sciences and Technology, Pakistan. Fatima Group does not own a share in this technology.

2

Fatima Group

- One of the largest business groups in Pakistan with diversified businesses in fertilizers, textiles, and sugar industries.
- Primary business includes manufacture of compound fertilizers- annual capacity 2.1 MTonnes
- Fatima Group has experience in undertaking large-scale green-field projects in Pakistan
- FFL was conceived at a cost of M USD 750 in December 2003

Fatima Group

Fertilizers

Pakarab Fertilizers

Fatima Fertilizer

Sugar

Fatima Sugar Mills

Trading

Reliance Commodities

Textiles

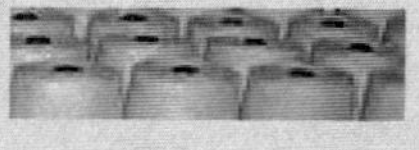

Reliance Weaving

Fazal Cloth Mills

Energy

Fatima Energy

3

Fatima Fertilizers Company

Vision, Mission & Core Values

Vision

To be a world class manufacturer of fertilizers and ancillary products, with a focus on safety, quality and contribution to national economic growth and development. We will care for the environment and the communities we work in while continuing to create shareholders' value.

Mission

- To be the preferred fertilizer company for farmers, business associates and suppliers through quality and service.
- To provide employees an exciting, enabling and supportive environment to excel in, be innovative, entrepreneurial in an ethical and safe working place based on meritocracy and equal opportunity.
- To be a responsible corporate citizen with a concern for the environment and the communities we deal with.

Core Values

Integrity: Our actions are driven by honesty, ethics, fairness and transparency
Innovation: We encourage creativity and recognize new ideas
Teamwork: We work collectively towards a common goal
Safety, Health and Environment: We care for people and communities around us
Customer Focus: We believe in listening to our customer
Excellence: We strive to excel in everything we do
Valuing People: We value our people as our greatest resource

4

Fatima Fertilizer Company Limited — Production

- **Final Products**
  - Urea — 1,500 TPD
  - Calcium Ammonium Nitrate (CAN) — 1,400 TPD
  - Nitro Phosphate (NP) — 1,200 TPD
  - Nitro Phosphorous Potash (NPK) — 1,000 TPD
- **Intermediary Products**
  - Ammonia — 1,500 TPD
  - Nitric Acid (NA) — 1,500 TPD
- **Off Sites & Utilities**
  - Power plant - 52 MW ISO Rating
  - Product Handling & Storage
  - 47 KM Natural Gas Pipeline
  - 12 KM Raw Water line & Pumping Station with own power
  - Railway Siding
- Debottlenecking
  - Ammonia — 1800 TPD

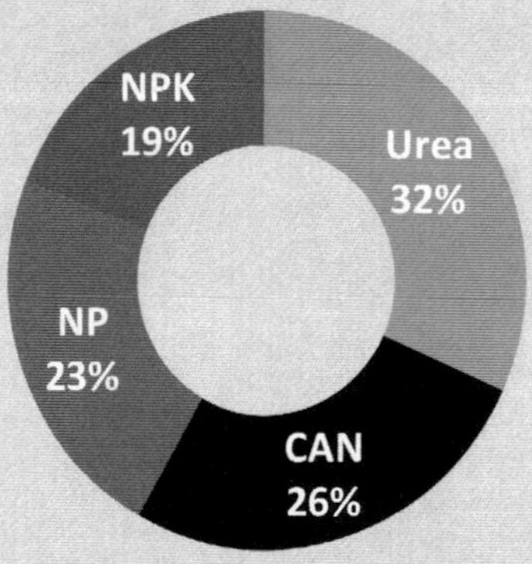

Per Annum Production Contribution

5

Fatima Fertilizer Company Limited — Major Milestones

- Apr. 2011 — NP Plant Production
- Apr. 2010 — NA Plant Production
- Mar. 2010 — Urea Plant Production
- Feb. 2010 — Ammonia Plant Production
- Nov. 2009 — CAN Plant Production
- Oct. 2009 — Ammonia Furnace 1st Fire
- Financial Closure achieved and LC's established — 26th Nov. 2006
- Construction & Commissioning Contract with CNCEC — Sep. 2006
- Contract with Sojitz signed — 27th Jun. 2006
- Ground Breaking Ceremony — Apr. 2006
- Gas Supply / Purchase Agreement Signed — 12th Jul. 2005
- Gas Allocation — 4th Sept. 2004
- Date of incorporation — 23rd Dec. 2003

6

Fatima Fertilizer Company Limited — FFL Site Safety

PSM

Lagging Indicators

Target 0.27

Total Recordable Injury Rate (Year to date)

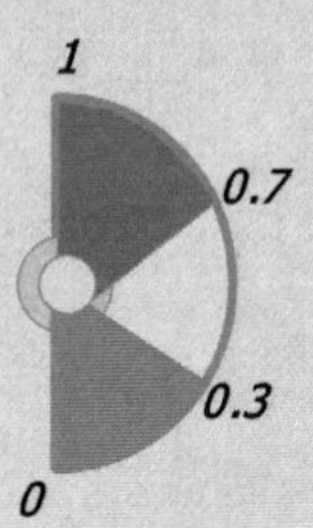

FFL 0.00
Cont. 0.087
Comb. TRIR = 0.061

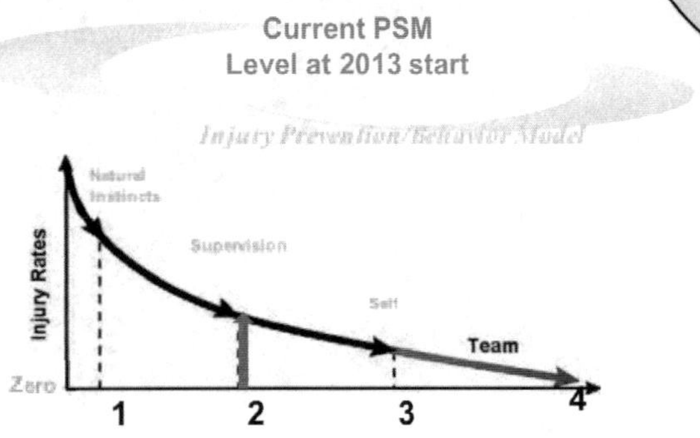

- SAFE WORKING HRS FFL = 6.80 Million
- SAFE WORKING HRS CONT. = 13.89 Million
- SAFE WORKING HRS COMBINE = 20.7 Million

Current 93
Target 80

7

Fatima Fertilizer Company Limited — Future Projects

Debottlenecking

- Ammonia DBN
- Phase 1 completion in Dec 2015 - 1500-->1650 TPD
- Phase 2 completion in 2017 - 1650-->1800 TPD

Midwest Fertilizers Company Project

- Ammonia 1500 TPD
- NA 1500 TPD
- UAN1500 TPD

8

Objective

- To make the Urea manufacturing process inherently independent of non-renewable feed sources (Natural Gas, Coal, etc)
- Sustainable and environment friendly production with minimum net energy requirement
- Minimizing carbon emissions in the atmosphere - Goal of regulating authorities: achieving 90% recovery of CO_2 with maximum 35% increase in electricity cost [1]
- Carbon Sequestration

9

Challenges

- Clean Hydrogen Source
- Clean Energy Source
- Flue Gas Separation into Carbon dioxide and Nitrogen

10

Proposed Solution

- SUST-E Urea Manufacturing Process
- Al-steam reaction for combined production of Hydrogen and Energy
- Membrane Separation of Flue Gases from Thermal Power Plant, Cement Industry, etc.
- Replace Front-end (Steam reforming)
- Renewable energy e.g. CSP for additional energy requirements

$$2Al + 3H_2O \rightarrow Al_2O_3 + 3H_2 + Heat$$

$$\text{Flue Gas} \rightarrow CO_2 + N_2$$

$$N_2 + 3H_2 \rightarrow 2NH_3 + Heat$$

$$2NH_3 + CO_2 \rightarrow NH_2COONH_4 \leftrightarrow CO(NH_2)_2 + H_2O$$

11

Scope

- Providing a workable design for the new synthesis route
- Investigation of technical feasibility of 7000 TPD Urea and 4000 TPD Ammonia production
- Investigation of economic feasibility of 7000 TPD and 4000 TPD production
- Providing direction for further research
- Tools – MATLAB, ASPEN PLUS

Process Development

Process Research & Innovation | Pilot Studies | Technology Transfer and Manufacturing | Process Intensification & Microprocess Engineering | Product Design

12

The Sust-E Urea Process

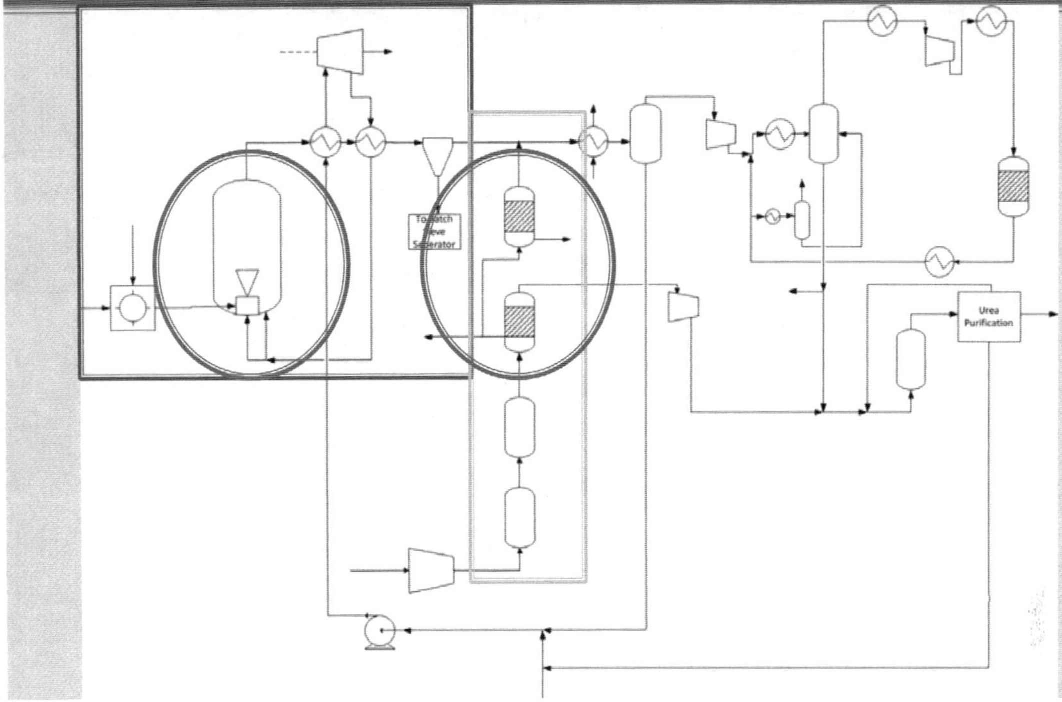

13

Raw Materials Chosen

- Aluminum ingots (grinding tool based on Watanabe's work) [2]
- Flue Gas from coal based plants
- Water
- Energy source – Al/H2O reaction

14

Aluminium- Steam Reactor

- $2Al + 3H_2O \rightarrow Al_2O_3 + 3H_2$ above 480 °C [3-6]
- Reactants are **fed** at only 600 °F (316 °C) and steam is fed at 150 psig via nozzle
- The Reactor is a combustor (in plug flow)
- **15442** tubes of 10" dia and height of 4ft, arranged in 16 floors, 34 rows of 29 tubes each
- Temperature of 2900 K achieved(> Al melting temperature) [4-5]
- The residence time is **60 ms** and conversion is 95%.[4-5]
- Heat of reaction **1100-1250 MW**, Heat Loss = 405-680 MW
- ΔH > **540 MW**

15

Membrane Separation

- Coal based flue gas
- Two stage, co-current process using PVAm membrane [7-8]
- Nitrogen purification using Carbon Molecular Sieve Membrane (Celo550) [9]
- Multi-component model used to for designing using MATLAB
- Membrane Area (m^2): PVAm Stage1= 2594122, Stage2= 627118
- Celo550 Area ~ 27,922,908.05 ft^2
- 94.6 % pure CO_2 & 94.5 % pure N_2 recovered after membrane separation.
- Celo 550 counter-current membrane separator gives 99.99% pure nitrogen.

16

PFD for Membrane Separation

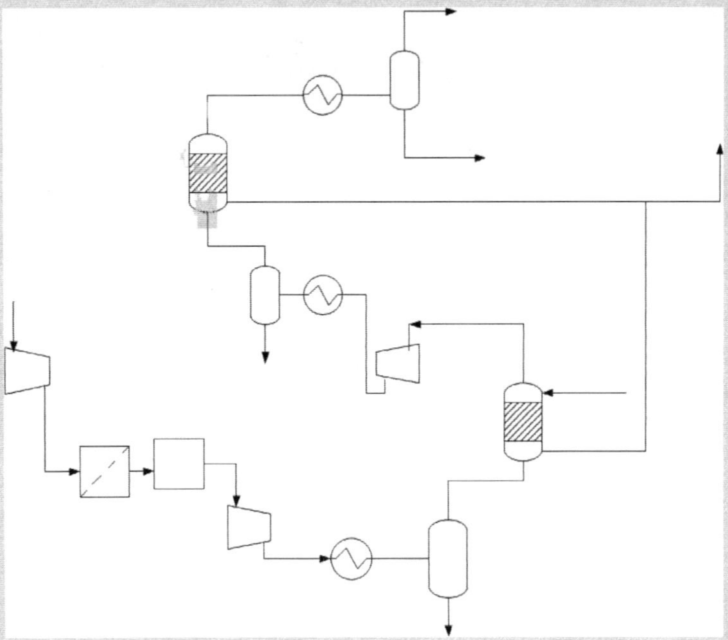

17

Process Integration

- Process Integration for reduction of energy and capital consumption.
- Reaction heat used to heat up feed as well as produce excess electricity
- Water vapor from exhaust steam turbine condensate used as sweep gas

18

Consumption and Production

Urea (99.8%)	7000 TPD
Ammonia (including inventory)	4000
Aluminium	6670
High Purity Alumina	11973
H_2O	4265
H_2	710
N_2	3290
CO_2	5119
Excess N_2	30008
Flue gas	39852 **(1206331.27 Nm^3/hr)**

19

Energy Efficiency

- Released by exothermic reaction: 1100-1200 MW
- Electricity production: 315 MW including 129 MW required for membrane separation
- Requirement for Al-Steam Feed: 260-280 MW
- Total major plant electricity requirements (grinding, vacuum systems, etc): 174.3 MW
- Surplus: 140.7 MW

*Turbine and generator efficiencies have already been accounted for

20

Cost Feasibility

- Total Plant Investment: $ 138 billion (at current cost of Alumina) – will be reduced by:
  - Increase in supply of alumina refractory after establishment of Sust-E Urea Technology
  - More efficient reactor design with better heat recovery and lower surface area
- Highly expensive bi-product (High Purity Alumina)
- Payback time: 2 yrs
- ROR = $ 69 billion/yr (can be achieved at $32/kg of alumina (34% of current cost) and $0.20/kg of Urea)

21

Sustainability

- Inherently independent of non-renewable resources - capable of running on solar power
- 0.73kg CO_2/kg Urea consumed, hence CO_2 emissions reduced.
- Feed Aluminum can be produced using renewable energy sources
- Eliminates requirement for catalysts (and hence emissions related to their production)

22

Conclusion

- The process is technically capable of achieving process requirements (purity, temperature, pressure) for ammonia and urea manufacture
- Economically feasible
- Energy Efficient

23

Areas for Further Research

- Pilot Studies
- Optimization of reactor design for reduced heat loss and surface area (and hence cost)
- Cost Models for Aluminum and High Purity Alumina nano-particles after establishment of Sust-E Urea plants
- Scale-up of electrolysis cells using non-consumable cathodes for Aluminum production
- Higher efficiency membranes

24

References

- [1] Favre E, "Carbon dioxide recovery from post-combustion processes: Can gas permeation membranes compete with absorption?"
- [2] Watanabe M, Ximeng J, Ryuichi S, "Method for generating hydrogen utilizing activated aluminum fine particles." US patent. US 2006/0034756 A1; 2006
- [3] Petrovic J, Thomas G, "Reaction of Aluminum with Water to Produce Hydrogen – A Study of Issues Related to the Use of Aluminum for On-Board Vehicular Hydrogen Storage", U. S. Department of Energy
- [4] Foote, J. P, Thompson, B. R, Lineberry, J. T, "Combustion of Aluminum with Steam for Underwater Propulsion"
- [5] Franzoni F, et al., Combined hydrogen production and power generation from aluminum combustion with water: Analysis of the concept, International Journal of Hydrogen Energy (2010),
- [6] Golovitchev, V. I, Al/H_2O Combustion, Report No. 2, Combustion Engine Research Centre, Chalmers University of Technology
- [7] Hussain, A., A single stage membrane process for CO_2 capture from flue gas by a facilitated transport membrane, Journal of Separation Science & Technology 47 (2012) 1857-1865.
- [8] Hussain, A., Hägg, M.-B., A feasibility study of CO_2 capture from flue gas by a facilitated transport membrane, Journal of Membrane Science 359 (2010) 140-148.
- [9] Campo M. C, Magalhães F. D, Mendes A, Separation of nitrogen from air by carbon molecular sieve membranes

25

Questions?

26

381792 Production of Bio-Oil from Lignin Extracted from Kraft Black Liquor By the Sunmembrane Process

Abstract:

1. Introduction

It is essential for the pulp and paper industry to utilize as much of the forest raw material as possible to achieve the sustainability goals and to increase the profit. The fibres that are liberated during the kraft pulping process are used for pulp production, while the rest of the components from the wood are mainly incinerated. Lignin is one major by-product from the kraft pulping process. It has a high heating value, and has not a large market outside internal boiler fuels.

Today, many researchers are focused on finding alternative bio-based and sustainable resources that can replace the fossil oil for production of liquid fuels used for vehicles. Lignin, which is already extracted from the wood in the kraft process, is an appropriate raw material for this purpose since it is available in large quantities. Using this lignin for production of renewable fuels does not only have an environmental advantage, but it also increases the income for the kraft pulp mills.

Scientists have proposed several methods for the conversion of kraft lignin into a liquid fuel, e.g. pyrolysis [1], palladium-catalysed depolymerisation [2] and base-catalysed depolymerisation [3]. However, these methods are all dependent on an efficient extraction process for lignin from the kraft black liquor.

Lignin can be extracted from the kraft black liquor by either acid precipitation [4], or by membrane filtration [5]. Membrane filtration has the advantage, compared to the acid precipitation technology used today, that the negative impact on the pulp mill operation is negligible, that no chemicals are added to the black liquor, and that the sodium sulphur balance in the pulp mill is not impaired. We have developed two processes for the production of a bio-oil from kraft lignin. Both processes are based on a membrane process, called the SunMembrane process. The first configuration allows production of a solid lignin. This solid lignin can then be fed into a pyrolysis unit for production of pyrolysis oil. By the use of the second process, a precursor for green gasoline is directly produced in the process with integrated base-catalysis, without production of an intermediate solid product.

2. Separation of Lignin from the Black Liquor

The process is based on membrane separation of the components in the black liquor. The first part of the process comprises several steps, and separates the lignin from the rest of the components in the black liquor. These steps can include one or more of the following processes: ultrafiltration (UF), fine ultrafiltration (FUF), nanofiltration (NF) and diafiltration (DF).

The UF step is an optional pre-filtration step. This step removes larger compounds in the black liquor, e.g. polymeric hemicelluloses and larger lignin molecules. In some applications these compounds have to be removed. However, in most applications, this step is not necessary.

The FUF/NF step isolates the lignin from the water and cooking chemicals. This is achieved by retaining the lignin and other organic compounds of relatively high molecular weight while the water, the cooking chemicals and other low molecular weight compounds can pass through the membrane without any significant retention. The stream that has passed through the membrane is sent to the recovery cycle for the cooking chemicals, while the stream with high lignin concentration is further processed with DF.

Water is continuously added to the stream with high lignin concentration during DF. By continuously adding water at the same rate as the permeate is withdrawn from the process, washout of the cooking chemicals from the lignin-rich solution is achieved. The DF permeate is directed to the recovery cycle while the stream with high lignin concentration is either concentrated further, or sent directly to next step.

3. Production of a Lignin Powder

If the aim of the process is to produce a lignin powder, the lignin in the solution with high lignin concentration is precipitated. In the precipitation, acid is added to the lignin-rich solution in order to decrease the pH. By decreasing the pH, the lignin starts to precipitate and form particles. Due to the previous recovery of the major part of the cooking chemicals, only a small amount of acid is needed compared to if the precipitation would be performed on the black liquor directly.

After the precipitation, the lignin slurry is filtered through a filter cloth in order to separate the lignin particles from the liquor. The filter cake of lignin is then washed before it is dried. The dried lignin powder can finally be fed into different downstream processes, e.g. a pyrolysis unit for production of a bio-oil.

4. In-situ Conversion of Lignin by Base Catalysis

Depolymerisation of the lignin molecules is needed to convert the lignin into a bio-oil. By smart integration into the membrane-based separation process, this can easily be done at a pulp mill. This process comprises, in addition to the previously described membrane process, a depolymerisation reactor and a separation step.

The depolymerisation is performed as a base-catalysed reaction, and is placed directly after the NF step. During the depolymerisation, partial deoxygenation and desulphurization also occurs. The main catalyst in this reactor is the sodium hydroxide that is already present in the black liquor. However, to improve the conversion of lignin into an oil fraction, a heterogeneous catalyst is also added.

The lignin molecules that are fully depolymerised are extracted from the liquor afterwards, and can be sold as a precursor for production of gasoline and diesel, or be used internally at the pulp mill to replace fossil fuel, e.g. in the lime kiln. The part of the lignin that is not converted into a bio-oil is recovered in the recovery boiler at the pulp mill.

5. Summary

Two different routes for production of a bio-oil have been presented. Both processes (patent pending) are currently under evaluation in lab-scale. The plan is to build a demonstration plant next year for evaluation of both alternatives for the production of bio-oil in a continuously operating plant.

Acknowledgments

The Swedish Foundation for Strategic Environmental Research, MISTRA, and Sweden's Innovation Agency, VINNOVA, are gratefully acknowledged for their financial support.

References

1. Ben H, Ragauskas A J, *NMR characterization of pyrolysis oils from kraft lignin*, Energy Fuels, 25 (2011) 2322-2332.
2. Liguori L, Barth T, *Palladium-nafion sac-13 catalysed depolymerisation of lignin to phenols in formic acid and water*, J. Anal. Appl. Pyrolysis, 92:2 (2011) 477-484.
3. Vigneault A, Johanson D K, Chornet E, *Base-catalysed depolymerization of lignin: separation of monomers*, Canadian J. Chem. Eng., 85:6 (2007) 906-916.
4. Wallmo H, *Lignin extraction from black liquor: Precipitation, filtration and washing*, Chalmers University of Technology (2007) doctoral thesis.
5. Arkell A, Olsson J, Wallberg O, *Process performance in lignin separation from softwood black liquor by membrane filtration*, Chem. Eng. Res. Des., (2014) In press.

387649 Upgraded Bio-Oil Production from Sweet Sorghum through 2-Stage Hydrothermal Liquefaction

Abstract:

Hydro-Thermal Liquefaction (HTL) technology is a promising method to convert high moisture lignocellulosic biomass into biofuels and platform chemicals at moderate temperature (280-370 °C) and pressure (10-25 MPa). Soluble sugars and benzene-type chemicals from sweet sorghum lignocellulose macromolecules (hemi-cellulose, cellulose and lignin) hydrolysis further decompose and cross react into solid residues, which significant reduce crude bio-oil yield. The main objective of this study was to investigate the two-stage liquefaction of sweet sorghum to selectively produce sugar mixtures and upgraded lignin derived bio-crude oil. In the first stage, biomass was treated below 280 °C to produce soluble sugar mixtures that were selectively hydrolyzed from cellulose and hemi-cellulose. Sugar stream contained mixtures of glucose, xylose, arabinose, 5-hydroxymethyfurfural (5-HMF) and levoglucosan. Solid residues from 1st stage liquefaction contained more than 70 wt % lignin. In the second stage, lignin containing solid residues was thermally treated along with hydrodeoxygenation (HDO) upgrading process at 350 °C for 1 h with 5 wt % Ru/C. The upgraded bio-oil yield reached to 63 wt % and contained cyclopenten-1-one, ethyl-phenol, hexadecanoic acid and octadecanoic acid. The moisture and O/C ratio of upgraded bio-oil were 5 wt % and 0.1 respectively, which could meet the fuel specifications of petroleum crude oil for effective co-processing into drop-in biofuels and platform chemicals.

367906 Optimizing and Improving Dilute Acid Pretreatment of Lignocellulose Biomass to Derive Sugars C5 and C6 for Production of Fuels and Biochemcials

Abstract:

Lignocellulose biomass is heterogeneous in nature have different ratios of xylans and cellulose sugars , and requires processing to closely homogeneous fractions and consistency for its application in the production of variety of fuels and biochemical. The breakdown of biomass is a key first step in the conversion of cellulosic sugars ,and this process has posed the great challenge to the economically viability of recovering sugars and its application for fuels and biochemicals.

This presentation will review Sweetwater Energy's overall process of optimizing dilute acid pretreatment conditions of various heterogeneous biomass feedstocks . Importance of optimizing and methods of pretreatment conditions [time , temperature , pressure and particle size] will be reviewed. Intent is to generate homogeneous, uniformly exploded and cooked, consistent particle size post treatment. Deriving a uniformly sized pretreated feedstock enabled and improved enzyme saccharification yields . In addition, specific challenges in controlling inhibitors sugar syrup will be discussed in details . Finally in technology scale up and accelerating the path to process commercialization and finally direct applications of SWE sugars in ethanol , biochemical and advanced fuels, will be reviewed .

Abstract for the AIChE Annual Meeting (16 to 24 November, 2014; Atlanta)

Process Technology Development for Biomass to Detergent Alcohol – Hurdles and Best Practices

Cawas Cooper, Sr. Project Manager, Codexis, Redwood City, CA (Phone: 650-421-8339)

Process technology development is a critical component for all projects to successfully transition from concept to commercialization. An industrial case study will be presented for a scale up of integrated biotechnology process from lab to a 1,500 liter demonstration plant involving international multifunctional team members.

Codexis' CodeXyme® 4X cellulase enzyme was used to produce cellulosic sugars from non-food biomass, while the CodeXol® detergent alcohol fermentation process technology – which includes Codexis' proprietary microorganism strain – converted these cellulosic sugars to detergent alcohols. We believe this was the world's first successful large scale effort to produce commercially relevant detergent alcohols from a cellulosic biomass feedstock.

The presentation will describe how key technical as well as non-technical components progressed during development and how multiple hurdles were overcome through best practices over the project duration.

BIOETHANOL FROM SWEET SORGHUM: A DECENTRALIZED PRODUCTION APPROACH

Anuradha Mukherjee, Rob Whiteley, Josh Ramsey, and Danielle Bellmer
Oklahoma State University, Stillwater, OK

E-mail: anuradha.mukherjee@okstate.edu , rob.whiteley@okstate.edu, , josh.ramsey@okstate.edu, danielle.bellmer@okstate.edu

Abstract

This work aims to present implementation of renewable fuel production from sweet sorghum on a farm scale. High concentrations of sugar in stalks, simple downstream technology, and minimum upkeep of feedstock make sweet sorghum an ideal candidate for decentralized energy production. The Alcohol Separation Unit (ASU) at Oklahoma State University (OSU) serves as both a research and demonstration prototype unit, first of its kind to study bioethanol production from fermented sweet sorghum liquid feedstock. This pilot plant has been built to establish a robust downstream separation strategy crucial for on-farm bioethanol production. The study investigates four key aspects of the process – equipment and instrumentation, safety, operation and maintenance, and finally economics.

The Alcohol Separation Unit has a two-column distillation set up for concentrating ethanol, with associated heat exchangers, pumps, and storage units. Each unit has been designed to work with the sweet sorghum liquid feedstock. This pilot plant was commissioned in Fall 2013. A total of 5000 gallons of sweet sorghum feed containing 5 wt% ethanol have been successfully processed at the pilot plant and 190 proof ethanol was produced in October 2013. The total mass balance closes at 92% and ethanol purities of 193.6 proof at 74°F with rectifier column conditions of 0.8 psig and 171°F were established during processing.

Overall economics of the pilot plant indicate fixed capital costs at 403,000 USD. Major unit operation equipment (columns, heat exchangers, storage tanks) makes approximately 50% of the costs. Major energy usage at the pilot plant for ethanol production is steam, which accounts for 75% of total operating costs of 68,000 USD.

1. Introduction

Nearly all gasoline sold in the United States include ethanol blended in various percentages starting at 10%. So far this volume has primarily been produced from corn-based feedstock - in 2012 bioethanol production in United States was 13,300 million gallons from these starch-based sources [1]. Non-corn cellulosic production volumes between 8 – 30 million gallons of cellulosic biofuels are expected only in 2014, based on current EPA estimates [2]. Despite a potential decrease in 2014 mandate for renewable fuels being considered in the United States, the total volume of biofuels produced will still be a staggering quantity and feedstock diversification will continue to be a heavily investigated research area.

Amongst alternative non-corn feedstock options sorghum crops are increasingly being considered a viable feedstock option for energy production. Sorghum crops require relatively lower inputs that their starch counterparts. With an estimated fertilizer input of 80:50:40 N:P:K kg/ha and water input of 8000 m^3 sweet sorghum is one of the lowest input crops [3, 4]. Grain, forage, and sweet sorghum all provide various options for fuel development albeit from different conversion routes. Sweet and forage sorghum are both suitable crops for cellulosic biomass conversion, although sweet sorghum has a significant quantity of juice in its stalks also suitable for direct liquid fermentation [5, 6]. Sweet sorghum in particular can be used in multiple ways – juice pressed from its stalks can be fermented, bagasse left in the field can be converted either via liquid or solid state cellulosic fermentation processes to fuel – this makes sweet sorghum extremely productive from a energy crop standpoint. Sweet sorghum has been estimated to produce 4000 liters per ha at half the cost of sugarcane ethanol. The advantages of sweet sorghum provide incentive for its use as an energy feedstock that has been demonstrated. In addition to its use as a fuel, sweet sorghum is also used as a precursor to food products and animal feed items [7].

Overall, sweet sorghum provides a low nutrient/fertilizer option, is easy to process, and can be converted to an array of end products. For a fermentation-based process, immediate utilization of sucrose upon expression or ensilage [8], pose as technical challenge. Upon storage or ensilage a significant portion of the sugars are lost. Additionally transporting the pressed juice, or cut stalks to a central facility has been estimated to be expensive and large facilities for sweet sorghum have been found not to be cost effective [9]. As suggested by researchers at the School of Biosystems and Agricultural Engineering and discussed by Kundiyana *et al* [10] using sweet sorghum at smaller scale decentralized processes removes the storage and transportation hurdle by fermenting the sweet sorghum in-field. Their study found even with minimal pH, temperature control and minimal nutrient addition, up to 95% sugar to ethanol conversion efficiencies could be achieved. For the overall success of a small-scale on-farm unit though, in addition to the in-field fermentation step, the downstream separation step to separate ethanol from the dilute post-fermentation process stream is crucial. Without a robust, economical, and safe process the success of on-farm bioethanol production can be compromised.

The Alcohol Separation Unit at Oklahoma State University is a research and demonstration prototype unit, the first of its kind to study bioethanol production from fermented sweet sorghum liquid feedstock. This pilot plant has been built to investigate and recommend a robust downstream separation strategy crucial, as discussed, for on-farm bioethanol production. Research activity at the pilot plant will translate into recommendations for farmers in terms of construction, operation, maintenance, and economics. The study investigates four key aspects of the process – equipment and instrumentation, safety, operation and maintenance, and finally economics. This includes design, set-up, and operation of the pilot-plant in addition to analytical requirements for product characterization.

2. Materials and Methods

2.1 Alcohol Separation Unit: Design Criteria

Main considerations for the design of the Alcohol Separation at ASU are:

1. Feed rate of 250 gal/h, for continuous operation of a steady state process servicing a 500 acre farm. A 30 ton/acre sweet sorghum yield is expected, and juice yields of 4000-6000 gal/acre.
2. Ethanol concentration of 6-10 vol% in fermented liquid sweet sorghum can be processed.
3. Final ethanol product of 95 vol% or 190 proof ethanol at 15 gal/h is expected. Ethanol recoveries are at least 90%.
4. Process designed to operate at atmospheric pressure.
5. A two-column process with a beer column and rectifier is used.
6. Feed to the distillation column enters at 150°F.
7. Steam is available onsite at maximum pressure of 150 psig.
8. Process has extensive data collection, control, and monitoring capability

2.2 Process Layout and Equipment Details

The Alcohol Separation Unit has a two-column distillation set up for concentrating ethanol, as seen in Figure 1, a common feature in ethanol separation units small or industrial. Details about and explanation of technology selection have been discussed in Mukherjee's work [11].

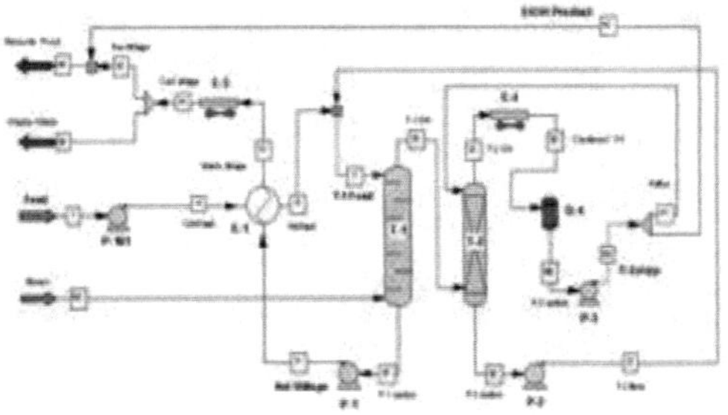

(a)

(b)

Figure 1: (a) Ethanol Simulation in ChemCAD (b) Alcohol Separation Unit: Storage and Process Area

In theory, the column can be thought of as a single unit. From Figure 1 vapors from the beer column enter the bottom of the rectifier and contents from the bottom of the rectifier are added back into the beer column. However physical size/height restrictions limit the construction of the distillation unit as one column. This two-column set up also allows use of packing in the rectifier, resulting in higher overall process efficiencies. In addition, solids are restricted to the beer column, where trays suitable for high solid loading process streams are used.

Material of construction of all equipment, purchased or fabricated, is stainless steel 304 or 316. Low pH values of bio-based ethanol processes streams make them moderately corrosive and material resistant to acidic conditions, such as stainless steel is preferred [12]. The bioethanol distillation facility is equipped with a beer column (T-1), and a rectifier (T-2), as indicated in Figure 1. Associated equipment to ensure steady state function of the pilot plant include an air-cooled heat exchanger (E-4), a feed pre-heater (E-1), stillage cooler (E-5), storage tanks (Tk-101 and Tk-102) as well as pumps (P1, P2, P3, and P4) and a condensate receiver tank (D-4). Each custom built unit including the distillation columns and the air-cooled condenser were pressure tested before installation and are rated for 180 psig @ 350°F.

Both beer column and rectifier are fabricated from 12 inch SA-312 TP304 pipe. Total height of the beer column is approximately 39 ft and the rectifier is 29 ft. Twenty three standard Sulzer cartridge SVG valve trays have been used in the beer column, each spaces at 15 inches The rectifier has a combination of four valve trays and 14 ft of M752Y packing (14 inch HETP). In the beer column, temperature measurement ports have been provided at Tray 1, 5, 7, 12, and 22. On the rectifier, trays 1, 2, and 3 have a liquid withdrawal port as well as temperature measurement capability. The tray-packing combination in the rectifier is an industry standard [13, 14], and has been adopted for on-farm design – solid entrainment to the packing is avoided, by presence of the trays. Additionally, fusel alcohol removal is also facilitated from the rectifier by virtue of the trays.

A custom designed induced draft air-cooler from GEA Rainey is used as the condenser. A three row finned tube bundle with a 1-inch slope containing 10 fins per inch on tubes is used to facilitate condensation. The tube bundle itself is 3 ft x 4 ft in dimension. A ¾ HP motor is used to power the fan. Two heat exchangers are present on the feed return line – the stillage cooler, E-5, a BOL-725 Thermal transfer air cooler, and the feed pre-heater, E-1, a USSC 1036 4-pass shell and tube heater. The shell and tube heater (20 tubes per pass, 4 pass, 0.375" OD x 0.02" wall) and air cooler are arranged in series to ensure the stillage temperature is close to 120°F, cool enough to store in polypropylene tanks, Tk-102 or Tk-101.

Either of two polypropylene tanks, each 5000 gallons in volume, can store sweet sorghum fermented juice feed. A spill protected storage area 40 ft x 24 ft x 1 ft high has been constructed around the tanks. Each tank has an Integral Flange Mounted Opening (IMFO) for easy cleaning and complete emptying of liquid content. A 24 inch manway at the top of the tank for observation, a 2 inch U-vent, and three 2 inch nozzles are some of the features of the tank. Both tanks are stored separately on the south side of the building, as seen in Figure 1 (b). Both tanks are interconnected to facilitate transfer from one tank to the other, and provide mixing of contents. Four Goulds 1.5 HP, 1MF1S5B4 centrifugal pumps are used for the feed, stillage, rectifier bottoms, and condensate return streams.

A Lattner 20 HP vertical boiler is used to provide steam to the distillation system. The steady state steam rate is 250 lb/h. This steam rate is achieved in stages only after steam system, and steam piping blowdown, in increments of 50 lb/h. The boiler is set at a maximum of 140 psi, with a 5 psi cutoff – in each cycle the pressure reaches 135 psi. The boiler trips at 140 psi. Steam is injected into the bottom of the beer column under Tray 1. The feed is introduced at the top of the beer column over Tray 23. The ethanol product stream is recombined with the stillage stream, and sent back to the tank area for storage. This is primarily due to Alcohol and Tobacco Tax and Trade Bureau (TTB) permitting requirements [15] for the facility, wherein a secured location is required for high purity ethanol storage. Therefore, ethanol product is recombined and stored in the tank area, which is secured by fence, and monitored by security cameras. In addition ethanol content and composition of the feed in Tk-101 and Tk-012 are monitored on a regular basis.

2.3. Instrumentation

As listed in the design criteria, process has an extensive data collection, control, and monitoring capability. Temperatures, flows, and pressures are recorded at multiple locations, and the process is fully automated to run during steady state operation. Data is collected as 4-20 mA signals from pressure or differential pressure transmitters, and flow meters. Two junction boxes are present at the base of T-1 and T-2 where all wiring terminates at the data collection device. Data from the junction boxes is transferred to the control room via Cat5e ethernet cables.

Six ports are available on T-1 with temperature measurements on Tray 1, 5, 12, 17, and 23. Similarly six ports are also available on T-2 with three on Tray 1, 2, and 3, and three at lower, mid, and upper points in along the packing, shown in Figure 2 (a). Process liquids and steam temperatures are also measured at each heat exchanger inlet and outlet, column inlet and outlet, as well reflux drum and storage area tanks. Thermowells fabricated from 1/4 inch copper tubes were used along with T-type thermocouples at point of measurement. In-field pressure measurements are provided in addition to transmitted data signals for both temperatures and pressures. This helps evaluating whether the process is running according to protocol, monitoring and reacting quickly to process upsets, in addition to providing a layer of redundancy.

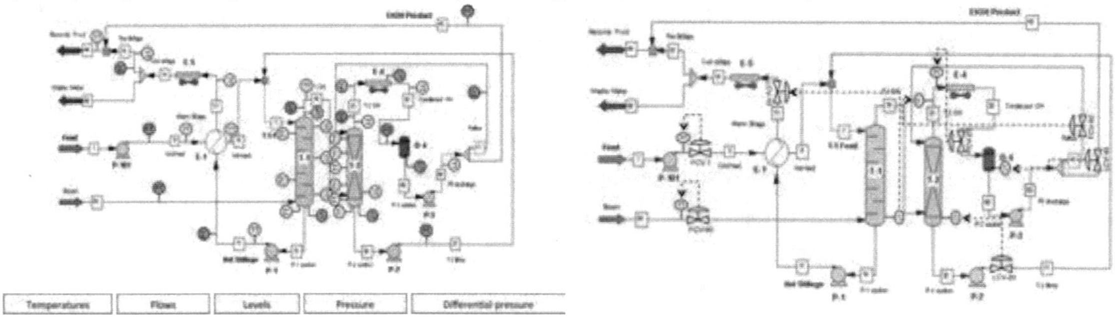

(a) (b)

Figure 2: (a) Instrumentation at ASU (b) Control strategy at ASU

Steam pressure (PT-90), T-1 and T-2 column pressures (PT-10 and PT-20 respectively), reflux drum pressure (PT-40), as well as differential pressure measurements across the column (dPT-10 and dPT-20) are recorded using Rosemount 1151 type pressure/differential pressure transmitters. In-field pressure measurements include heat exchanger inlet and outlets, column top pressures, reflux drum pressure, and discharge pressure measurements at all pumps.

A total of seven flows are measured at ASU, which include feed, stillage, T-1 bottoms, T-2 bottoms, reflux, distillate, and steam, also indicated in Figure 2 (a). There are four highly accurate coriolis meters in use for feed (FT-1), stillage (FT-60), distillate (FT-50), and T-2 bottoms flow measurement. The other meters are orifice meters couples with differential pressure transmitters and flow is calculated from the measured delta-P across the orifice. Process levels are measured with a wet-leg differential pressure approach. Silicone oil (specific gravity = 1.1) is used as the fill fluid on the low-pressure side of the transmitter. Sump levels on T-1 and T-2 as well as D-4 levels are measured using differential pressure. Tank Tk-101 and Tk-102 levels are measured using electric liquid level indicators. There is a dial for height display, and a 4-20mA current signal can be wired into the control program.

There are eight control loops at the pilot plant - control variables, manipulated variables, along with set point for each control loop are described in Table 1 and highlighted in Figure 2 (b). Overall, content coming into the system (steam and feed) is controlled in a flow control loop. Levels are maintained by controlling outflow from the columns. Pressure in E-4 and subsequently column pressure is maintained by keeping the liquid level in its tubes. Reflux flow into the column is set up as a flow control loop or a temperature cascade loop for T-2 top temperature.

Table 1: Control Strategy at Alcohol Separation Unit

Control Loop	Air-to-close/open	Stem	Cv	Tag	Service
OSU-CV1	ATC	B-linear	2.00	PCV-20	Overhead Pressure/E-4 Level
OSU-CV2	ATC	F-linear	0.32	FCV-40	Reflux/T-2 Overhead Temperature
OSU-CV3	ATO	B-linear	2.00	FCV-1	Feed flow
OSU-CV4	ATO	B-linear	2.00	LCV-10	T-1 level
OSU-CV5	ATO	H-linear	0.13	LCV-20	T-2 level
OSU-CV6	ATO	B-linear	2.00	FCV-90	Steam flow
OSU-CV7	ATO	I-linear	0.08	LCV-40	D-4 level

Data acquisition from the process is via two National Instruments (NI) cDAQ 9188 chassis, and input modules. Ethernet based DAQ hardware provided a 50 percent cost savings on materials, which we would have otherwise spent on collecting measurement data. A NI 9203 module is used for 4-20 mA current acquisition. This provided flexibility in process measurement as highly accurate coriolis flow sensors, level transmitters, and differential pressures were all coupled with the DAQ hardware. Thermocouple temperatures are acquired directly using NI 9213 module. Output signals to control valves are generated using the NI 9265 modules. The control strategy is implemented using Labview programming. Critical process information including levels, pressures, temperatures, and flows are displayed, complete with an alarm system for high or low level occurrences on the graphical user interface, as displayed in Figure 3 (a) and (b). Control valve tuning, level, flow, temperature control capabilities have also been built into the control system using PID algorithm provided in NI's Control and Fuzzy logic toolkit.

(a)

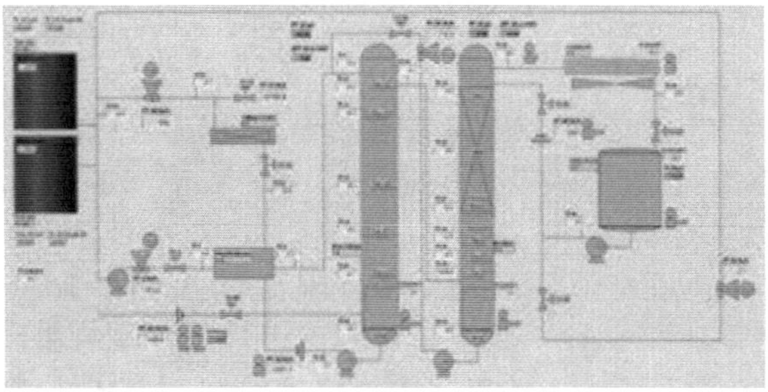

(b)

Figure 3: (a) Critical process information on an easy to follow process flow diagram (b) Main process control window for the Alcohol Separation Unit

Program development, initial testing of control loops, controller tuning, and initial commissioning activities were accomplished in three months of summer 2013. The intensity of all start-up and commissioning activities, it has to be emphasized can be expected at a level that of any chemical process or plant.

2.4 Safety Considerations at the Alcohol Separation Unit

Before the design was finalized and construction activities on-site began, a detailed Process Hazard Analysis (PHA) was conducted. As with any chemical process, safety is the highest priority right from design to operation. This exercise not only helped identify safety concerns in ASU's design, but as a group various possible solutions were identified. Over-pressurization, firefighting, storage of alcohol, spills, and monitoring were acknowledged as concerns regarding the facility and safeguards were included into the design to address these concerns.

As required by the Alcohol and Tobacco Tax and Trade Bureau (TTB), any ethanol produced on site must be gauged and stored under lock and key. Therefore, to tackle this primary concern, ethanol content of feed tanks is measured on a regular basis - this provides an accurate estimation of ethanol content in the fermentation broth. The storage tanks are in a fenced area, and are monitored with security cameras. To eliminate the storage of high purity ethanol the process operates in redistillation mode where ethanol produced in the process is mixed back into stillage and pumped back to storage area. When not in operation, process liquids are expected only in the storage area. Spills in this area will be restricted with a dyke 1 ft tall, built around the storage tanks. The volume of the diked area is large enough to accommodate contents of the largest tank inside it, 5,000 gallons in our case, as recommended in guidelines set by the National Fire Protection Agency (NFPA) in NFPA 30 [16].

Detailed relief calculations with three over-pressurization scenarios were examined to address over-pressurization concerns - (a) fire at the base of columns, (b) runaway steam boiler, and (c) loss of electricity and failure of air-cooler. Two rupture disk assemblies, on the basis of these calculations, were selected to protect columns and associated equipment from over-pressurization. Fire fighting measures were analyzed using the NFPA 11 standard [17]. For process area spillage, AR-AFFF requirements included contents in the T-1 and T-2 sumps, reflux drum, and air cooler.

2.5 Steady State Operations and Commissioning:

Liquid sweet sorghum juice was procured from Delta Renewables Inc. – a total volume of 4890 gallons was transferred to the storage tanks Tk-101 via a 2 in inlet valve using the transfer pump on the delivery truck. The liquid was allowed to circulate in the tank using feed pump P-101 for half an hour before samples were taken for feed ethanol determination.

An initial filtration step was set up for removal of large sized leafy and solid particles. Mesh filters (constructed from lining holed buckets with air conditioner filters) were installed in Tk-102 manway and liquid from Tk-101 was diverted through these filters, and into Tk-102, and back into Tk-101. Feed in Tk-101 was recirculated for at least half an hour prior to processing.
Steam was let into the system through the steam line into T-1 till target of 250 lb/h was reached. Vents on T-2 overhead line and D-4 was closed once both distillation columns were allowed to fill up with steam. A liquid level in the beer column, T-1, the rectifier, T-2, and condensate drum, D-4 respectively develops upon condensation of vapors in the system. Upon reaching the 20%

level on D-4, the collected liquid was sent back to T-2, as reflux, at approximately 100 lb/h. Feed was added to the column in increments till a steady state value of 1700 lb/h was reached.

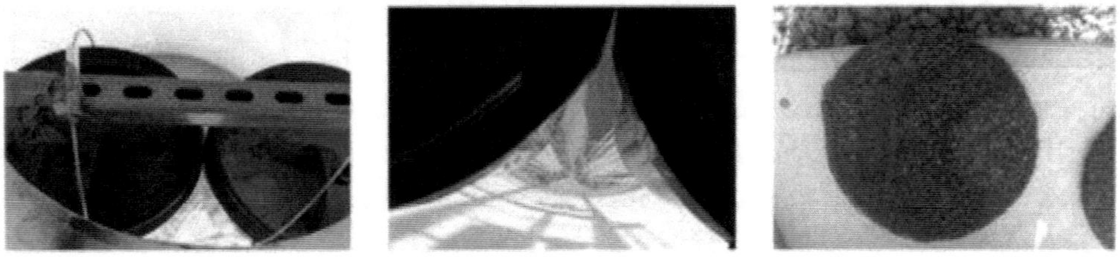

Filtration setup Filterate Retentate

Figure 4: Filtration Setup Details at ASU

Headspace gases from the air cooler/condensate drum were diverted through a sparger assembly – carbon dioxide gases trapped in the fermented sweet sorghum juice still retained in the distilled product were removed by dissolving in water. Some ethanol is lost in this process, however over-pressurization of air cooler and condensate drum are avoided. A constant pressure of 0.8 psig was maintained in the reflux drum. As ethanol is concentrates in the column, temperature separation was observed along the columns. Levels in T-1 and T-2 were maintained between 10 to 15%. Maintaining low levels in columns assists in a faster shut down process as equipment is emptied out quickly. Product ethanol concentration was checked on-site with hydrometers. 193.6 proof ethanol concentration t at 74°F (translates to 190 proof at 60°F) was recorded as product from ASU.

2.6 Analytical Sample Analysis

An Aminex HPX-87H column (Bio-Rad, Sunnyvale, CA, USA) at 60°C was used for off-site HPLC analysis. Sample flowrate was 0.6 mL/min and a 0.01 N H_2SO_4 solution was used as eluent and refractive index detection (1100 series, Agilent) was used for quantification. All samples were examined in triplicate. Samples with high ethanol concentration (T-2 samples) were diluted to 1:100 and feed and stillage samples (from T-1) were diluted to 1:10. Ethanol standards were prepared from 200 proof laboratory grade ethyl alcohol (AL200-500). A calibration standard with fructose 20.002 g/L, glucose 19.870 g/L, sucrose 20.074 g/L, and acetic acid 19.78 g/L was prepared for non-ethanol constituent analysis. Dilutions of 1:1, 1:2, 1:4, 1:10, and 1:20 were used for calibration samples.

2.7 Economic Analysis

Capital and installation costs of constructing a farm scale unit were documented. These costs represent actual spending on equipment and installation. Installation costs include labor and contractual services (not including engineering salaries of the PI and graduate student), piping, hardware, fittings material cost, and construction equipment rental costs. Operating costs include operation boiler steam and electricity. Electricity costs also includes demand from all pumps (P1, P2, P3, P4, and P101 at 1.5 HP, E4 and E-5 fan motor at 2 and 1.5 HP respectively).

3. Results

3.1 Process Operation

3.1.1 Mass Balance

Feed and steam flows account for all mass entering the system. Stillage and product flows account for all mass exiting the process. Mass flows shown in Table 3 were recorded at azeotropic product formation - 190 proof at 60°F, mentioned earlier. This mass balance closes at 92% for total material.

Table 3: Total Mass Balance over ASU

Total Flows (lb/h)	Mass In	1752	Mass Out	1610
Stream	Feed	Steam	Stillage	Product
Sensor	FT-1	FT-90	FT-60	FT-50
Flowrate (lb/h)	1500	252	1557	53

With the exception of FT-90 all sensors accounting for the flow are high accuracy Micromotion flow devices. An orifice meter is recommended and used for steam flowrate measurement along with calibrated differential pressure transmitter. The data recorded has some inherent scatter (for each flowrate), which is possibly contributing to the difference.

3.1.2 Process Flows and Temperatures

Steam flowrate was maintained at 250 lb/h, at an average delivery pressure of 110 psig, as seen in Figure 5 (a). Typically boiler pressure swings between 100 and 130 psig, reflecting cold make up water addition cycles in the boiler, also seen in Figure 5 (a). The cut-off pressure for the boiler is 140 psig. We found that the boiler could maintain a maximum steam flowrate of 250 lb/h without tripping. Feed was introduced in 200 lb/h increments till 1700 lb/h was achieved, as seen in Figure 5 (b).

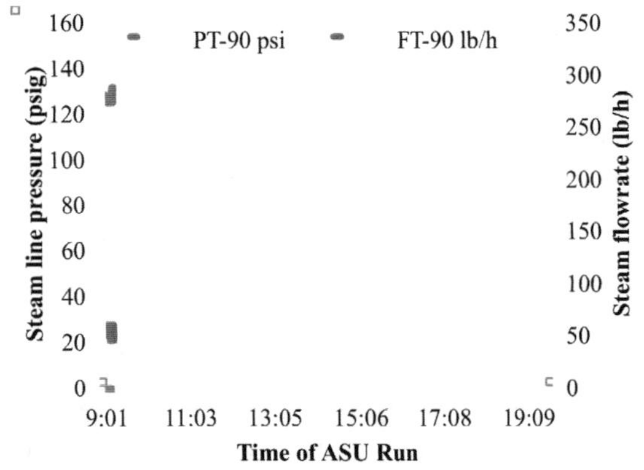

(a)

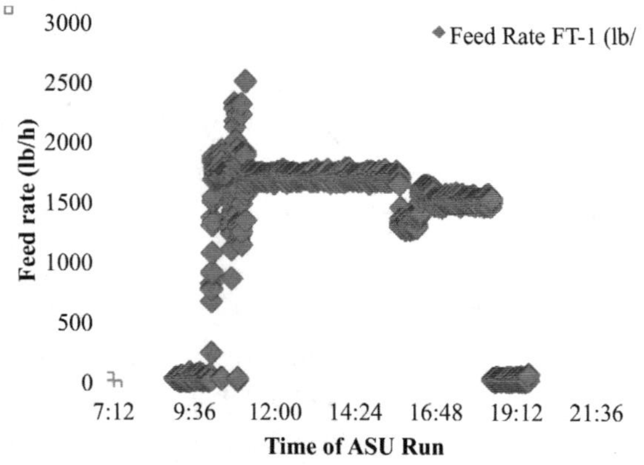

(b)

Figure 5: Flows entering the system (a) Steam (b) Feed. Also shown in 4 (a) is steam boiler pressure behavior

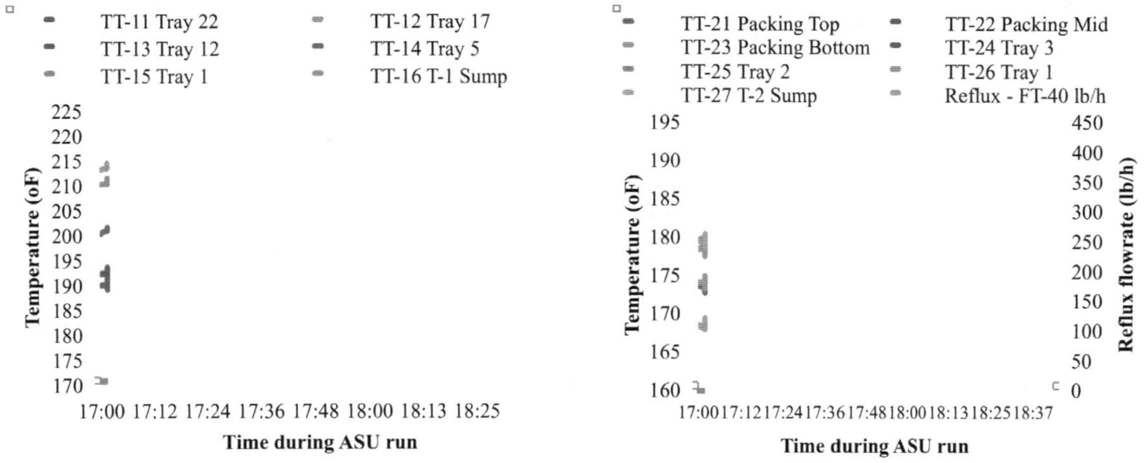

(a) (b)

Figure 6: Temperature profile of column (a) T-1, Beer column (b) T-2 Rectifier

As ethanol is concentrates in the column, temperature separation was observed along the columns as shown in Figures 6 (a) and (b). The bottom temperature of approximately 216°F and a T-1 top temperature of 191°F were observed at product specification. T-2 top and bottom temperatures were 171°F and 181°F.

3.2 Analytical Results

The initial ethanol concentration on delivery was determined to be 4.25 wt% using HPLC analytical technique.

Table 2: Fermented sweet sorghum feed composition on arrival

Name	Sucrose	Glucose	Fructose	Glycerol	Acetic Acid	Ethanol
wt%	1.30	0.18	1.64	0.47	0.48	4.25
Std dev	0.25	0.01	0.16	0.04	0.04	0.25

Apart from using hydrometer readings, which provide accuracy to 0.2 proof, HPLC protocols were also used to verify ethanol concentrations in various liquid samples taken throughout the pilot plant during operation. Table 4 compares results from hydrometer tests as well as HPLC results for ethanol samples drawn downstream of E-4 at three time intervals. Temperature corrected hydrometer results for ethanol volume concentrations agree with ethanol concentrations established with HPLC within 0.5%.

Table 4: Ethanol product concentration: Hydrometer and HPLC comparison

Time (pm)	Measured Proof (°F)	Temperature at proof measured (°F)	Corrected Proof at 60°F	Hydrometer Results Vol%	HPLC Results Vol%
3:23	190.0	77.7	185.7	92.9	93.03
3:36	191.0	78.1	186.8	93.4	93.03
6:13	193.6	74.0	190.0	95.0	95.64

Feed, stillage, T-1 and T-2 bottoms product, T-2 tray 3, and sparger samples were also collected and are tabulated in Table 5. Ethanol was not found in stillage stream, leaving the process. For feed ethanol at 4.5 vol% the ethanol content found on tray 3 on T-2 was 43.4 vol%. The ethanol content in rises in the liquid as we traverse the column height, as intended in distillation.

Table 5: Ethanol concentration at various process streams in ASU

Process Stream	**Feed**	**Stillage**	**Product**	**T-2 Tray3**	**T-2 Btms**
vol%	4.49	0.00	96.50	43.35	28.43
Std Dev	0.05	0.00	0.03	3.06	

In order to fulfill TTB requirements of tracking ethanol concentrations experiments were conducted to establish the variation of ethanol content in the feed storage tanks over time. In addition to ethanol sugars (sucrose, fructose, glucose) and fermentation products, namely acetic acid, and glycerol were also measured in HPLC experiments. Average ethanol concentrations of feed was approximately 5wt% and increased slightly from the time sweet sorghum shipment was received, 10th Oct 2013, to process run on Nov 4th 2013. This can be attributed to continued fermentation inside storage tanks after delivery. Unfermented sugars were still present in the fermented broth and accounted for 2.4 wt% of the solution.

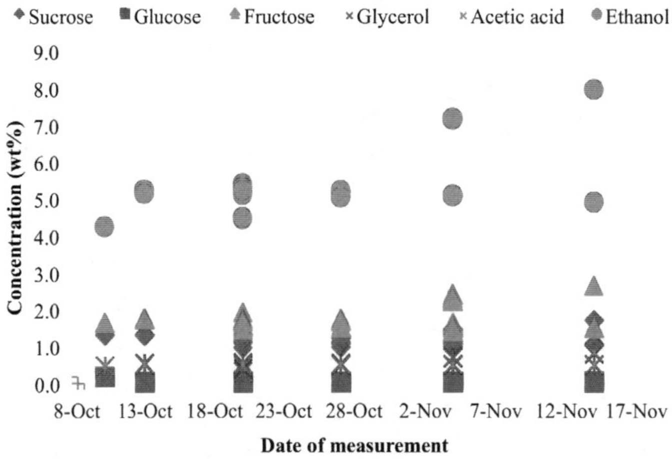

Figure 7: Feed characterization: Level of sugars, ethanol, and fermentation by products over time

Stillage and ethanol product are recombined and recycled back to the storage area. The contents are diluted with water equivalent of steam added into the process. Average ethanol concentration of Tk-102 was found to be 3.4wt%. Overall, stratification in tanks was not very significant.

GC experiments were conducted to estimate levels of dissolved CO_2 in sweet sorghum feed based on Liu *et al*'s estimation method [18]. Additionally, CO_2 level in the waste stream (recombined product and stillage from process stored in Tk-102) was also measured to see if sparging the process streams had caused CO_2 reduction. Carbon dioxide levels in feed samples from Tk-101 were 0.062 g/L and from Tk-102 was 0.03 g/L. A 52% reduction in carbon dioxide levels is seen. We expect the level of CO_2 to decrease with time and with the progress in fermentation occurring in the feed.

3.3 Economics of Small Scale Ethanol Separation

Capital and installation costs of constructing a farm scale unit are tabulated in Table 6. These costs represent actual spending on equipment and installation spending. An initial total fixed capital cost of 251,000 USD was estimated in 2009. These values were based on industry quotes and internet pricing of readily available items. A 49% cost factor was used for installation costs. In reality, the total expenditure was 38% higher than initial estimates. Installation costs include labor and contractual services (67,400 USD not including engineering salaries of the PI and graduate student), piping, hardware and associated material cost (38,000 USD), and construction equipment rental costs (16,000 USD).

Table 6: Fixed capital and installation costs of ASU and comparison with initial estimates

Costs	Initial Estimate	Actual Cost	%Difference
Instrumentation (Control and Measurement)	60,400	74,300	19
Column Body and Internals	49,500	95,590	48
Boiler	16,000	33,800	53
Heat Exchangers	2,700	41,000	93
Pumps	3,700	2,400	(54)
Storage Tanks	36,500	28,300	(29)
Equipment Cost	168,800	275,390	39
Installation Cost	82,700	127,600	35
Fixed Capital Costs	251,500	402,990	38

Main contributors to the operation of a steam stripping ethanol separation facility are from operation of the boiler, which account for 75% of operating costs, seen in Table 7. Total annual operating costs for a plant operating for 45 weeks is 68,300 USD. Electricity costs includes demand from all pumps (P1, P2, P3, P4, and P101 at 1.5 hP, E4 and E-5 fan motor at 2 and 1.5 hP respectively).

Table 7: Operational costs for ASU

Direct Operating Costs - excluding raw materials	**Usage**		**Unit Cost**		**Annual Cost $/yr**
Utilities	Utility use		$/unit		
Electricity (distillate, bottoms, reflux and feed pumps, condenser, air cooler blowers)	42877	kW-hr	0.09	$ / kW-hr	3,859
Boiler fuel cost	678	SCFH	9.78	$/1000 SCFH	50,827
Boiler water cost	67	gph	2.99	$ / kgal	1,531
Total Utility Costs					*56,217*
Maintenance and repairs (3% of fixed capital)					*12,090*
Total Annual Operating Expenses ($/yr)					**68,307**

3.4 Safety Considerations

A total relief rate of 1,877 lb/h of an ethanol-water mixture is expected from the first scenario (fire at base of column) at a Maximum Allowable Working Pressure (MAWP) of 50 psig. Liquid inventory included sump liquids for both T-1 and T-2, all liquid in trays (23 trays with 2 inches of liquid) and packing (14ft of packing with 3.5% liquid hold up and 4 trys), reflux drum level, and air cooler tubes liquid hold up. The pilot plant was thus equipped with two union type rupture disk assemblies, each with a rated capacity of 1251 lb/h air (relief rate is 5,329 lb/hr for ethanol-water mixture both calculated at 50 psig and 277°F) providing adequate relief capacity in case of over-pressurization.

Solutions of ethanol with even 80% water are flammable [19] and burn with a near smokeless flame (depending on the concentration of the ethanol in the liquid). Ethanol fires containing more than 10% ethanol need an Alcohol Resistant (AR) Aqueous Film Forming Foam (AFFF) [20]. The real concerns for fire at the pilot plant are in the process area, where high purity ethanol, up to 95 vol% can be found. The storage area contains only dilute concentrations of ethanol (less than 10 vol%), not considered flammable.

Liquid accumulation from T-1 and T-2 sumps, reflux drum, and air cooler liquid in the process area (of dimensions 20 ft x 11 ft) would account for a liquid layer of 0.84 inches. For the storage area, flammable liquid content is 6.5 vol% (design ethanol content in sweet sorghum liquid) of the 10,000 gallons present in both tanks. Calculations showed the height of flammable liquid in spills in process and storage areas was lower than the 1 inch contained spill height. This indicates the pilot plant is an exceptional case for fire response since the spill and flammable substances volumes are low. In addition the City of Stillwater fire department is not equipped with AR-AFFF foam, and it was not economical to maintain an inventory for such small quantities of alcohol. According to the EERC it is "important to understand all you can safely do is contain the incident and let the fire run its course. Knowing when to let this happen is an important component of safety" [21]. After discussions with the Fire Marshall for OSU it was established the best response in case of fire was indeed to let the fire run its course.

4. Conclusion

The Alcohol Separation Unit successfully served as a research and demonstration unit for bioethanol separation (190 proof product) from fermented sweet sorghum feedstock. Detailed records were maintained to document true operating costs and energy usage for feeds varying between 6-10 vol % ethanol with the ASU. This operational pilot plant can serve as the benchmark for development of sweet sorghum bioethanol processes. Updated and efficient equipment were used to set up rigorous measurement and control strategies making ASU one-of-a-kind multi functional bioethanol production unit. Start-up and shut down procedures have been established. Procedures were also established for on-site and laboratory analytical procedures for ethanol product determination. This demonstrative research strengthens existing body of sweet sorghum research.

5. References

1. Gelman, R., *2012 Renewable Energy Databook*, 2013, National Renewable Energy Laboratory.

2. EPA, *EPA Proposes 2014 Renewable Fuel Standards, 2015 Biomass-Based Diesel Volume*, 2013, Office of Transportation and Air Quality

3. Rao, S., et al., *Sweet sorghum for biofuel and strategies for its improvement* 2009: International Crops Research Institute for the Semi-Arid Tropics.

4. Reddy, B.V., et al. *Sweet sorghum as a biofuel crop: Where are we now.* in *Proceedings of sixth international biofuels conference*. 2009.

5. Rooney, W.L., et al., *Designing sorghum as a dedicated bioenergy feedstock.* Biofuels, Bioproducts and Biorefining, 2007. **1**(2): p. 147-157.

6. Turhollow, A.F., E.G. Webb, and M. Downing, *Review of sorghum production practices: Applications for bioenergy.* ORNL/TM-2010/7. Oak Ridge Natl. Lab., Oak Ridge, TN, 2010.

7. *ICRISAT: Sweet Sorghum Provides Fuel and Food*, 2008.

8. Linden, J.C., et al., *Preservation of potential fermentables in sweet sorghum by ensiling.* Biotechnology and Bioengineering, 1987. **30**(7): p. 860-867.

9. Worley, J.W., D.H. Vaughan, and J.S. Cundiff, *Energy analysis of ethanol production from sweet sorghum.* Bioresource Technology, 1992. **40**(3): p. 263-273.

10. Kundiyana, D.K., et al., *Influence of temperature, pH and yeast on in-field production of ethanol from unsterilized sweet sorghum juice.* Biomass and Bioenergy, 2010. **34**(10): p. 1481-1486.

11. Mukherjee, A., *On farm ethanol dewatering: Detailed design of distillation unit*, 2009, Oklahoma State University: United States -- Oklahoma. p. 146.

12. Osterman, K., *Stainless Steels: Cost-effective Materials*, in *Ethanol Producer Magazine*2012, BBI International: North Dakota.

13. Jacques, K.A., et al., *The alcohol textbook: a reference for the beverage, fuel and industrial alcohol industries* 2003: Nottingham University Press.

14. Summers, D.R., *Advances in Distillation Equipment and Applications*, in *AIChE Annual Meeting* 2006: San Francisco.

15. TTB, *Title 27: Alcohol, Tobacco and Firearms*, in *Part 19—Distilled Spirits Plants*, A.a.T.T.a.T. Bureau, Editor.

16. NFPA, *NFPA 30: Flammable And Combustible Liquids Code*, 2012, National Fire Protection Association.

17. NFPA, *NFPA 11: Standard For Low-, Medium-, And High-Expansion Foam*, 2010, National Fire Protection Association.

18. Liu, K., et al., *Fermentative production of ethanol from syngas using novel moderately alkaliphilic strains of Alkalibaculum bacchi.* Bioresource Technology, 2012. **104**(0): p. 336-341.

19. Hristova, M., D. Damgaliev, and D. Popova, *ESTIMATION OF WATER-ALCOHOL MIXTURE FLASH POINT.* Journal of the University of Chemical Technology and Metallurgy, 2010. **45**(1): p. 19-24.

20. White, D., *R-AFFF Best for Ethanol Fires*, in *The Chief* 2008.

21. Butters, T., *Ethanol: A New challenge for Fire Fighters*, in *VCOS News* 2008.

387396 Anhydrous Bio-Ethanol Production: Development and Optimization of Distillation and Dehydration Steps

Abstract:

Bio-ethanol production is a rapidly growing market due to the possible uses of the anhydrous product in energy generation. This production however, is weighed down by energetic demands and environmental effects.

This paper examines the distillation and dehydration steps of a bio-ethanol production plant complementing a sugar mill. The distillation step is optimized by performing rigorous energy analysis and heat integration on the process utilizing the resources made available by the sugar mill, while pervaporation is used in the dehydration step due to the formulation and use of a pervaporation module integrated into a process simulation engine, several schemes and membrane materials are assessed and compared in light of their energy demands and membrane area required.

The two steps are then combined and the optimization is conducted on the resulting process.

374159 Determination of the Reaction Equilibrium and Kinetics of the Hydroamination of β-Myrcene with Morpholine in a Liquid-Liquid-Two-Phase-System

Abstract:

Myrcene is obtained from renewable raw materials such as pine plants and offers the possibility to replace petroleum-based synthesis routes without being in competition with the food production. One example for the possibility of replacing a petroleum-based product is terpenyl amine, which can be produced by hydroamination of myrcene with morpholine. Terpenyl amine is already industrially used in paints or inks.

In this study terpenyl amine is investigated by the hydroamination reaction of myrcene and morpholine and the goal is to set up an experimentally validated process model. The reaction requires a homogeneous platinum catalyst and heat supply to increase the reaction yield. A liquid-liquid-two-phase-system, composed of an organic phase and an aqueous phase, is used to achieve high selectivity in combination with a low catalyst loss. While the organic phase consists only of the reactant myrcene, the catalyst in addition to the second reactant morpholine is dissolved in the aqueous phase. Since the catalyst is practically not soluble in myrcene and myrcene does not dissolve in water, the reaction takes place only at the interface between the organic and the aqueous phase. After the reaction the product terpenyl amine partitions in the organic phase. By the use of this liquid-liquid-two-phase-system the loss of catalyst can be minimized.

For the design of an industrial relevant process, knowledge of the reaction equilibrium and the reaction kinetics is required. Furthermore, the dependency of the reaction equilibrium and the reaction kinetics on temperature has to be measured. The dependence between reaction equilibrium and temperature is experimentally determined in a glass reactor and can be described by the Van't Hoff equation. In addition the calculation of the reaction enthalpy is possible via the Van't Hoff equation. For the measurement of the reaction kinetics, a double stirred cell is used. This ensures a stable interfacial area while simultaneously mixing the two individual phases without intermixing them. Thus, the reaction kinetics can be concluded from the measurement of the concentrations of reactants and product at different temperatures over time. By using an Arrhenius approach the experimentally determined reaction rate coefficients are described as a function of the temperature. Hence, the activation energy and the pre-exponential factor are determined. Applying these two approaches, a process model for the calculation of the reaction equilibrium and the reaction kinetics can be developed. Additionally, this process model allows a simulation of an industrial process. With these results of the simulation, both the size of a reactor and the residence time can be defined for the industrial scale to meet production requirements.

Acknowledgement:

The project „Sustainable Chemical Synthesis (SusChemSys)" is co-financed by the European Regional Development Fund (ERDF) and the state of North Rhine-Westphalia, Germany, under the Operational Programme „Regional Competitiveness and Employment" 2003 - 2013.

383235 Air Dispersion Modeling with Multi-Criteria Decision Approach for Sustainability

Abstract:

With the advancement of computer simulation software, air dispersion modeling is widely used to estimate or predict the concentration of air pollutants or toxicants emitted from various sources for different purposes (e.g., air permitting and regulatory compliance). A preliminary air permitting study can identify the air permitting programs necessary for different manufacturing plants or refineries. In this work, air dispersion modeling software AERMOD is utilized in a preliminary air permitting study with maximum process throughputs from a manufacturing plant for possible emission sources. Initial results show that the concentration of one of the pollutants exceeding the regulation standard.

To reduce the air pollution and comply with the regulatory requirements, the plant needs to consider several options to control the concentration within the air permitting limit. A multi-criteria decision-making approach, the analytic hierarchy process (AHP), is utilized with multi-level hierarchical structure of objectives, criteria and alternatives. Based on the philosophy of the triple bottom line of sustainability (economic, social and environmental), several sub-criteria are defined and the pertinent data are derived by using a set of pairwise comparisons. These comparisons are used to calculate the relative performance measures of alternative options and the results illustrate the benefits of the multi-criteria decision approach.

Multi-objective Optimization for Air-quality Monitoring Network Design

Min Chen and Qiang Xu

Dan F. Smith Department of Chemical Engineering
Lamar University, Beaumont, TX 77710, USA

Abstract

For an industrial zone where petrochemical and chemical plants are heavily populated, the design of an effective air-quality monitoring network (AQMN) is very important for local environmental and industrial sustainability. Given the situation that emission profiles from chemical plants may dynamically changes with respect to time due to their capacity expansion/reduction, raw material and product portfolio variation, or manufacturing technology advancement, the design of a regional AQMN should be generally sensitive and cost-effective to emission changes from all of the plants in the region. In this paper, a general methodology with multi-objective and deterministic optimization for AQMN design/redesign is developed.

In this methodology, Gaussian dispersion model is firstly employed to create spatial pollution distributions according to historical meteorological conditions and updated emission source profiles in a studied chemical industrial zone. Next, an MILP model is developed to optimally determine the best design of an AQMN, which allows the relocation of old monitoring stations and the addition of new monitoring stations to maximize the number of detections for potential pollution violations. Certainly, such a design should also minimize the total budget cost with considerations of old monitor relocation, new monitor construction, and land use at different prices in different locations, as well as satisfy many other spatial and temporal constraints. The Pareto frontier of pollution-violation detection times vs. total budget cost will be provided for decision makings in the final solution. The proposed methodology can be used to design a new AQMN from the scratch or to reconfigure a current AQMN to monitor regional air quality more effectively. Multiple case studies are employed to demonstrate its efficacy.

MANCHESTER
1824
The University of Manchester

AIChE

Title: Optimal Design and Integration of a Process Site Waste Heat Recovery System
Paper Number: 382091

Authors: Gbemi Oluleye, Megan Jobson & Robin Smith

The Challenge!

AIChE

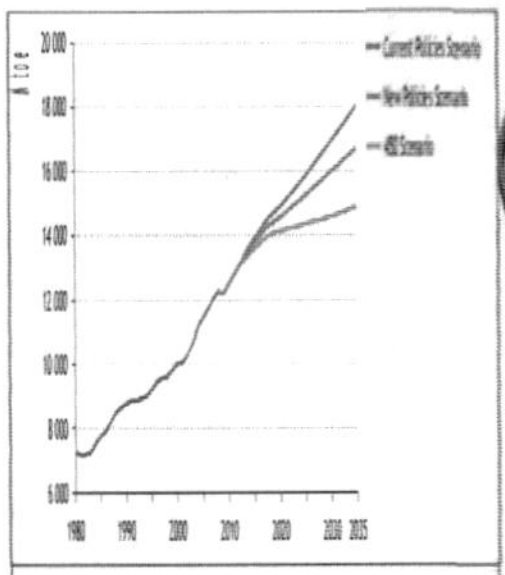

Increasing energy demand and depleting reserves

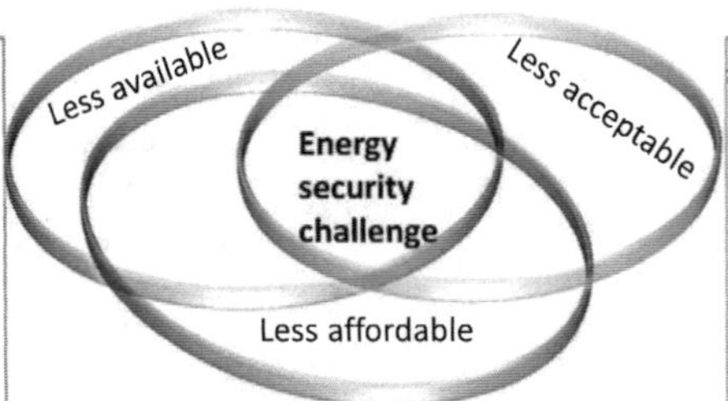

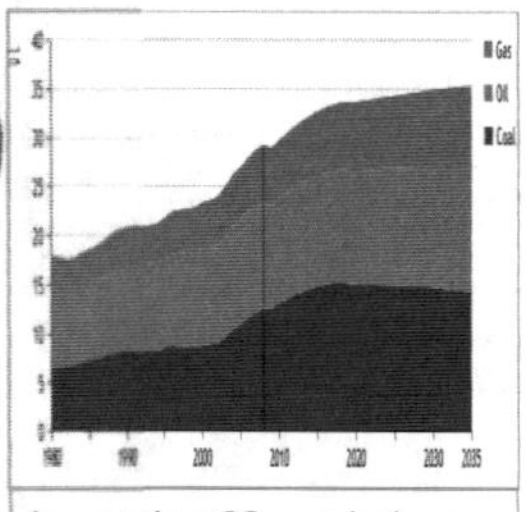

Increasing CO_2 emissions

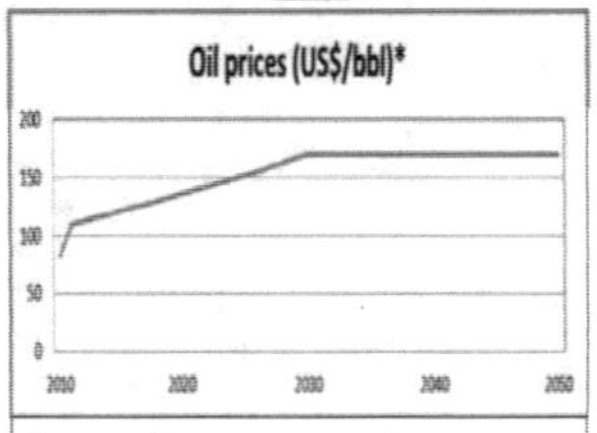

Increasing energy prices

Yet energy is still being wasted!

AIChE

CONTENTS :

1. Introduction
2. Background
3. Objectives
4. Methodology
5. Case study
6. Conclusions and future work

AIChE

1. Introduction

Introduction

- A considerable amount of energy is wasted at medium to low temperatures in the process industries
  - ✓ In the United Kingdom, heat wasted
    - ➢ At least 40% of the energy content of fuel[1]
    - ➢ The heat wasted can satisfy thermal demand of England
  - ✓ In the United States
    - ➢ Two-thirds of input energy for electricity generation is lost as heat during conversion processes[2]
    - ➢ the heat wasted can satisfy domestic demand for heating

"A major drawback is the location of industries"

- There could be potential in the heat wasted to reduce emissions, costs and improve energy efficiency

[1] Ammar,Y. et al 2012. Low grade thermal energy sources and uses from the process industry in the UK. Applied energy,pp.3-20
[2] Rattner A. S., Garimella S. Energy harvesting, reuse and upgrade to reduce primary energy usage in the USA. Energy 2011; 36(10): 6172-83.

Introduction

- Optimization strategies have been introduced to design energy systems
  - ✓ An efficient and systematic decision-making approach to ensure capital energy trade-offs are captured is provided

Optimisation →
- Design Optimisation: finding the nominal operating load of a system
- Operational Optimisation: finding the optimal operating point
- Retrofit Optimisation: addition of new features to an old system to improve the operating point

This work involves design optimisation for the waste heat recovery system and retrofit optimisation for integrating the system into an existing site

AIChE

2. Background

AIChE

Background

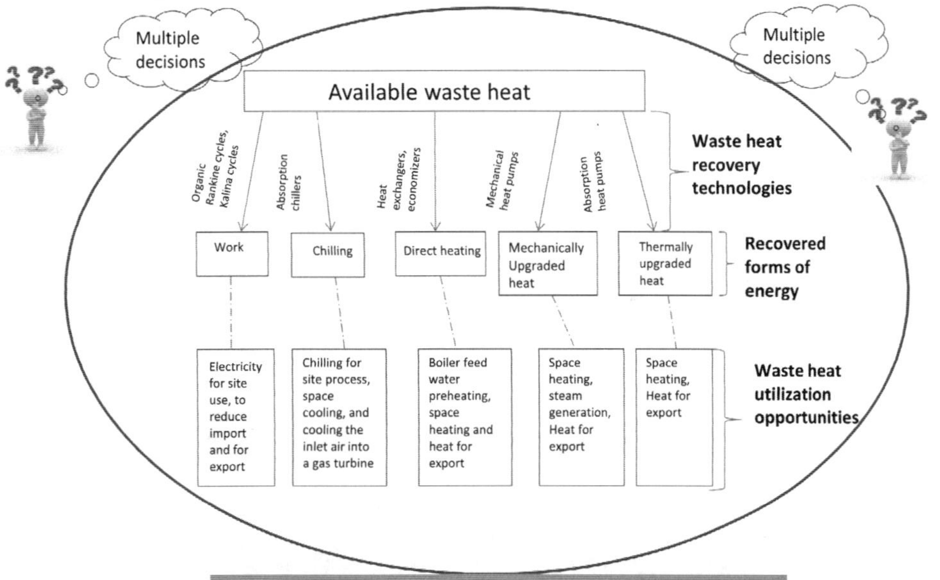

There are diverse sources of waste heat in a process site

Background 1 (Waste Heat Sources)

AIChE

- Waste heat is available from site processes
  - The pinch concept has been introduced
    - ✓ Thermodynamically feasible energy targets (including heat and power targets) are set
    - ✓ Maximum energy recovery can be guaranteed

"What happens to the residual heat when a site is pinched?"

- Waste heat also available from site cogeneration system designed to satisfy site energy demand
  - ➢ Condensate from utility system
  - ➢ Fired heater exhaust (limited by acid dew point)
- Conventional sinks
  - ➢ Heat rejected to cooling water
  - ➢ Heat rejected to air through the stack

Heat from different units available over a wide temperature range

Background 1 (Waste Heat Sources)

AIChE

- Systematic way developed to represent the available waste heat sources in a process site
  - The waste heat source profile[3]
    - ✓ Shows the energy flow variation with temperature
    - ✓ Temperature shifted by ΔT_{min} for heat recovery

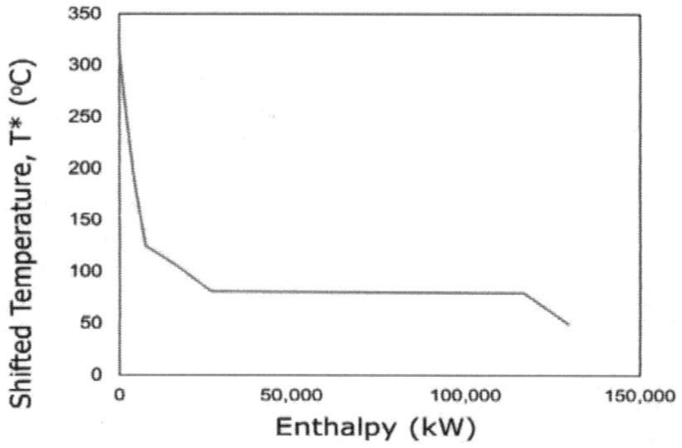

The kinks on the profile represent the supply and target temperature of the waste heat sources

[3] Oluleye G., Jobson M., Smith R., Perry S.J., 2014, Evaluating the potential of a process site for waste heat recovery, Chemical Engineering Transactions, 39, 1069-1074 DOI:10.3303/CET1439179

AIChE

Background 2 (Waste Heat Recovery Technologies)

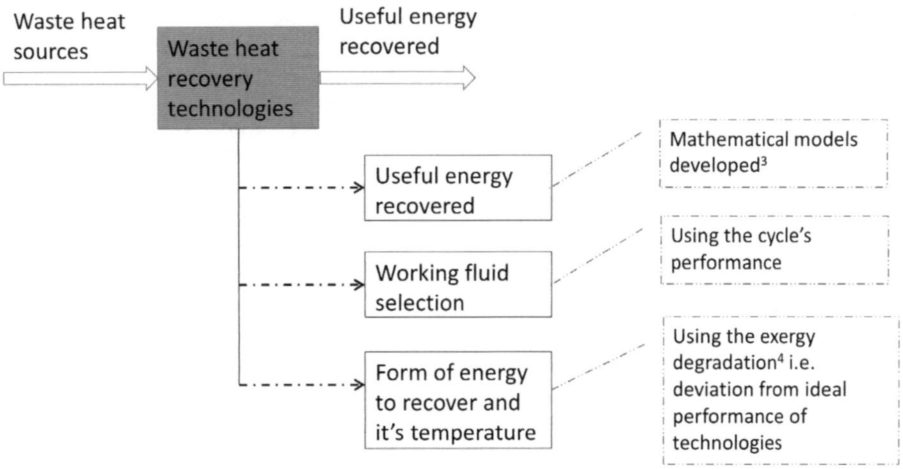

[4] Oluleye G., Jobson M., Smith R., 2014, Exergy analysis for the design of site waste heat recovery systems, Applied Thermal Engineering (Submitted)

L287c – 11 2014 AIChE Annual Meeting Centre for Process Integration © 2014

AIChE

Background 2 (Useful energy recovered)

- Simple models proposed to evaluate the useful energy recovered

$$Performance_{ideal} = \text{f}\ (Technology\ temperatures)$$

$$Performance_{real} = Factor_{Technology}\ \text{X}\ Performance_{ideal}$$

$$Factor_{Technology} = \alpha_{technology}(Performance_{ideal}) + \beta_{technology}$$

- The ideal performance is synonymous with the ideal efficiency for an Organic Rankine cycle
- The Factor accounts for irreversibility's in the technology components and working fluid non-ideal behaviour
- Model can be regressed against rigorous simulations
- Validated in Oluleye *et al.* 2014[3]
- Model does not account for the usefulness of the heat sources

[3] Oluleye G., Jobson M., Smith R., Perry S.J., 2014, Evaluating the potential of a process site for waste heat recovery, Chemical Engineering Transactions, 39, 1069-1074 DOI:10.3303/CET1439179

L287c – 12 2014 AIChE Annual Meeting Centre for Process Integration © 2014

AIChE

Background 2 (Exergy Output factor)

- Exergy output factor
  - ➢ Includes 'usefulness' of the heat source (waste heat)

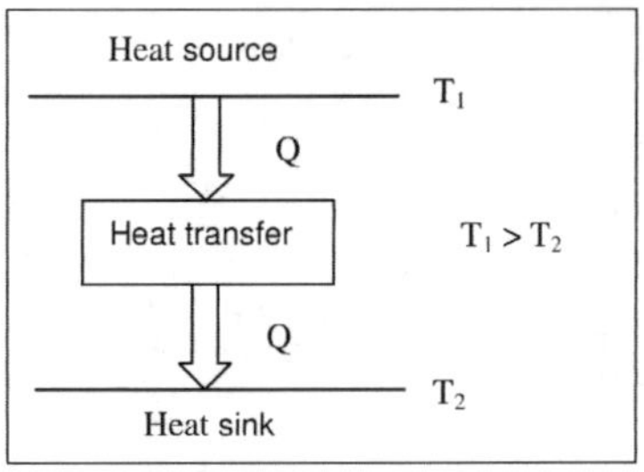

$$Exergy\ Output\ Factor = \frac{Output\ exergy\ tranfer\ by\ heat}{Input\ exergy\ transfer\ by\ heat}$$

$$Exergy\ transfer\ by\ heat = 1 - \frac{T_{ambient}}{T_i}$$

- ➢ Real performance is corrected for by the exergy output factor

$$Performance_{real*} = (Factor_{Technology}\ \text{x}\ Performance_{ideal})\ \text{x}\ Exergy\ Output\ Factor$$

The real performance is corrected for by the exergy output factor to account for the usefulness of the heat source so as to predict what quality of heat to use to recover work, chilling, heat etc.

L287c – 13 2014 AIChE Annual Meeting Centre for Process Integration © 2014

AIChE

Background 2 (Useful Energy Recovered)

- Different forms of energy can be recovered from industrial waste heat
  - ➢ Determine the best quality of heat to use to generate work, chilling, direct heat, mechanically upgraded heat and thermally upgraded heat
  - ➢ Heat recovery temperatures
- Using the exergy degradation associated with heat recovery technologies
  - ➢ measures deviation from ideal performance of technologies

$$Exergy\ degradation = \frac{Performance_{ideal} - Performance_{real}}{Performance_{ideal}}$$

- ➢ Ideal performance is the maximum possible thermodynamically

The rule is to reduce the exergy destruction accompanying the energy utilization/conversion process

L287c – 14 2014 AIChE Annual Meeting Centre for Process Integration © 2014

AIChE

Background 2 (Waste Heat Recovery Technologies)

- Organic Rankine cycles: Work generation

$$W_{net} = Q_{available} \times Factor_{ORC} \times \left(1 - \frac{T_{condenser}}{T_{evaporator}}\right)$$

$$T_{available} > T_{evaporator} + \Delta T_{min}$$

$$Factor_{ORC} = \alpha_{ORC} \times \left(1 - \frac{T_{condenser}}{T_{evaporator}}\right) + \beta_{ORC}$$

$$\text{Exergy degradation} = \left(\left(Factor_{ORC}\right) \times \left(\frac{1 - \frac{T_{ambient}}{T_{evaporator}}}{1 - \frac{T_{ambient}}{T_{available}}} \right) \right)$$

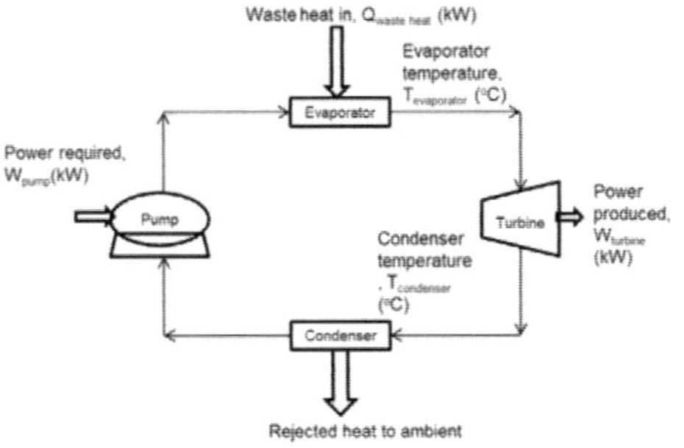

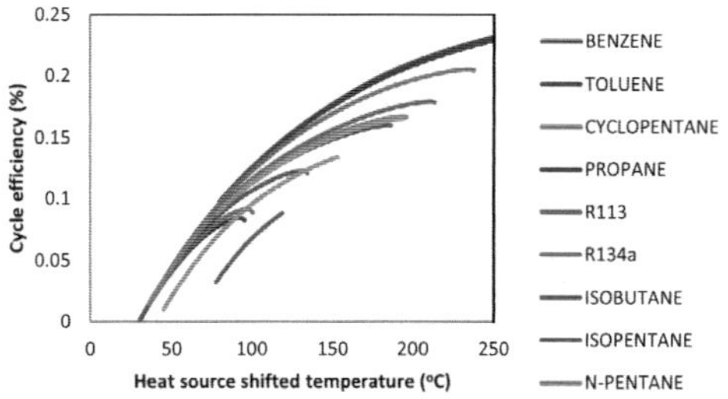

AIChE

Background 2 (Waste Heat Recovery Technologies)

- Absorption chillers: Chilling provision

$$Q_{chilling} = (Q_{available} + W_{pump}) \times Factor_{AbC} \times \left(1 - \frac{T_{condenser}}{T_{generator}}\right)\left(\frac{T_{evaporator}}{T_{condenser} - T_{evaporator}}\right)$$

$$Factor_{AbC} = \beta_{AbC} / (COP_{ideal} - \alpha_{AbC})$$

$$\text{Exergy degradation} = \left(\left(Factor_{AbC}\right) \times \frac{\left(\frac{1 - \frac{T_{ambient}}{T_{generator}}}{1 - \frac{T_{ambient}}{T_{available}}}\right)}{\left(\frac{1 - \frac{T_{ambient}}{T_{evaporator}}}{1 - \frac{T_{ambient}}{T_{chill}}}\right)} \right) \times \left(\left(Factor_{VCC}\right) \times \left(\frac{1 - \frac{T_{ambient}}{T_{sink}}}{1 - \frac{T_{ambient}}{T_{condenser}}} \right) \right)$$

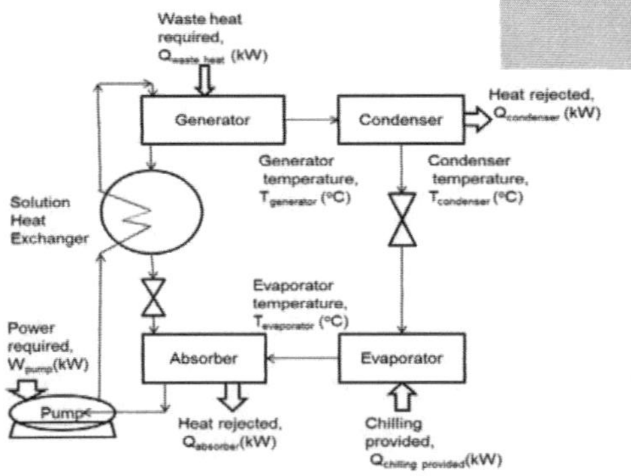

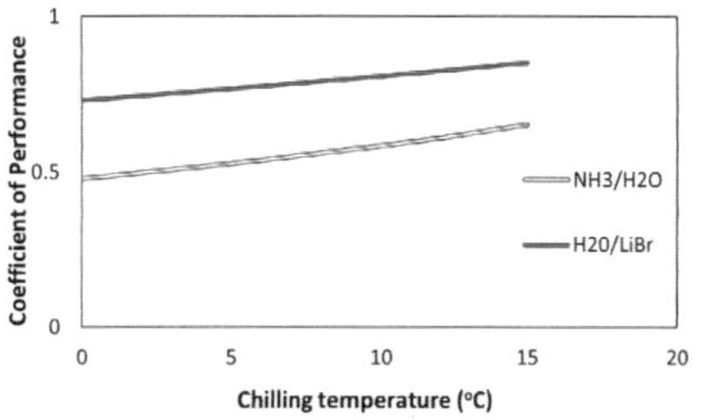

AIChE

Background 2 (Waste Heat Recovery Technologies)

- Absorption heat pumps: Thermally upgraded heat

$$Q_{upgraded} = (Q_{available} + W_{pump}) \times Factor_{AHP} \times \left[1 + \left(1 - \frac{T_{condenser}}{T_{generator}}\right)\left(\frac{T_{evaporator}}{T_{condenser} - T_{evaporator}}\right)\right]$$

$$Factor_{AHP} = \beta_{AHP} / (COP_{ideal} - \alpha_{AHP})$$

$$\text{Exergy degradation} = (Factor_{AHP}) \times \left(\frac{\left(\frac{1 - \frac{T_{ambient}}{T_{sink}}}{1 - \frac{T_{ambient}}{T_{condenser}}}\right)}{\left(\frac{1 - \frac{T_{ambient}}{T_{available}}}{1 - \frac{T_{ambient}}{T_{generator}}}\right)}\right)$$

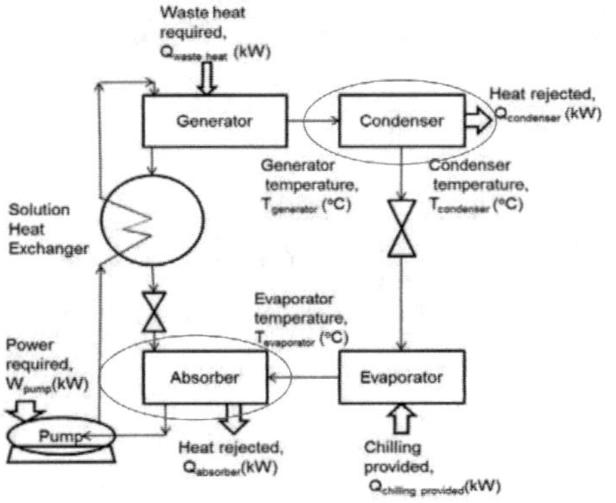

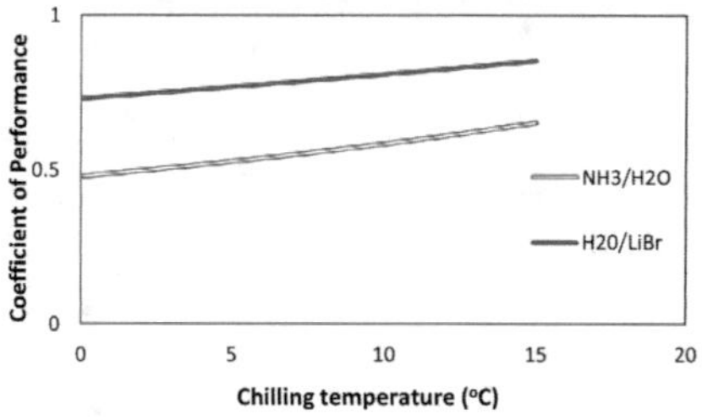

AIChE

Background 2 (Waste Heat Recovery Technologies)

- Mechanical heat pumps: mechanically upgraded heat

$$Q_{upgraded} = (Q_{available}) \times Factor_{MHP} \times \left[\left(\frac{T_{condenser}}{T_{condenser} - T_{evaporator}}\right)\right]$$

$$Factor_{MHP} = \alpha_{MHP} \times \left(\frac{T_{condenser}}{T_{condenser} - T_{evaporator}}\right) + \beta_{MHP}$$

$$\text{Exergy degradation} = (Factor_{MHP}) \times \left(\frac{1 - \frac{T_{ambient}}{T_{sink}}}{1 - \frac{T_{ambient}}{T_{condenser}}}\right)$$

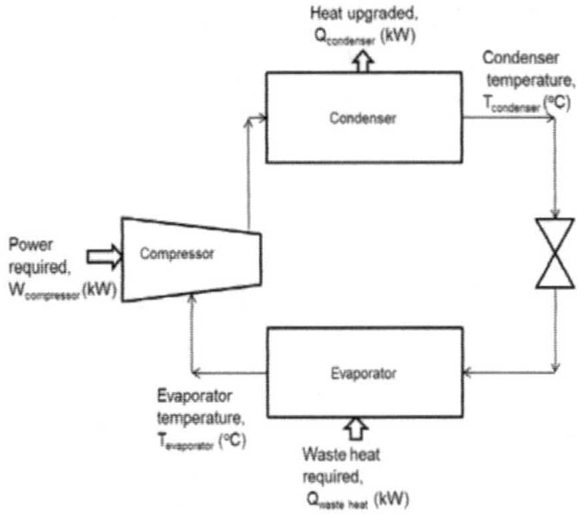

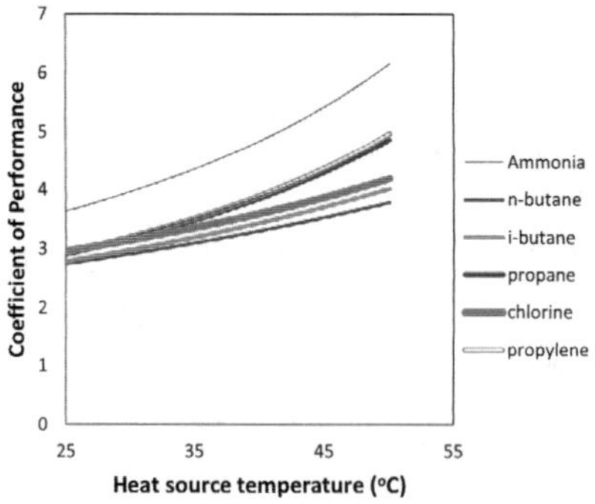

Background 2 (Waste Heat Recovery Technologies)

- Heat exchangers: direct use of heat

$$Q_{recovered} = (Q_{available})$$

$$T_{available} > T_{sink} + \Delta T_{min}$$

$$\text{Exergy degradation} = \left(1 - \left(\frac{1 - \frac{T_{ambient}}{T_{sink}}}{1 - \frac{T_{ambient}}{T_{available}}}\right)\right)$$

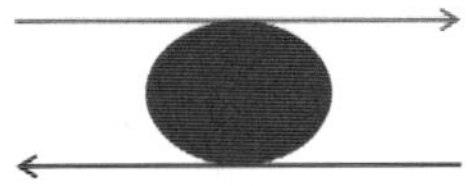

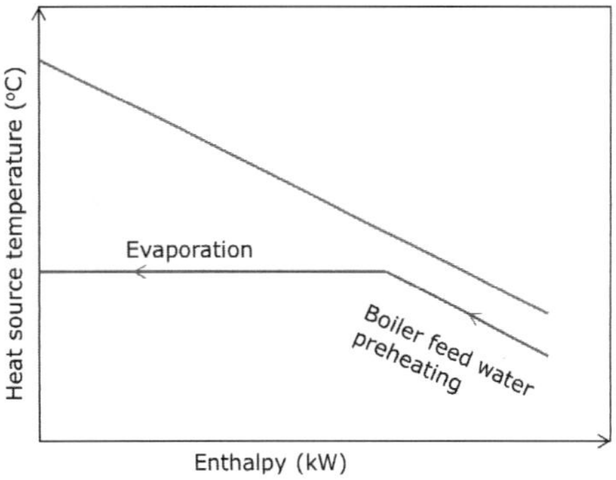

Background 3 (Waste Heat Utilisation Opportunities)

- ➢ Diverse opportunities exist to use the recovered energy
- ➢ Grouped according to the end-users; on-site and off-site

Recovered Energy	On – site opportunities	Off – site opportunities
Electricity	• Reduce import on-site • Satisfy site electricity demand	• Electricity export • Charging port for electric vehicles
Chilling	• Space cooling • Gas turbine compressor inlet air chilling • Process chilling	• Space cooling
Heat	• Boiler feed water preheating • Space heating	• Heat export to new buildings • Heat export to existing buildings • Heat for agriculture

AIChE

Summary: Background

- Waste heat available over a wide temperature range from different sources
- Different forms of energy can be recovered from industrial waste heat
  - ✓ Form of energy to recover and it's temperature determined using exergy degradation associated with technologies
- Various opportunities exist to use the recovered energy from industrial waste heat
- The operational limits of the waste heat recovery technologies and physical limitations of a site dictate what waste heat it is physically possible to recover

AIChE

3. Objectives

AIChE

Objectives

This work is based on the premise that waste heat recovery can improve the energy security of process sites

- Identify and evaluate options for recovering useful energy from industrial waste heat
- Design process site waste heat recovery systems
- Integrate the waste heat recovery system with the existing site cogeneration system
- Investigate the benefits of waste heat recovery

L287c – 23 | 2014 AIChE Annual Meeting | Centre for Process Integration © 2014

AIChE

4. Methodology

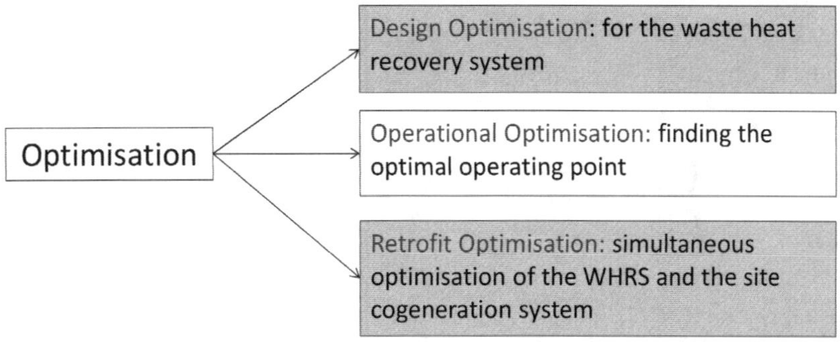

L287c – 24 | 2014 AIChE Annual Meeting | Centre for Process Integration © 2014

Optimisation-based Design of WHRS

- Optimisation strategies provide an efficient and systematic decision making approach

An objective function	Predictive Model	Decision variables
(Scalar quantitative performance measure)	(describing system behaviour) • Equations • Inequality constraints	(degrees of freedom) • Operational • Structural

- Optimisation framework applied to design waste heat recovery systems

Form of energy recovered, heat recovery temperatures and utilization opportunities determined to satisfy the objective function and imposed constraints

Optimisation-based Design of WHRS

- Mixed integer approach is used (simultaneous optimisation of structural and operational variables)
- The strategy requires:

1. Creating a reducible superstructure that includes all technologies that can be potentially selected
2. Reducing the superstructure in order to select the design to satisfy the objective function
3. Showing the optimization output i.e. placement of technologies on the waste heat source profile
4. Evaluation of results

AIChE

Design Optimisation

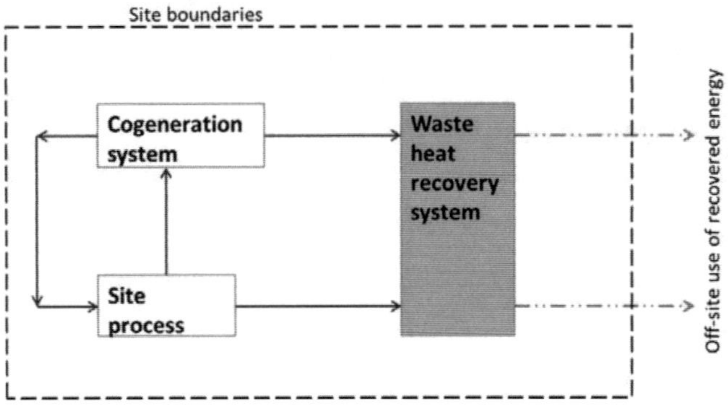

- Objective function

Max { Economic potential associated with heat export + Economic potential associated with electricity export }

AIChE

Retrofit Optimisation

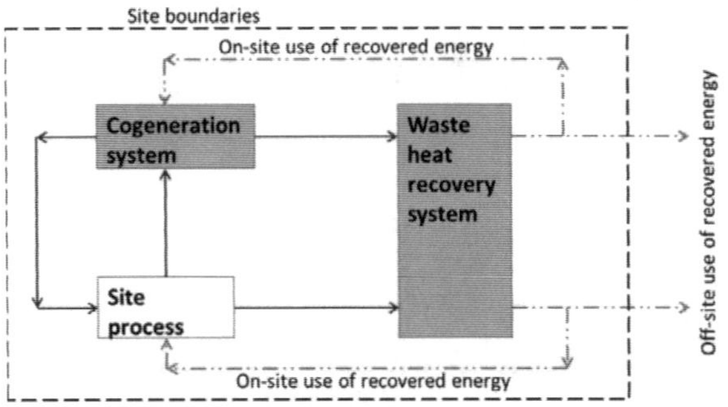

- Objective function

Max { Site operating cost before heat recovery − Site operating cost after heat recovery + Economic potential associated with heat export + Economic potential associated with electricity export − Total annualised cost of opportunities integrated with the cogeneration system }

Design involves optimising the waste heat recovery system and the site cogeneration system

Optimisation Framework

AIChE

- Variables

Integer variables
- On/off modes of technologies i.e. selection of technologies and associated utilisation opportunities

Continuous variables
- Recovered energy associated with utilisation opportunities
- Cogeneration system fuel flow
- Flow rate of steam to steam turbines
- Site power import and export
- Condensate return temperature

- Constraints imposed, for example
  - Energy and mass balances
  - Implication constraints for technology sizes
  - User constraints can also be imposed
- ❑ Optimisation outputs
- ❑ Results/ Design evaluation:
  - Change in site energy efficiency,
  - Change in site CO_2 Emissions
  - Change in site operating costs

L287c – 29 2014 AIChE Annual Meeting

Summary: Methodology

AIChE

- Optimisation framework developed to design new waste heat recovery systems for existing process sites
- Optimisation framework also developed to optimise the waste heat recovery system and site cogeneration systems to ensure a holistic approach to energy system design
- Design will be evaluated based on efficiency, economics and emissions

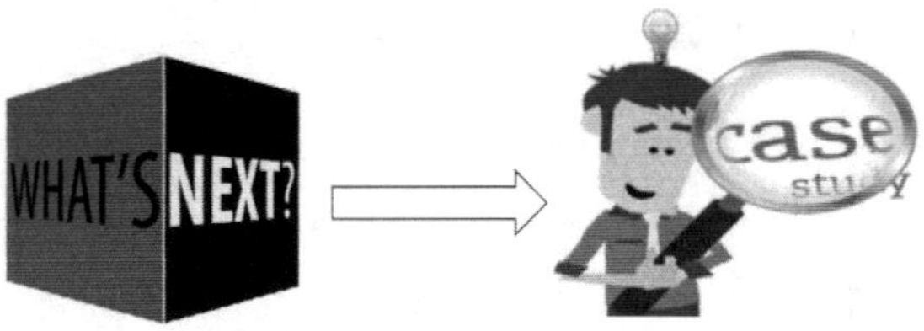

L287c – 30 2014 AIChE Annual Meeting

AIChE

5. Case Study

Case Study

AIChE

- Petroleum refinery[5]
  - 7 processing units: hydrotreaters (naphtha, kerosene and diesel), vacuum distillation unit, platformer, visbreaker and fluidised catalytic cracking unit
  - Cogeneration system (gas turbine, boilers, steam turbines etc.)
  - Refrigeration system
  - Cooling water system
  - Fired heaters
  - Site energy efficiency is 50%
  - Neighbourhood demand for heat: 2.4 MW
  - Power export allowed

Energy Prices (DECC, IAG 2012)

- Industrial electricity price: 13.4 p/kWh
- Industrial natural gas price: 3.2 p/kWh
- Domestic electricity price: 16.7 p/kWh
- Domestic natural gas price: 5.0 p/kWh
- Grid emission factor: 0.485 kg/kWh
- Natural gas emission factor: 0.193 kg/kWh

Capital costs

- Organic Rankine cycle [6]: 3510 £/$kW_{electricity}$
- Absorption chiller[6]: 374 £/$kW_{chilling}$
- Discount rate: 15%
- Operating hours: 8760

5. Fraser and Gillespie (1992) The application of pinch technology to retrofit energy integration of an entire oil refinery. Trans IChemE, vol. 70, Part A
6. Oluleye G., Jobson M., Smith R., 2014, A hierarchical approach for evaluation of waste heat utilization opportunities, Chemical Engineering Transactions, 39, 1093-1098 DOI:10.3303/CET1439183

Case Study

AIChE

- Design a waste heat recovery system for this refinery site
- Waste heat utilisation opportunities
  - ➤ Power generation for export
  - ➤ Hot water generation for export
  - ➤ Power generation for use within the site
  - ➤ Steam generation
  - ➤ Chilling for site process
  - ➤ Chilling the inlet air into the gas turbine
  - ➤ Boiler feed water preheating

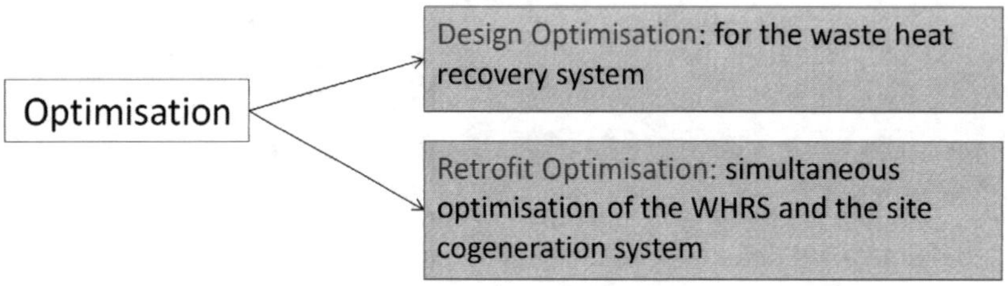

Case Study (Methodology Overview)

AIChE

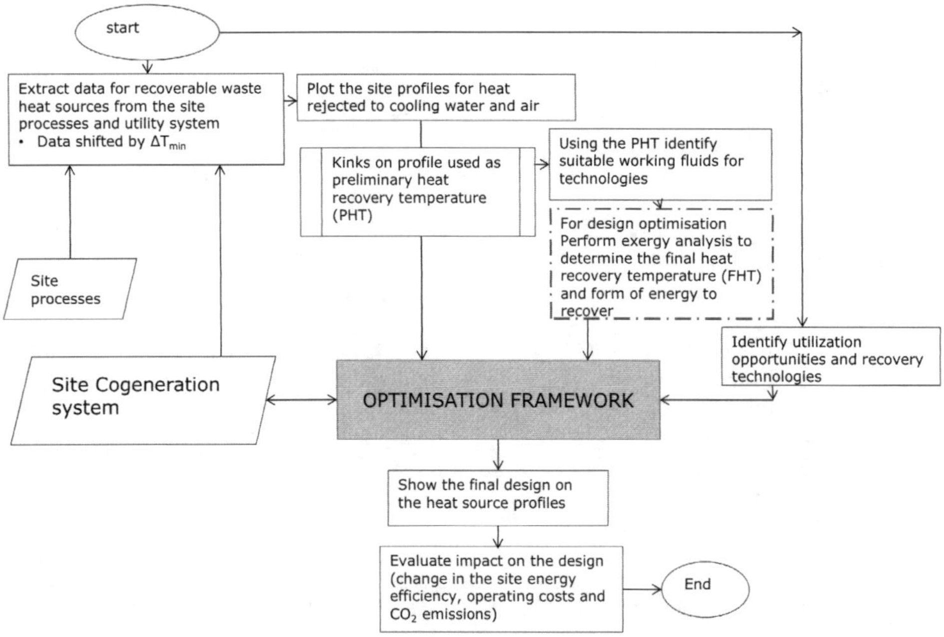

Case Study

AIChE

❖ Heat rejected to cooling water

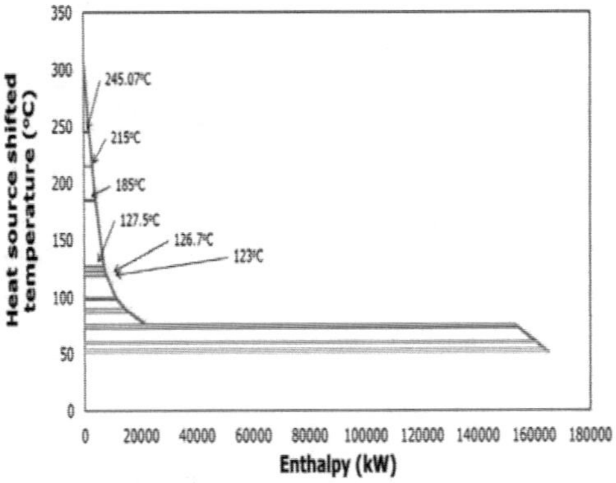

Kinks on profile used as preliminary heat recovery temperatures

❖ Heat rejected to air

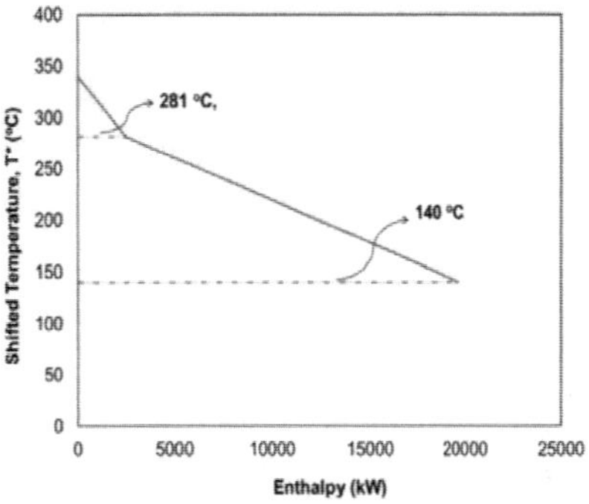

Organic Rankine cycle
- Above 80°C: benzene
- Below 80°C: cyclopentane

Absorption chiller
- Lithium Bromide/water

Mechanical Heat pump
- Ammonia

Case Study

AIChE

❖ Exergy analysis

➢ Select final heat recovery temperature and recovered form of energy to minimize exergy degradation associated with technologies

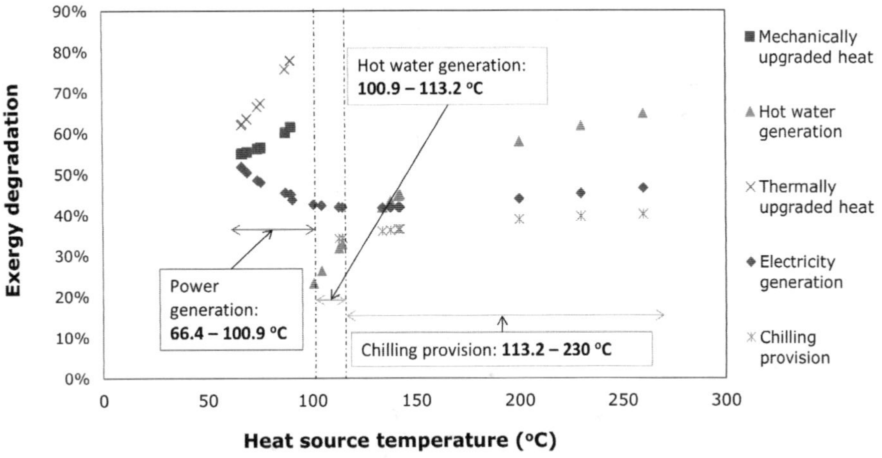

Case Study (Design without integration)

AIChE

❖ Waste Heat Recovery System Design

✓ Design for heat that was rejected to cooling water

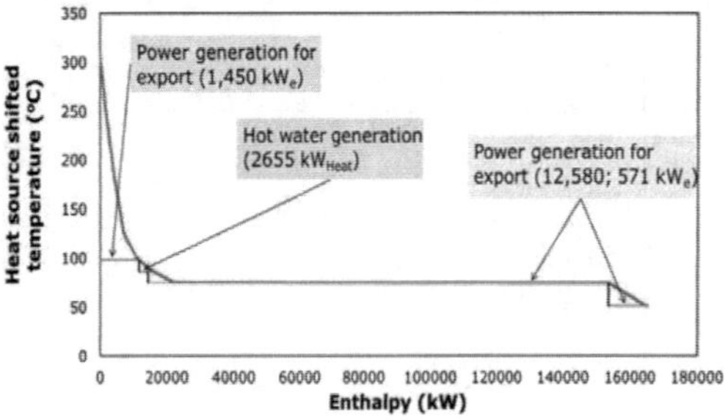

✓ Design for heat that was rejected to air

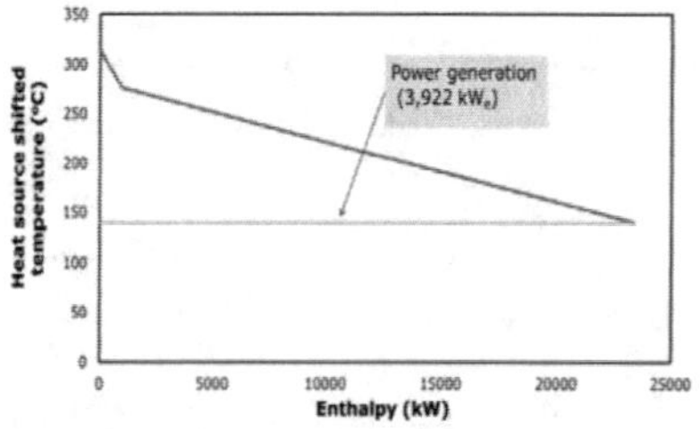

❖ Design Evaluation

Technologies required
- 4 Organic Rankine cycles
- 1 heat exchanger

❖ Total useful energy recovered
- ➢ 7.8% increase in site energy efficiency
- ➢ 11% of available waste heat recovered

❖ Potential to reduce emissions
- ➢ Reduces site emissions by 9.8%

❖ Economics*
- ➢ Reduces the site operating cost by 19.6%

*heat distribution cost to buildings not included in this analysis

Case Study (Design with integration)

AIChE

❖ Waste Heat Recovery System Design

✓ Design for heat that was rejected to cooling water

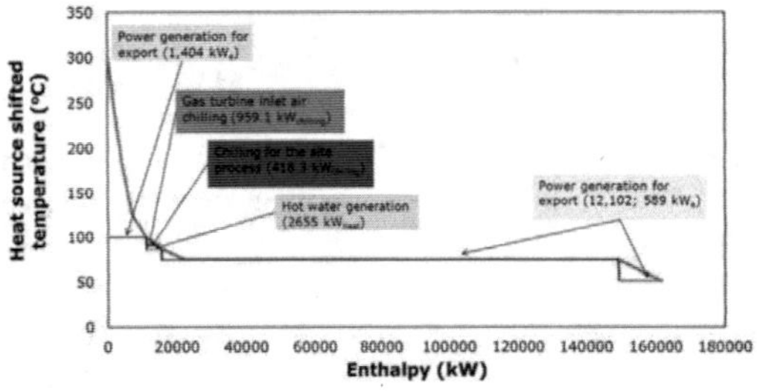

✓ Design for heat that was rejected to air

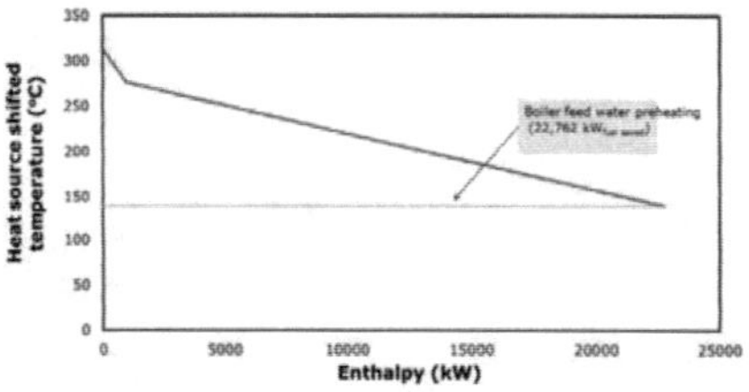

❖ Design Evaluation

Technologies required
- 3 organic Rankine cycles
- 1 heat exchanger
- 2 absorption chillers
- 1 economiser

❖ Total useful energy recovered
- ➢ 26 % increase in Site energy efficiency
- ➢ 22% of available waste heat recovered

❖ Potential to reduce emissions
- ➢ Reduces the site emissions by 17.7%

❖ Economics*
- ➢ Reduces the site operating cost by 30.38%

*heat distribution cost to buildings not included in this analysis

Summary: Case Studies

AIChE

Design basis	Useful energy recovered	Reduction in site CO_2 emissions	Reduction in site operating cost*	Technologies selected
Design without integration	11%	9.8%	19.6%	• 4 organic Rankine cycles • 1 heat exchanger
Design with integration	22%	17.7%	30%	• 3 organic Rankine cycles • 1 heat exchanger • 2 absorption chillers • 1 economiser

- Design with integration considers interactions between the waste heat recovery system and the site cogeneration system
  - ✓ Both systems optimised simultaneously resulting in better savings
  - ✓ Allows for steam generation and BFW preheating
  - ✓ Allows for use of power generated within the site
  - ✓ Savings in primary fuel consumption
  - ✓ Waste heat produced from cogeneration system is reduced
  - ✓ A holistic approach to design and optimisation has more benefits

L287c – 39 2014 AIChE Annual Meeting Centre for Process Integration © 2014

There is potential in Industrial waste heat:

AIChE

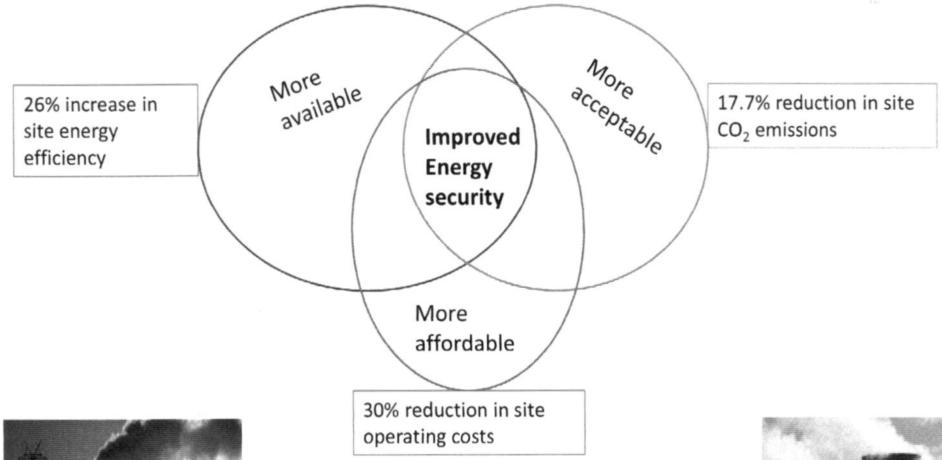

Don't waste heat!

L287c – 40 2014 AIChE Annual Meeting Centre for Process Integration © 2014

Summary: case study

AIChE

- Methodology developed to identify the potential of industrial waste heat and design a site waste heat recovery system
- System design evaluated based on efficiency, emissions and economics
- More benefits from integrating the waste heat recovery system with the site cogeneration system
- Benefits depend on the demand for recovered energy

AIChE

7. Conclusions and future work

Conclusions

AIChE

- ❑ Waste heat recovery can indeed improve the energy security of a process site since efficiency is increased, costs and emissions reduced
- ❑ Waste heat from both the site process and utility system should be taken into account
- ❑ Mathematical model of waste heat recovery technologies presented
- ❑ Exergy analysis introduced to determine the form of energy to recovery and the temperature of heat to use
- ❑ The methodology can be applied to any establishments producing waste heat
- ❑ A holistic approach to design should be encouraged

Future work

AIChE

- Take into account the economics of heat distribution for off-site use of recovered heat
- Develop a multi-period optimisation framework to account variations is heat availability, energy prices and ambient temperature
- Perform sensitivity analysis to determine the major drivers for economic recovery of waste heat in process sites.

AIChE

Thank You ☺

The financial support from Process Integration Research Consortium (PIRC) is sincerely acknowledged

From Data to Miniplant Concepts: An Approach to Accelerate Process Development

Alejandro Merchan-Restrepo[a], Duc Hoang-Minh[a], Günter Wozny[a], Harvey Arellano-Garcia[b]

[a] *Chair of Process Dynamics and Operation, Technische Universität Berlin, Sekr. KWT-9, D-10623 Berlin, Germany*

[b] *School of Engineering & Informatics, University of Bradford, West Yorkshire, BD7 1DP, United Kingdom*

Abstract

In this work, a hierarchical approach to process design and development starting from data to miniplant concepts is presented. It includes the experimental application and practical insights gained from the adaption of a developed optimal model-based design framework for an improved chemical process development. For this purpose, special attention is paid to the explicit modelling of all experimental steps so as to prepare, implement, and analyse experiments in order to have a realistic definition of the numeric design problem and to formally include experimental restrictions and sources of uncertainties in the problem formulation. The proposed framework will be discussed based on the quantification of the reaction kinetics in multi-phase systems using the example of the hydroformylation of long-chain olefins. The developed framework represents a systematic approach to support the development of chemical processes from laboratory experiments to the optimization of complete processes by investigating the design space and quickly screening alternatives.

The case study is concerned with the hydroformylation process of 1-dodecene, a homogeneous catalyzed reaction via transition metals. The main goals for this process focus on an economical catalyst recycling and a high selectivity regarding the linear aldehyde product n-tridecanal. For this purpose, the hydroformylation process in a DMF-decane thermomorphic solvent (TMS) system with a Biphephos-modified rhodium catalyst is investigated. The postulated reaction network was proposed by (Kiedorf et al. 2013) and successfully applied in (Müller et al. 2013).

Literature cited

G. Kiedorf, D. M. Hoang, A. Müller, A. Jörke, J. Markert, H. Arellano-Garcia, A. Seidel-Morgenstern, C. Hamel. Kinetic of 1-dodecene hydroformylation in thermomorphic solvent system using a rhodium-biphephos catalyst. Chemical Engineering Science (2013), http://dx.doi.org/10.1016/j.ces.2013/06.027.

D. Müller, D. M. Hoang, V. A. Merchan, H. Arellano-Garcia, Y. Kasaka, M. Müller, R. Schomäcker, and G. Wozny. Towards a novel process concept for the hydroformylation of higher alkanes: Mini-plant operation strategies via model development and optimal experimental design. Chemical Engineering Science (2013), http://dx.doi.org/10.1016/j.ces.2013.05.022

388384 Batch-to-Continuous Protocol – a Guideline for Lab-Scale Chemistry

Abstract:

Batch-to-continuous protocol – a guideline for lab-scale chemistry

M. Padaro, M. A. Liauw, ITMC - RWTH Aachen University,

Aachen, Germany

Introduction

Traditional lab-scale syntheses are typically developed by chemists, and carried out as batch reactions. When an industrial application becomes interesting, there often is a move towards continuous operation, be it because of higher productivities, more constant product quality, or higher safety. The transition batch-to-continuous is particularly difficult if crucial aspects were not considered during batch optimization.

A systematic protocol, based on a sound understanding of the reactive system, is desirable that will facilitate the implementation of a continuous operation. This would also resolve many scale-up issues.

Approach

In the project SusChemSys at RWTH Aachen, a systematic protocol is developed as a guideline to lab-scale chemistry. For a given reaction system, the batch protocol is assessed. Individual steps of the protocol, such as dosing, mixing, heating or reacting, are depicted as a material flow through spatially separated elements (see e.g. [1], [2]). This structure is now evaluated against selected criteria outlined below. Where a potential problem is identified, the guideline gives suggestions (many taken from literature) for solving this problem. After their implementation, the evaluation is resumed.

Results

The list of criteria encompasses issues like solids handling, corrosion, order of reactant addition, heat management and phase behavior. As an example, the *solids* issue may be resolved by varying the temperature or the solvent, by adding a co-solvent or an immiscible second liquid phase that sweeps up precursors to solids formation. All these points are backed by references where this approach has been successfully applied. Obviously, the *corrosion*issue will strongly benefit from including a corresponding material database in the protocol environment. Of course, many other criteria have to be addressed: mixing, viscosity, pressure dependences and pressure drop, separation.

It is crucial to acknowledge that all elements in the synthesis structure are closely related to one another. If e.g. an undesired reaction is identified, it may be decided to avoid it by adding an inhibitor to the reactants. That inhibitor may however lead to corrosion upstream of the reaction, or to solids formation downstream. Ideally, the synthesis chemist is assisted in taking the decisions by an expert system that reconciliates all these cross-correlations efficiently. As a limiting case, the protocol yields the insight that a given synthesis cannot be run continuously in an economically viable manner.

As a case study, the approach is demonstrated for one or two examples (e.g. [3]). It is finally discussed how process development engineers may be interested in supporting the development of this protocol that promises a smoother transition from lab-scale batch syntheses into continuous operation.

Acknowledgment

Financial support by SusChemSys is gratefully acknowledged. The project "Sustainable Chemical Synthesis (SusChemSys)" is co financed by the European Regional Development Fund (ERDF) and the state of North Rhine Westphalia, Germany, under the Operational Programme "Regional Competitiveness and Employment" 2007 - 2013.

References

[1] J. A. Arizmendi-Sánchez, P. N. Sharratt, "Multilevel Phenomenological Modelling Approach to Support the Evaluation and Generation of Intensified Processes", *European Symposium on Computer Aided Process Engineering – 15*, Eds. L. Puigjaner, A. Espuña (2005).

[2] H. Freund, K. Sundmacher, "Towards a methodology for the systematic analysis and design of efficient chemical processes: Part 1. From unit operations to elementary process functions", *Chem. Eng. Process* **47**(2008) 2051-2060.

[3] C. B. Minnich, L. Greiner, C. Reimers, M. Uerdingen, M. A. Liauw, "Bridging the gap: A nested-pipe reactor for slow reactions in continuous flow chemical synthesis", *Chem. Eng. J.* **168** (2011) 759-764.

387006 Introduction of an Organophilic Membrane System in an Ethanol Production Process Based on Lignocellulosic Biomass

Abstract:

Ethanol production from lignocellulosic biomass is a promising path to produce this alcohol for the purposes of clean energy generation. However, the production of ethanol via this route represents a challenge in terms of energy consumption and achievable purity, especially in the separation stages of the process. The concentration of ethanol after the fermentation stage ranges from 5 to 10% w/w depending on the available substrate in the fermentation broth and the strain used [1]. Conventional separation techniques such as distillation are widely used to separate ethanol from the fermentation broth, but the achievable concentration of the distillate by using this separation method may not be high enough for biofuel standard purity (>99%). Membrane-based separation techniques have been widely investigated for industrial applications as they are less energy intensive [2]. One of the most widely employed techniques for the separation of alcohols is pervaporation, and in particular using hydrophilic membranes to dehydrate ethanol beyond the azeotropic point reached during the distillation [3].

This work aims to implement an organophilic membrane module between the simultaneous saccharification and co-fermentation (SSCF) stage and the separation stages (distillation combined with hydrophilic membrane separation) to intensify the production of ethanol by adding a new unit that will allow a reduction of the overall energy demand of the process. The new unit will also increase the productivity of ethanol during the fermentation stage by reducing the inhibitory effects of this alcohol and the associated impurities on the fermenting microorganisms. The combination of fermentation and hydrophilic membrane system has been widely studied showing promising results in terms of ethanol production, but the impact of this combination on the overall process has so far not been considered [4,5].

The overall process configuration consists of a SSCF reactor, an organophilic membrane module and a separation section consisting of either distillation and/or a hydrophilic membrane process. The permeate stream from the organophilic membrane is sent to the subsequent separation stage to purify the ethanol. The retentate stream (mainly reducing sugars and other large molecules that do not permeate) is recycled back to the SSCF reactor to further the ethanol production by using all the available substrate.

In order to consider the relative merits of the configuration, a dynamic model is developed of the overall process taking into account each individual processing step. Whilst distillation modelling is generally straight-forward, membrane modelling is not due to the intrinsic characteristics of the membrane properties. This relates in particular to the organophilic membrane as the impact of the numerous impurities from the fermentation broth (e.g glucose, xylose, acetic acid, glycerol and xylitol) on the performance of the membrane is not clearly understood, and hence very difficult to model.

This work introduces a mathematical model of an organophilic membrane PERVAP 4060, provided by Sulzer, which is to be used to remove ethanol from the fermentation broth taking into account the effects of the impurities on the performance of the membrane. The organophilic membrane used in this work is a composite membrane consisting of a very thin separation layer (0.5 – 5 μm) on top of a porous support, coated on a polymer fleece. A number of experiments have been conducted at different feed concentrations, temperatures and permeate pressures to determine the model parameters which have been estimated using gOPT in gPROMS.

The optimal design and operation of the overall process is then considered for a given lignocellulosic biomass feedstock based on different alternative plant configurations, including with and without the organophilic membrane and with and without a hybrid separation stage, showing that a) the introduction of the organophilic membrane improves the overall performance of the process in terms of reduced energy consumption and improved product yield, and b) the use of a hybrid separation stage also improves the overall performance.

References

[1]. Baia FW, Anderson WA, Moo-Young M. (2008). Ethanol fermentation technologies from sugar and starch feedstocks. Biotechnology Advances, 26: 89–105.

[2]. Kaewkannetra P, Chutinatea N, Moonamarta S, Kamsanc T, Chiud TY. (2011). Separation of ethanol from ethanol–water mixture and fermented sweet sorghum juice using pervaporation membrane reactor. Desalination, 271. pp: 88–91.

[3]. Wei P, Cheng LH, Zhang L, Xu XH, Chen HL, Gao CJ. (2014). A review of membrane technology for bioethanol production. Renewable and Sustainable Energy Reviews, 30: 388–400.

[4]. Fan S, Xiao Z, Zhang Y, Tang X, Chen C, Li W, Deng Q, Yao P. (2014). Enhanced ethanol fermentation in a pervaporation membrane bioreactor with the convenient permeate vapour recovery. Bioresource Technology, 155: 229–34.

[5]. Gaykawad SS, Zha Y, Punt PJ, van Groenestijn JW, van der Wielen LAM, Straathof AJJ. (2013). Pervaporation of ethanol from lignocellulosic fermentation broth. Bioresource Technology, 129: 469–76.

371828 Comparative Study of MFI and SAPO-34 Zeolite Membrane Reactors for Propane Dehydrogenation

Abstract:

Propane dehydrogenation (PDH) reaction is an attractive process to meet the increasing demand of propylene and polypropylene. PDH is an equilibrium-limited endothermic reaction, and the use of permselective membrane reactors (MRs) is thought to be advantageous for intensifying the process. Hydrogen-permselective zeolitic membranes are potentially useful for PDH MRs. Compared to Pd-alloy dense membranes, zeolite membranes have high thermal stability and chemical resistance. The main focus of this work is to compare and understand the effects of zeolite membranes with differing H_2 permeation characteristics, on the MR performance in PDH reactions. In particular, we discuss in detail the effect of PDH reaction enhancement using MRs constructed with MFI (medium-pore) and SAPO-34 (small-pore) zeolitic membranes.

MFI and SAPO-34 zeolite membranes were synthesized on the inner surface of porous α-alumina tubes. Binary permeation data showed that SAPO-34 membranes exhibited much higher H_2/C_3H_8 selectivity = 20 but slightly lower H_2 permeance = 1.4×10^{-7} mol m^{-2} s^{-1} Pa^{-1} at 650°C, compared to the MFI zeolite membrane (H_2/C_3H_8 selectivity = 7.2 and H_2 permeance = 1.7×10^{-7} mol m^{-2} s^{-1} Pa^{-1}). The MFI and SAPO-34 membrane tubes were packed with a Na_2O doped Cr_2O_3/Al_2O_3 catalyst. The operation parameters studied in this work include reaction temperature, space velocity, sweep flow rate, sweep flow direction, H_2/HC ratio and H_2O/HC ratio. At 650°C, both zeolite MRs effectively exceeded the equilibrium conversion. In comparison to MFI MR, the propane conversion and propylene selectivity in SAPO-34 MR were significantly higher, due to its high H_2 selectivity which allows for smaller permeation of HCs during the reaction. It is found that the enhancement of the propane conversion in MR depends strongly upon the H_2 permeation characteristics of the membrane. We then discuss the optimal operation of the PDH MR system. Rapid catalyst deactivation and regeneration are important considerations for commercial PDH processes. To avoid the rapid activity decrease, hydrogen was added as diluent to the propane feed. This provided significantly more stable catalyst activity. PDH was also carried out in the presence of H_2O, and its effects on reducing coke formation and catalyst deactivation are described. Furthermore, the MRs were operated in the counter-current and co-current sweep modes. It is observed that the counter-current mode leads to higher propane conversion but lower propylene selectivity. This finding is discussed in terms of competing reaction and transport effects.

Intensification of Emulsion Copolymerization – Development and Application of Mathematical Model for Nonlinear Model Predictive Control

R. Pokorný, A. Zubov, P. Matuška, M. Kroupa, J. Kosek

Department of Chemical Engineering, Institute of Chemical Technology Prague, Technická 5, 166 28 Praha 6, Czech Republic, Tel.: 224 35 32 96, E-mail: Juraj.Kosek@vscht.cz

Emulsion polymerization represents well-known and mature technology, in which products are often manufactured in semi-batch reactors according to a fixed recipe schedule. Although this tight and robust recipe is finalized only after a period of product and process development, it does not allow for batch-to-batch changes in order to further increase the process efficiency. However, such intensification can be achieved by employing nonlinear model predictive control (NMPC), which allows time-varying feeds and reactor temperature to run the reaction faster, operate the reactor closer to its limits, and still keep the designed product specifications (such as molecular weight distribution, copolymer composition, viscosity, etc.). In the case of emulsion polymerization, the most critical constraint is usually the cooling capacity, which significantly varies from batch to batch due to the fouling of the reactor walls.

This contribution deals with the development of the mathematical model for an emulsion polymerization case study. The developed model is based on the solution of differential and algebraic equations, which describe the polymerization process in terms of mass and enthalpy balances, polymerization kinetics (employing the method of moments), thermodynamics, and by population balances describing nucleation of polymer particles. Because of the application of the model in the on-line controller, focus was kept on the model speed and robustness. The simulation results were validated by available experimental data from our project partners – University of Warwick, University of Hamburg, and BASF.

In the cooperation with Cybernetica AS, the model is currently being implemented in the Cybernetica CENIT software; the technology will be demonstrated on pilot plant scale for proof of process intensification concept.

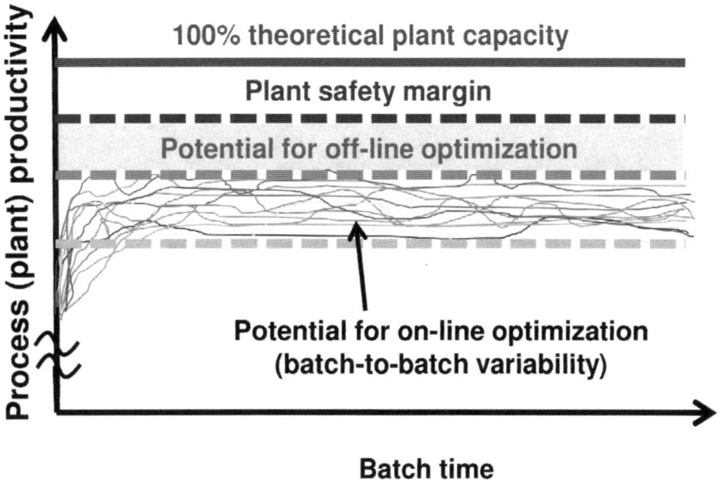

Fig. 1 The dependence of the plant productivity on time. The green box shows potential for off-line recipe optimization, the yellow box shows potential for on-line optimization due to batch-to-batch variations in process parameters

An MINLP Approach for Synthesis and Design of Multi-stream Heat Exchangers Network

Xiong Zou, Yongqiang Li, Bing Wang, Hong-Guang Dong*, Jinqu Wang

School of Chemical Engineering, Dalian University of Technology, Dalian 116024, China

Nowadays, Multi-stream heat exchangers (MHEX) have been paid more attention to in recent years for its high efficiency and low capital. In general, the MHEXs are more compact than regular two-stream heat exchangers, this means larger heat exchanging area per unit volume and less external surface exposing in environment accompanied with MHEXs, therefore, the technology option of MHEXs is believed to bring profit in both equipment capital and operation cost. However, the design of MHEXs and their networks are much difficult because of intricate channel configuration, complicated flow arrangement, complex heat-transfer paths as well as multiple potential thermal matches among cold and hot streams.

Usually, equiment design and network synthesis are dealt and solved separately. Many methods have been proposed over the years for thermal design of multistream plate pin exchangers. See Das & Ghosh [1] for an extensive review of these methods. Most MHEXs network synthesis approaches are expanded from the two-stream heat exchangers based methodologies and could be also further divided into two categories: Pinch Technology(M. Picon-Nunez et al.[2]; Xing Luo, Wilfried Roetzel and co-workers [3-4]) and stage-wise superstructure-based optimization (Yee and Grossmann[5]; M.M. Faruque Hasan et al. [6]). However, the methods in terms of "Pinch Technology" are incapable of realizing the thermal matches spanning temperature interval and the number of equipment as well as utilities are unable to be synthesized simultaneously. Furthermore, this method cannot weigh equipment investment and utilities simultaneously, which will be achieved in mathematical programming method based on stage-wise superstructure. A series of improvements on original stage-wise superstructure are proposed in recent years and actually ameliorate the probabilities of multi-stream heat exchanging matches, but still not sufficient for the optimal match structure and streams splitting and mixing by spanning stages are still not realized thoroughly.

MHEXs are widely used in energy-intensive industrial and cryogenic processes, such as air separation/liquefaction, rectisol process in ammonia plant, cryogenic section of ethylene plants, and liquefied natural gas (LNG). Due to the large difference between ambient and operating temperatures, heat exchange between the heat exchanger and the environment can not be neglected. Unfortunately, most technologies are designed under adiabatic hypothesis and do not consider the interaction between stream arrangement and network structure which have a serious impact on utility consumption and capital cost.

In this paper, we propose a framework for simultaneous multi-stream heat exchangers network synthesis and stream arrangement optimization with considering heat transfer with the surroundings. An improved State-Space superstructure based on hierarchy matching multi-stream heat exchanger (MHEX) process operator (PO) is proposed. The rigorous heat transfer calculation between adjacent streams is implemented through temperature synergistic effect in MHEXs. Moreover, arbitrary splitting and mixing of any stream is configured by mixers and splitters embedded in distribution network (DN). The mathematical modeling of the proposed superstructure is performed with mixed integer non-linear programming (MINLP). The following key details will be highlighted:

- The effect of thermal interactions (heat loss and/or heat leakage) on the performance of MHEX and networks has been considered through detailed stream arrangement and surface area calculation.

- Effective insulation materials are used to alleviate deterioration caused by heat loss or heat leakage. New objective function are developed to capture the trade-off between utility cost, capital cost and insulation materials cost.

-The performance analysis of a series of MHEXs network configuration which have never been included within previous superstructure.

The applicability of the systematic approach is demonstrated in two case studies: a high-temperature system in petrochemical industry and a cryogenic systems in ethylene industry.

Reference:

[1] P.K. Das, I. Ghosh, Thermal design of multistream plate fin heat exchangers—a state-of-the-art review, Heat Transfer Eng. 33 (2012) 284–300.

[2] M. Picon-Nunez, G.T. Polley, M. Medina-Flores, Thermal design of multi-stream heat exchangers, Appl. Therm. Eng. 22 (2002) 1643-1660.

[3] Xing Luo, Meiling Li, Wilfried Roetzel, A general solution for one-dimensional multistream heat exchangers and their networks, Int. J. Heat Mass Transfer 45 (2002) 2695-2705.

[4] Xing Luo, Xin Guan, Meiling Li, Wilfried Roetzel, Dynamic behavior of one-dimensional flow multistream heat exchangers and their networks, Int. J. Heat Mass Transfer 46 (2003) 705-715.

[5] T. F. Yee, I. E. Grossmann, Z. Kravanja, Simultaneous optimization models for heat integration-I. area and energy targeting and modeling of multi-stream exchangers, Comput. Chem. Eng. 14 (10) (1990) 1151-1164.

[6] M.M. Faruque Hasan, I.A. Karimi, H.E. Alfadala, H. Grootjans, Operational modeling of multistream heat exchangers with phase changes, AIChE J. 55(1) (2009) 150-171.

IMPACT OF SEED TYPE ON API POWDER PROPERTY CONTROL

Jason Sweeney[1], Brenda Remy[1], Jim Vernille[1], Steve Guzikowski[1], Shawn Pack[1], Lotfi Derdour[2], Elizabeth Borgeson[3],
[1]Chemical Development, [2]Drug Product Science and Technology, [3]Chemical Development Operations, Bristol-Myers Squibb Co.,
New Brunswick, NJ, 08903, USA

Session 12A04 Crystallization Process Development

Abstract

Powder property control is often an essential requirement in the production of Active Pharmaceutical Ingredients (API). Properties such as particle size can impact drug product efficacy as well as play an important role in the formulation process. Ideally, the API crystallization should be designed to consistently deliver the desired powder properties. Process parameters such as temperature, anti-solvent addition rate and seed characteristics can be used to control the crystallization. This talk discusses an API crystallization where nucleation and agglomeration play a significant role. In this case study, API seeds generated via air attrition (jet) milling and wet high shear milling were used to control the crystallization. The impact of API seed parent particle size and milling parameters on seed particle size and morphology were first evaluated. Seeds generated on the upper and lower particle size extremes were then evaluated in lab and pilot scale API batches to understand the impact of seed properties on the ability to control the growth rate relative to nucleation and agglomeration rates. The resulting API powder properties from these batches provided an understanding of the impact of seed type on the crystallization control strategy. Ultimately, the API particle size variability, manufacturing cost, and operational complexity were evaluated to select a seed type which minimized the quality and manufacturing risks.

Continuous-Flow Tubular Crystallization in Slugs Spontaneously Induced by Hydrodynamics

Mo Jiang[a], Zhilong Zhu[a], Ernesto Jimenez[a], Charles D. Papageorgiou[b], Josh Waetzig[b], Andrew Hardy[b], Marianne Langston[b], Richard D. Braatz[a]

[a] Department of Chemical Engineering, Massachusetts Institute of Technology
[b] Takeda Pharmaceuticals International Co., Process Chemistry Research & Development

Among the most inefficient and least reproducible steps in the manufacture of most active pharmaceutical ingredients are batch crystallizations, which motivate the development of new continuous-flow crystallization technologies [1]–[5]. A novel continuous-flow crystallizer design is described with the potential to provide improved control of crystal properties, improved process reproducibility, and reduced scale-up risk [6]. Liquid and gas are introduced into one end of the tube at flow rates selected to spontaneously generate alternating slugs of liquid and gas that remain stable while cooling crystallization occurs in each liquid slug. Mixing within each stable self-circulating slug is maximized by controlling the slug aspect ratio through specification of liquid and gas flow rates. The crystallizer is designed so that nucleation and growth processes are decoupled to enhance the control of each individual process. Coaxial or radial mixers combine liquid streams to generate seed crystals immediately upstream of the growth zone where nucleation is minimized, and crystal growth is controlled by the varying temperature profile along the length of the tube. The crystal size, shape, and distribution within each slug are monitored in real-time by employing in-situ video microscopy through a transparent wall of the tube. The slug-flow crystallizer design is experimentally demonstrated to generate large uniform crystals of L-asparagine monohydrate in less than 5 min. Our continuous-flow crystallizer design increases the process productivity by nearly two orders of magnitude compared to the existing batch crystallizers used in the pharmaceutical industry.

References

[1] A. J. Alvarez and A. S. Myerson, "Continuous plug flow crystallization of pharmaceutical compounds," *Cryst. Growth Des.*, vol. 10, no. 5, pp. 2219–2228, 2010.

[2] R. Vacassy, J. Lemaître, H. Hofmann, and J. H. Gerlings, "Calcium carbonate precipitation using new segmented flow tubular reactor," *AIChE J.*, vol. 46, no. 6, pp. 1241–1252, 2000.

[3] R. J. P. Eder, S. Radl, E. Schmitt, S. Innerhofer, M. Maier, H. Gruber-Woelfler, and J. G. Khinast, "Continuously Seeded, Continuously Operated Tubular Crystallizer for the Production of Active Pharmaceutical Ingredients," *Cryst. Growth Des.*, vol. 10, no. 5, pp. 2247–2257, 2010.

[4] M. Jiang, M. H. Wong, Z. Zhu, J. Zhang, L. Zhou, K. Wang, A. N. Ford Versypt, T. Si, L. M. Hasenberg, Y. E. Li, and R. D. Braatz, "Towards achieving a flattop crystal size distribution by continuous seeding and controlled growth," *Chem. Eng. Sci.*, vol. 77, pp. 2–9, 2012.

[5] M. Jiang, X. Zhu, M. C. Molaro, M. L. Rasche, H. Zhang, K. Chadwick, D. M. Raimondo, K.-K. K. Kim, L. Zhou, Z. Zhu, M. H. Wong, D. O'Grady, D. Hebrault, J. Tedesco, and R. D. Braatz,

"Modification of Crystal Shape through Deep Temperature Cycling," *Ind. Eng. Chem. Res.*, vol. 53, no. 13, pp. 5325–5336, 2014.

[6] M. Jiang, Z. Zhu, E. Jimenez, C. D. Papageorgiou, J. Waetzig, A. Hardy, M. Langston, and R. D. Braatz, "Continuous-Flow Tubular Crystallization in Slugs Spontaneously Induced by Hydrodynamics," *Cryst. Growth Des.*, vol. 14, no. 2, pp. 851–860, 2014.

Design of Kinetically-Controlled Enzymatic Reactive Crystallization

L. G. Encarnación-Gómez[1], A. S. Bommarius[1], and R.W. Rousseau[1]
[1] *School of Chemical and Biomolecular Engineering*
Georgia Institute of Technology
Atlanta, GA 30332-0100, USA

β-lactam antibiotics are among the most-used pharmaceutical compounds worldwide; they span approximately 65% of the antibiotic market and their production exceeds 45000 tons per year [1]. Industrial manufacture of these compounds mainly has been through chemical synthesis, which involves the utilization of multiple solvents and protection and de-protection groups that generate significant amounts of waste [2].

An alternate route for the manufacture of β-lactam antibiotics it is the enzymatic synthesis by pencillin G acylase (PGA). This enzyme can catalyze the synthesis of phenylglycine methyl ester (or other acyl donor compounds) and 6-aminopenicillanic acid (6-APA) to ampicillin. However, PGA also catalyzes the hydrolysis of phenylglycine methyl ester (primary hydrolysis) into phenylglycine and the hydrolysis of ampicillin into 6-APA and phenylglycine (secondary hydrolysis).

The present work focuses on the design of enzymatic reactive crystallizationin which ampicillin can be recovered as a crystalline product and thus protected from secondary hydrolysis. The design of such an operation requires knowledge of the kinetics of reaction and crystallizationand the solubility of all components in the system, including the dependence on pH and temperature.

Crystallization (nucleation and growth) kinetics were obtained using an online-monitoring system that provided the solute concentration in solution and the crystal chord-length distribution (CLD) in real time. Concentration vs. time data from multiple experiments were combined and used to estimate the rate of primary nucleation, secondary nucleation, and crystal growth. Since multiple experiments were used (different pH ramps, supersaturations, seed loadings), this parameter estimation routine allowed indentification of kinetic parameters that span multiple operational conditions. Crystallization kinetics were combined with the reaction kinetics developed by Youshko and Svedas [3] to develop a reactive crystallization model. Our results show the advantages of performing semi-batch operations and seeding the reactive medium. Seeding had the advantage of allowing crystallization of ampicillin at a low supersaturation (within its metastable region). Without seeding, ampicillin within the metastable region was subject to secondary hydrolysis which decreased the yield of the operation.

Preliminary experimental results show the possibility of performing the reaction at supersaturated conditions. However, the yield of ampicillin crystals was low because the reaction was unseeded, and ampicillin was subject to secondary hydrolysis for a longer period of time. Also, significant crystallization of phenylglycine occurred since PGME was constantly hydrolyzed by PGA. However, as the model suggests, phenylglycine crystallization can be suppressed and precipitation of ampicillin can be improved by performing a seeded-semi-batch reactive crystallization.

[1] R.P. Elander. Industrial production of B-lactam antibiotics. *Appl. Microbiol. Biotechnology.* **61**: 385-392. (2003)

[2] R. C. Giordano, M. P. A. Ribeiro, and R. L. C. Goirdano. Kinetics of B-lactam antibiotics synthesis by penicillin G acylase (PGA) from the viewpoint of the industrial enzymatic reactor optimization.
Biotechnology Advances. **24:** 27-41 (2006)
[3] M. I. Youshko and V. K. Svedas. Kinetics of amipicillin synthesis catalyzed by penicillin acylase from E. coli in homogeneous and heterogeneous systems. Quantitative characterization of nucleophile reactivity and mathematical modeling of the process. Biochemistry **65**: 1367-1375 (2000)

Continuous Crystallization of Proteins in a Tubular Plug-flow Crystallizer

P. **Neugebauer**[1], J.G. **Khinast**[1,2]

1 Graz University of Technology, Institute for Process and Particle Engineering, 8010 Graz, Austria

2 Research Center Pharmaceutical Engineering, 8010 Graz, Austria

Email for correspondence:
peter.neugebauer@tugraz.at
khinast@tugraz.at

In the past significant efforts have devoted to the development of large-scale crystallizers for the production of protein crystals due to the increasing demand of such crystals in the field of biocatalysis/biotranformations and for therapeutic applications [1;2]. Nevertheless, protein crystal growth in batch reactors is negatively influenced by either detrimental shear forces at high agitation rates or diffusion limitations caused by insufficient agitation [3].
To overcome these problems a novel continuous plug-flow crystallizer was developed, based on previous work on tubular crystallizers [4;5], using a crystallization agent solution to induce crystallization. Low flow rates reliably prevent the formation of amorphous precipitate and optimal temperature control leads to high yields within a residence time in the minute range. Minimization of the residence time distribution is achieved by the introduction of air bubbles, thereby generating a segmented gas-slurry flow. By implementing an online crystal size distribution (CSD) measurement device continuous monitoring of the CSD is feasible.

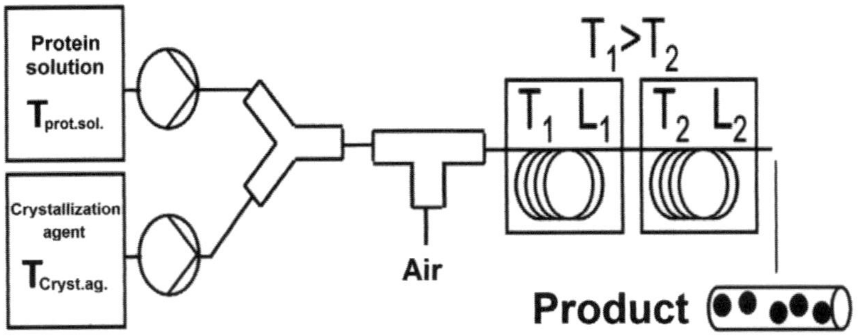

Figure 1: Schematic drawing of the tubular plug-flow crystallizer

[1] Margolin, Alexey L., and Manuel A. Navia. "Proteinkristalle als neue Katalysatoren." Angewandte Chemie 113.12 (2001): 2262-2281.

[2] Basu, Sujit K., et al. "Protein crystals for the delivery of biopharmaceuticals." Expert opinion on biological therapy 4.3 (2004): 301-317.

[3] Schmidt, Sebastian, et al. "Kristallisation für die Aufarbeitung von Proteinen." Chemie Ingenieur Technik 76.6 (2004): 819-822.

[4] Eder, Rafael JP, et al. "Continuously seeded, continuously operated tubular crystallizer for the production of active pharmaceutical ingredients." Crystal Growth & Design 10.5 (2010): 2247-2257.

[5] Eder, Rafael JP, et al. "Continuous sonocrystallization of acetylsalicylic acid (ASA): control of crystal size." Crystal Growth & Design 12.10 (2012): 4733-4738.

371601 Developing Crystallization Options for Manufacturing

Abstract:

Crystallization is a critical unit operation in the production of active pharmaceutical ingredients. The importance of controlling particle size, form, and supersaturation have led to extensive development around addition mode for antisolvent or reactive crystallizations. Three typical addition modes are forward-addition, reverse-addition, and co-addition. Different addition modes have different inherent risks, equipment requirements, and complexities. Co-addition crystallizations are typically successful where forward-addition and reverse-addition crystallizations fail; however, co-additions require additional equipment and control than other addition modes.

A case study is presented of a reactive crystallization which had undesirable crystallization behavior in forward and reverse modes, only maintaining supersaturation control with a co-addition strategy. In the process of understanding this behavior, new process options were developed which allowed a reverse addition to be performed with supersaturation control, providing manufacturing with multiple crystallization options.

389120 Methods for the Reduction of Cycle Time in Intermediate Crystallization Steps

Abstract:

In recent years great attention and research focus has been paid to improving the final crystallization step in process development. The final crystallization step will typically define final product attributes such as purity, yield and particle size and is critical for delivering material that can be effectively processed and formulated into a final product fit for the market.

While focus on the final crystallization remains important it is apparent that an opportunity exists for the improvement of many non-optimized intermediate crystallization steps. Intermediate steps differ from the final crystallization in that the intermediate product is usually transferred to the next step where it is dissolved and processed. In most cases the final product attributes are less important that the speed and efficiency with which the intermediate product can be delivered to the next step.

The challenge for intermediate crystallization development is that increasing the speed of crystallization typically means running at higher supersaturation levels. Such operation can result in undesired crystallization mechanisms excessive secondary nucleation, phase separation (oiling out) morphology changes. Such mechanisms often lead to common downstream bottlenecks such as slow filtration rates, poor flowability and caking during transportation.

In this presentation techniques will be presented that focus on reducing the overall cycle time of intermediate crystallization steps. The application of *in situ* measurement and imaging tools will be shown to facilitate the choice of process parameters that ensure the fastest crystallization possible without sacrificing the speed of downstream separation and delivery to the next processing step.

國立中央大學
National Central University

The Impact of Reaction and Crystallization of Acetaminophen (Paracetamol) on Filtration and Drying through in-Process Controls

Org. Proc. Res. Dev. **2013**, *17* (9), 1168-1178.

Department of Chemical and Materials Engineering
National Central University, Taiwan R.O.C.

Report: Hung Lin Lee
Advisor: Prof. Tu Lee
Date: 11/18/2014
E-mail: tulee@cc.ncu.edu.tw

AIChE
The Global Home of Chemical Engineers

No.300, Jhongda Rd., Jhongli City, Taoyuan County 32001, Taiwan R.O.C.

Chemical Engineering

3

- Reaction-related factors: conversion, impurity levels, yield
- Scale up & reproducibility
- Controllable process
- Impact on downstream operations, particle shapes, particle size distributions (PSD) and solid-state properties

The shape of snowflake is dependent on crystallization pathways!

Motivation

4

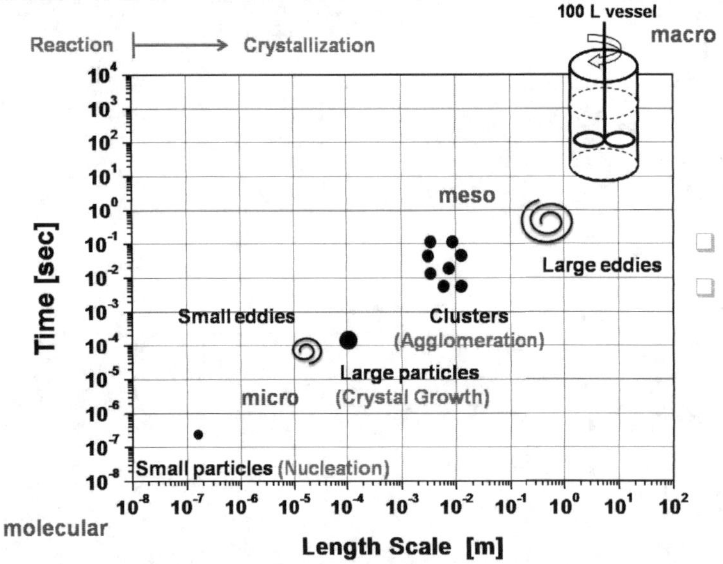

- **Bottom-up fabrication**
- **Desirable particle size distribution (PSD)**

The aim of our study is to provide an enabling platform to realize how reaction and crystallization affect the PSD, which in turn influence the downstream operations, filtration and drying.

Powder Technol. **2005**, *150* (2), 60-71.

Acetaminophen (Paracetamol)

5

- Active pharmaceutical ingredient (API)
- Analgesic (Pain Reliever)
- Antipyretic (Pain Reducer)
- First marketed in the US in 1950
- 24.6 billion tablets sold in 2008 (~ 2.6 billion US$ in sales)

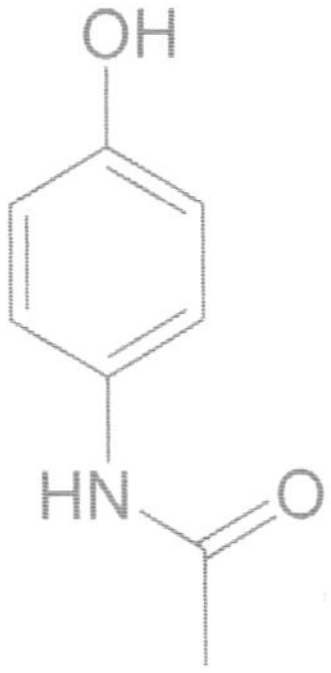

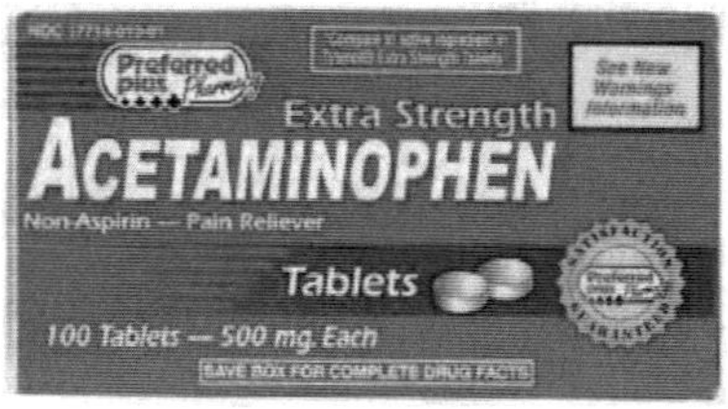

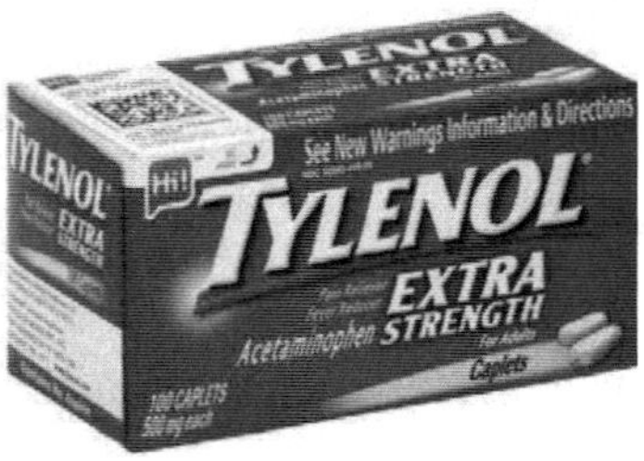

Outline

6

- Motivation
- Reaction & Crystallization of Acetaminophen
- Apparatus for In-Process Controls
- Unit Operations
  - Filtration
  - Drying
- Experimental Procedures
- Results and Discussion
  - In-Process Controls for Cases I, II and III
  - Crystallization Routes and Particle Size Distributions
  - Filtration and Drying
- Conclusions

Reaction & Crystallization

7

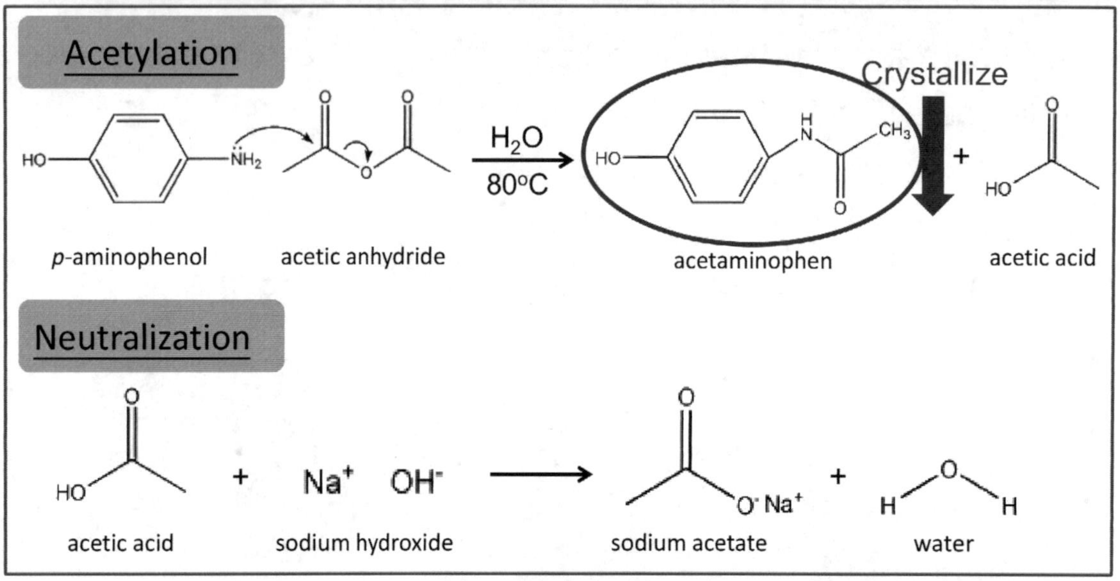

Crystallization → Filtration → Drying

Apparatus for In-Process Controls

8

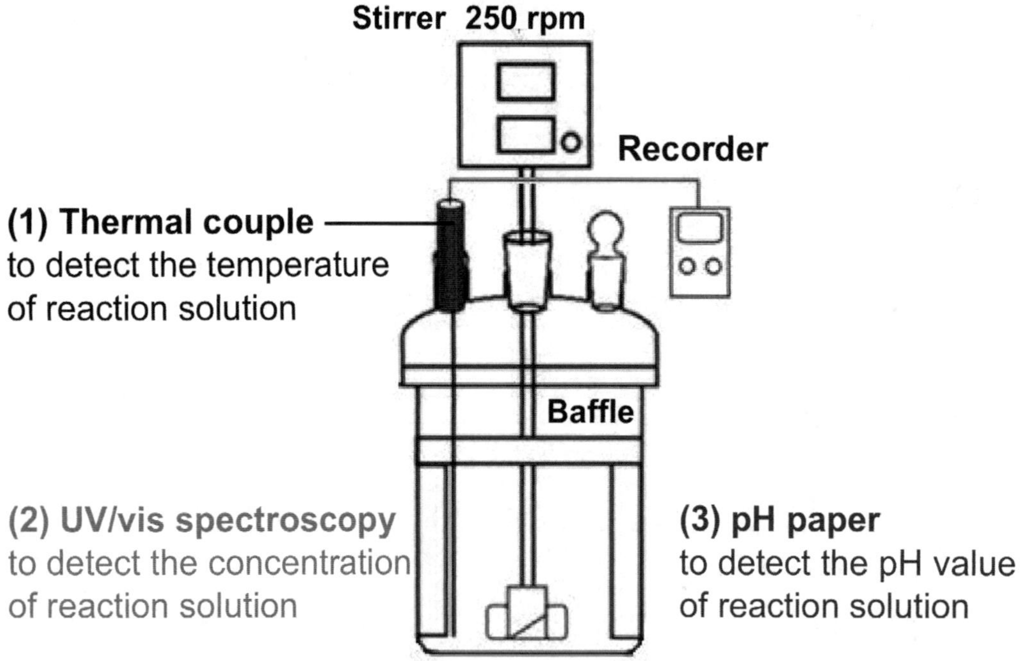

Unit Operations

9

Filtration

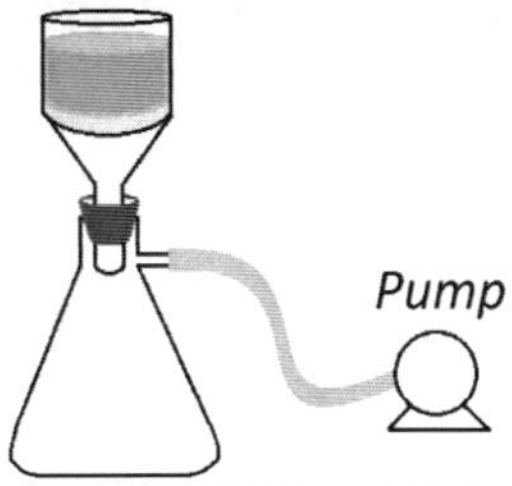

$$\alpha = \frac{\Delta p_c A}{\mu u m_c}$$

α = specific cake resistance (m/kg)
ΔP_c = pressure drop over cake (Pa)
A = cake area (m^2)
μ = viscosity of filtrate (Pa·s) ≈ viscosity of water
u = linear velocity of filtrate (m/s) = $dV/(dt*A)$
m_c = total mass of solids in the cake (kg)

→ ***To calculate specific cake resistance***

Drying

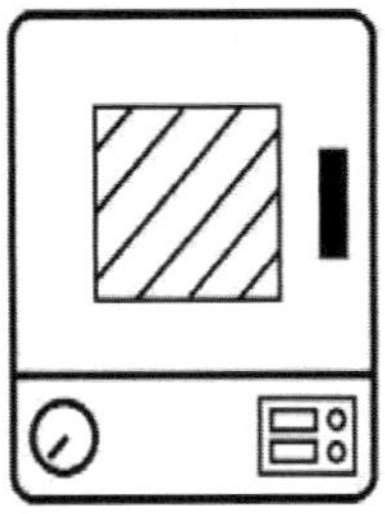

$$X = \frac{m_t - m_s}{m_s}$$

$$\rightarrow R = -\frac{m_s}{A}\frac{dX}{dt}$$

m_t: wet cake weight
m_s: dried solid weight (kg)
R: drying rate (kg/hr·m^2)
A: cake area (m^2)
X: free moisture (kg/kg solid)
t: time (hr)

→ ***To obtain drying curve***

Unit Operations of Chemical Engineering, McCabe et al., 7th Ed., **2005**.

Experimental Procedures

10

- Two engineering parameters: pH control and agitation mode

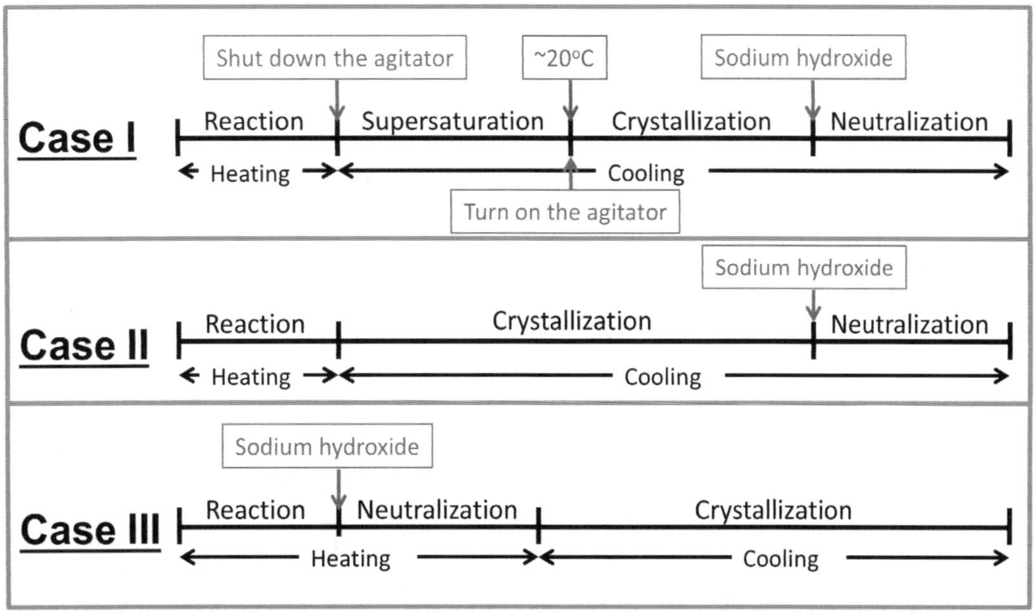

Results & Discussion

In-Process Control for Case I

12

Nucleation rate, *B*:

$$B = k_b (C - C^*)^b$$

Crystal Growth rate, *G*:

$$G = k_g (C - C^*)^g$$

Metastable zone width:

$$\Delta C_{max} = C - C^* = S - 1$$

where k_b = nucleation rate constant,
b = nucleation order,
C = acetaminophen concentration,
C^* = saturated conc. or solubility,
k_g = crystal growth rate constant,
g = crystal growth order,
S = supersaturation.

Org. Process Res. Dev. **2013**, *17* (9), 1168-1178.

In-Process Control for Case I

13

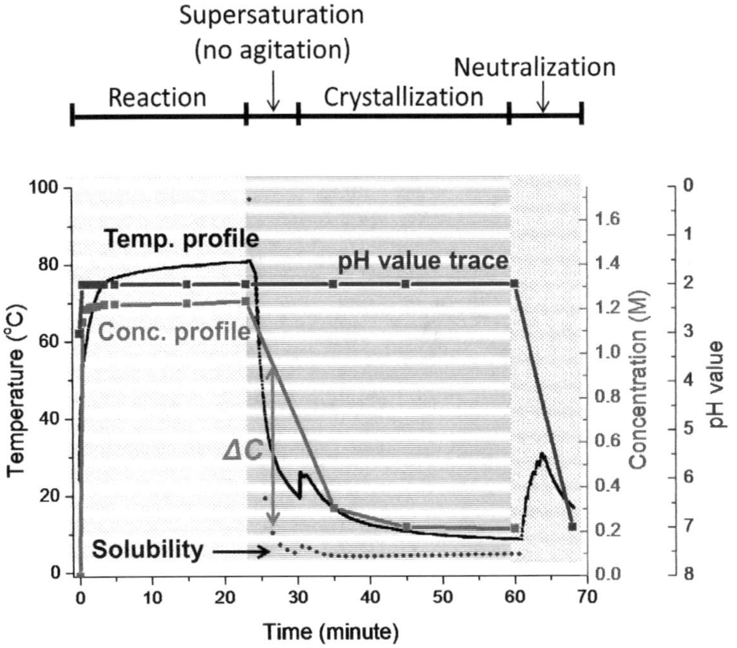

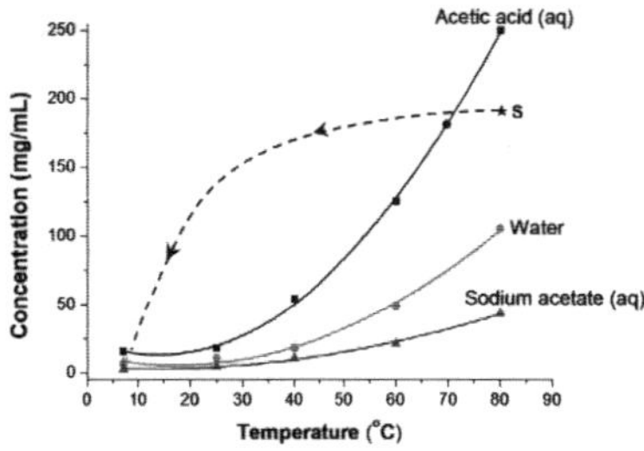

- **Poorer mixing**
  - → Broader ΔC
  - → Higher nucleation rate
  - → Insufficient growth

Org. Process Res. Dev. **2013**, *17* (9), 1168-1178.

In-Process Control for Case II

14

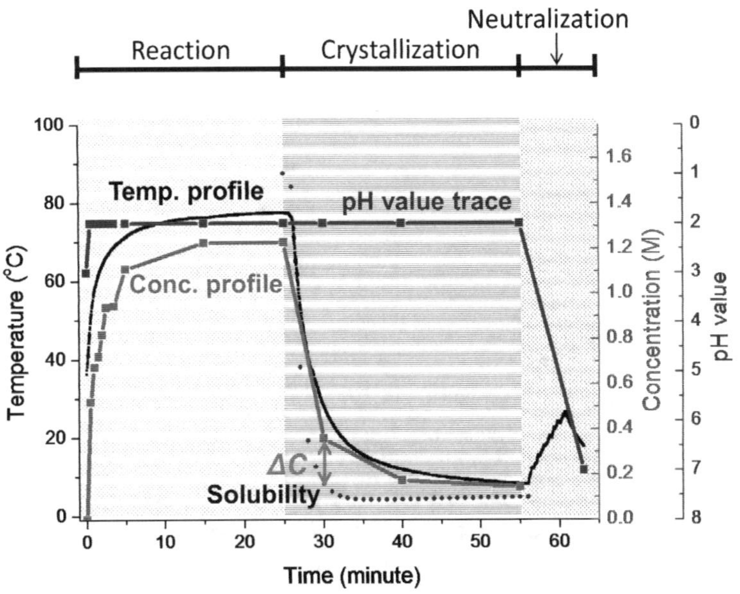

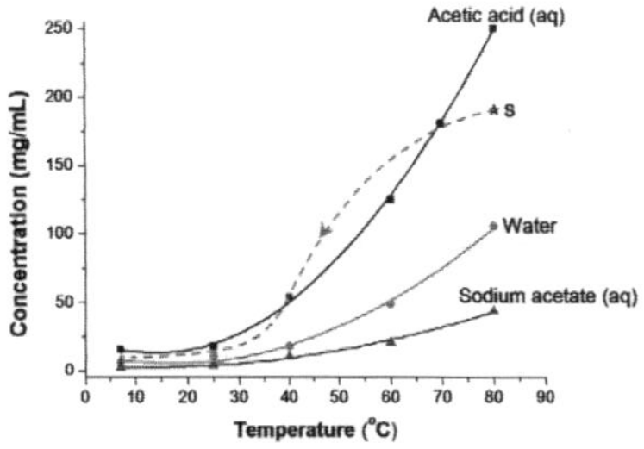

- **Good mixing**
  - → Efficient heat & mass transfer
  - → Smaller supersaturation

Org. Process Res. Dev. **2013**, *17* (9), 1168-1178.

In-Process Control for Case III

15

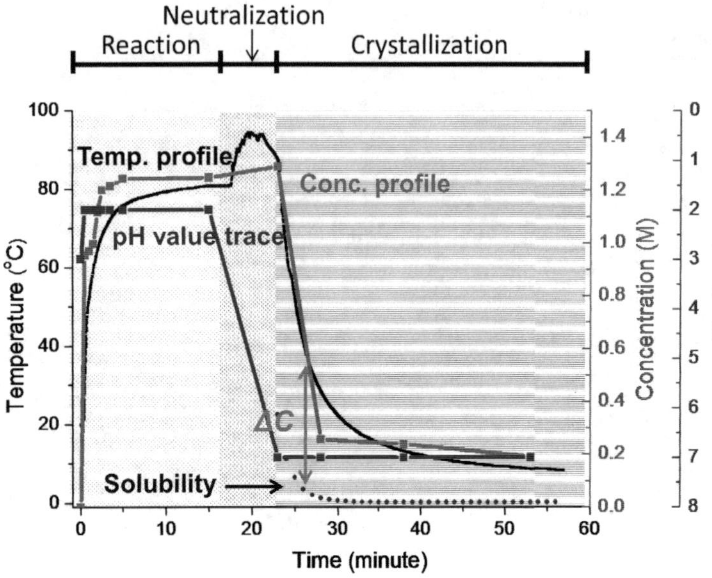

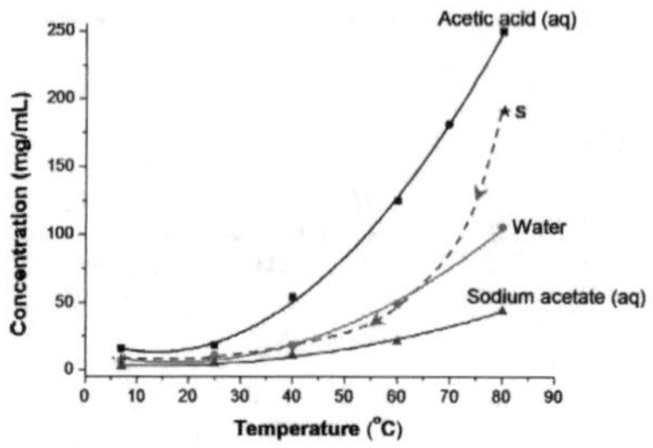

- **Good mixing**
- **Solution medium was changed**
  - → Ten times less solubility
  - → High nucleation rate

Org. Process Res. Dev. **2013**, *17* (9), 1168-1178.

Crystallization Routes

16

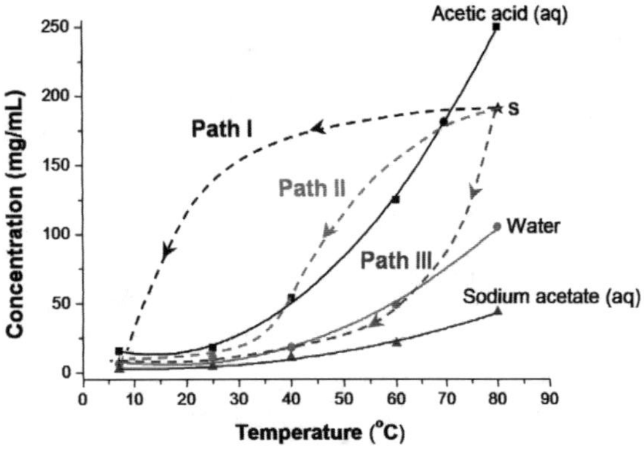

- **After reaction, starting point S was determined by UV/vis.**
- **Different crystallization routes,**
  - **•PSDs**
  - **•Operation behaviors**

Particle Size Distribution (PSD) by Sieving Analysis

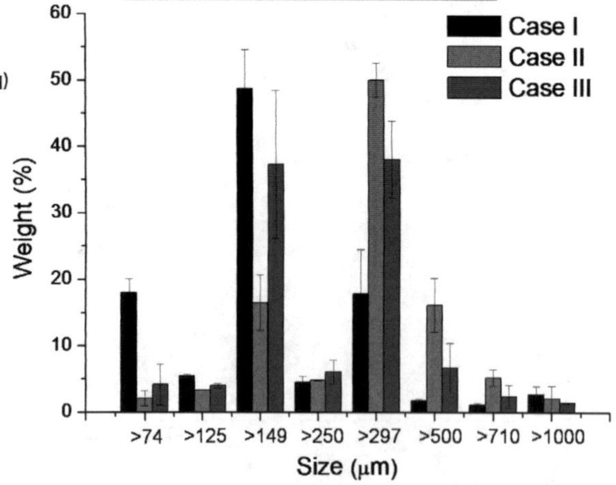

Particle Size (>297μm): Case II (red) > Case III (blue) > Case I (black)

Filtration

17

➢Specific cake resistances calculated by equation: $\alpha = \frac{\Delta p_c A}{\mu u m_c}$

Case I	Case II	Case III
8.80×10^4^		

➢Cake resistance, α : Case I > Case III > Case II

➢Case II had relative large voids; Cases I and III provided much less space between agglomerate particles in the cake

Drying

18

➢Drying Curve obtained by equation: $R = -\frac{m_s}{A}\frac{dX}{dt}$

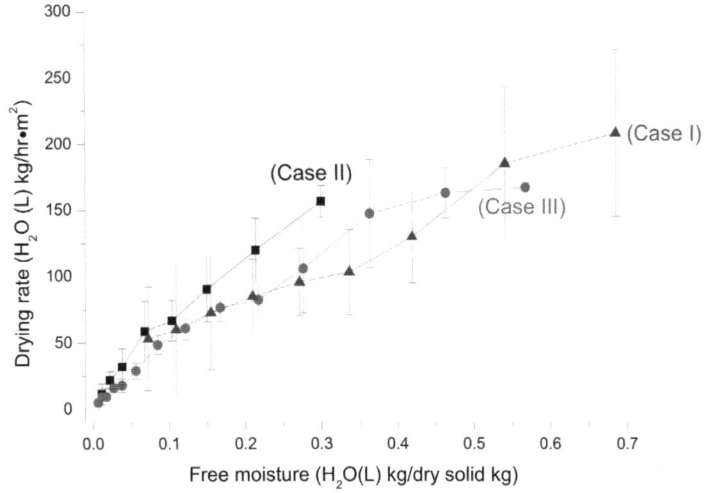

Lumping

➢Faster drying rate implies that it has larger pores and PSD

➢Lumping phenomenon is undesirable in process

Overall Mass Balance

19

- The weights of acetaminophen crystals produced and yields for all cases were about 14 g and 70 % respectively.
- The three cases were almost identical in reaction (molecular level).

	Solvent volume (mL)	Total weight of starting materials (g)	Residence time (sec)	Acetaminophen weight retained on filter paper (g)	Yield (wt%)	Ideal weight of acetaminophen (g)	Conversion (%)	Mother liquor volume (mL)	Overall weight of acetaminophen (g)
Case I	110.0	15.00±0.003	1800	13.99±1.71	67.3±8.27	20.78	85.25	84.67±6.03	17.70
Case II	110.0	15.00±0.007	1800	14.52±0.90	69.88±4.34	20.78	84.64	98±15.59	17.59
Case III	110.0	15.01±0.004	1800	14.85±0.35	71.49±1.69	20.78	88.82	83±5.29	18.46

Conclusions

20

- ❑ Crystallization is critical step for bottom-up fabrication.
- ❑ Acetaminophen particles produced in cases I, II, and III were nearly identical in purity and yield, but the PSD, drying, and filtration behaviors are different at higher length scales depending crystallization routes taken.
- ❑ Optimization for process performance depending on different usage was not discussed in this study.
- ❑ To design a robust and reproducible process, all processing and functional properties must be considered equally.

Acknowledgements

21

- My advisor – Prof. Tu Lee
- My lab partners
- My family
- National Science Council (Ministry of Science and Technology)
  - NSC 102-2221-E-008-071-MY2

Thanks for your attention!

393632 Turning the Big Idea into a Startup

Abstract:

When a university researcher or garage inventor is thinking about commercializing an invention, there are a number of early considerations that are not always obvious to first time technical entrepreneurs. This talk will focus on some of the early stage commercialization decisions and how those decisions may impact the startup company potential later. Some of the topics to be discussed include fundraising, teambuilding, and intellectual property management.

386885 Starting an Small Engineering Consulting Business – the 10 Year Anniversary of a Startup Company

Abstract:

A small engineering consulting business was started in 2004 with the modest objectives of addressing short-term business opportunities while remaining in a specific geographic location. Full-time commitment to this venture occurred for a seven-year period. Ten years and 154 engineering projects later the business continues to operate. Personal and professional lessons learned on business planning and strategies, accounting and financial tools, fixed asset investments, family involvement, and finally, the pros and cons of this career path versus the traditional non-ownership career path will be discussed.

387532 The Art of the Mass Balance: The Cultural Tools of Entrepreneurial Chemical Startups

Abstract:

The product cycle processes of tech based startups and chemical startups are inherently different. Tech startup thought leaders teach entrepreneurs to hit the ship date even if all the kinks aren't worked out. Later a patch can be uploaded by customers to fix bugs or add functionality at minimal cost to the startup.

Chemical startups differ significantly as processing plants are capital intensive and must incorporate safety, equipment lead times, and construction efforts. Adopting an entrepreneurial approach, chemical startups can deliver against the fast paced timeline while designing the flexibility required to commercialize a first of its kind technology. The tool for success in an entrepreneurial chemical startup is the mass balance.

Keeping score in an industrial startup is dependent upon the value and quality of the product. The disposition of the constituents in the product determines the cash flow possibilities in the operation cost and the needed investment to enable the conversion of the raw material feedstock mass into the valued finished product. The key to a successful mass balance is the validation of the instruments and test methods measure the chemical engineering entrepreneurial effort's success.

The Personal and Family Side of Starting an Independent Business

Starting a business has its special and unique technical challenges relating to deciding on a focus, designing a strategy for communication and outreach, identifying clients, obtaining financing, and many others. But unless you are a single adult, there is another factor and that is your family. If they have gotten used to a regular paycheck, travel being reimbursed, promotions coming on a regular business, and the occasional special recognition, this can all come to a screeching halt. The support and understanding needed from a family and external networks is critical for a new venture to succeed. The level of this support will also depend greatly on how large the imagined enterprise is envisioned to be.

These are some questions to consider:

1. Short term economic survival. How long can you survive without a regular paycheck? Have you built up sufficient savings in the event that your first clients are harder to find than you thought? If your spouse is not working, will they be able to work to cushion the cash flow? Do they want to? What is the market in the local area for the skill set they possess? Will the cost of child care be part of the equation? It is not unusual at all for someone starting an independent business to wait at least six months before receiving their first payment. More companies are moving to 90 day plus payment cycles. Is that tolerable? A discount on pricing and acceptance of a client's corporate credit card are ways of increasing short term cash flow at a small long term cost. There may also be short term training opportunities available through various organizations, but not necessarily at the same rate of pay that may be desired long term.

2. Health insurance. The degree to which this is of concern will depend on a large number of factors including age, state of residence, health history barriers to insurability, and whether a working spouse has insurance to which the new "independent" can be added. Regardless of the specific answers to these questions, the cost of covering health insurance must be considered when starting an independent business. One way of minimizing a major impact is to purchase low cost catastrophic insurance with a very high deductible, usually available in many areas without medical questions. And of course, AIChE offers group health insurance plans, but which come with holding periods before covering pre-existing conditions.

3. Spouses. No two people are created equal and the reaction of family members to such a venture, even when openly discussed ahead of time, can vary and change. What is the personal tolerance for going without a paycheck for an extended period of time (even assuming the points in "short term survival" have been addressed)? Is how the neighbors and your social circle view your situation important? If you are a male and your wife is working or needs to go back to work, are you willing to do household chores you've never done before? Some of these attitudes will depend upon whether your starting of an independent business was totally your choosing or as an indirect result of a reduction in force, a general layoff or closure/relocation of a business, or a personal performance issue. Is your spouse willing to cut back on some of the luxuries they may have enjoyed previously?

4. Children. This is a different issue than spouses and will greatly depend on their ages. What would your child's reaction be to someone in their school, possibly informed by one of your neighbors about your situation, saying something like "Your mommy or daddy is out of work...ha, ha". Will they start a fight? How well have you prepared them?

5. Support structure. In today's world, the stigma of losing a job is far less than it used to be due to numerous factors including mergers and acquisitions, change in business emphasis, and "forced" geographic moves to retain a job. Yes, there are still performance issues, but many times this can be a result of different approaches to a problem or opportunity vs. genuinely poor performance. It is

highly unlikely that anyone today will have only one job for their entire career. Since everyone, over some period of time, will experience the same kind of dislocation, there is virtually no barrier to discussing possible job opportunities with anyone. Today's professional networking tools such as LinkedIn are tremendous sources of connections and opportunities. Local professional engineering chapter meetings are another great source, especially if there is a strong desire to stay in the same geographic area.

It is important to remember that starting such a venture within a family structure has additional needs to consider beyond just hanging out the professional consulting shingle. Consider them all before taking the plunge and your venture has a much higher probability of success.

364573 Direct Synthesis of Hydrogen Peroxide in a Microreactor

Abstract:

The direct synthesis of hydrogen peroxide in a microreactor is a safe and efficient process. Conventionally, hydrogen peroxide is produced using the anthraquinone autooxidation process, which is rather complex and can only be performed cost-effectively on a large scale. As a result, hydrogen peroxide needs to be stored and transported to locations where is it used with all the risks involved. The direct synthesis of hydrogen peroxide in a single step over a palladium catalyst, on the other hand, allows for the production of ultra-pure hydrogen peroxide on a smaller scale, which can then be used directly without the need to add stabilizers, which can affect the processes in which it is used. The hydrogen peroxide direct synthesis, however, has two main issues. Firstly, the simultaneous use of hydrogen and oxygen in a conventional reactor can be dangerous due to explosion risks. Secondly, the catalysts used in the hydrogen peroxide direct synthesis also catalyze the decomposition of hydrogen peroxide once higher concentrations are produced.

In this presentation we will show the development of an efficient wall-coated catalytic microreactor system for the direct synthesis of hydrogen peroxide over a bimetallic gold-palladium catalyst. In this microreactor we can work safely with non-diluted mixtures of hydrogen and oxygen (50vol% each). At these reaction conditions, we can achieve hydrogen peroxide productivities which are an order of magnitude higher than at the dilute (4 vol% H_2 and O_2) conditions which are conventionally used. In addition to this leap in productivity, also the reaction selectivity improves dramatically from 30 to 70 %. This increase in selectivity we will explain using a kinetic study, which shows that a higher surface occupancy on the catalyst surface of the reactants at these higher partial pressures prevents the peroxide decomposition.

In an optimization study, we will show that using this technology, it is possible to produce 1 wt% H_2O_2 at a high selectivity (>80 %) in a single step. It is possible to produce up to 5 wt% of hydrogen peroxide in a single step, however, in that case the peroxide decomposition rate becomes high and the overall selectivity drops to only about 25%. In the final part of the presentation, we will present our current research in which we attempt to combine a high productivity with a high selectivity by taking the process intensification one step further: directly combining the hydrogen peroxide production with its consumption in a fine chemical selective oxidation. The direct usage of the peroxide in the same microreactor keeps its concentration low, thereby preventing its decomposition.

379831 Implementation of a Process Intensification Approach in the Production of an API Intermediate

Abstract:

Organic synthesis is essential for the production of an important class of pharmaceuticals and its implementation on an industrial scale is a great challenge for the modern pharmaceutical industry. The majority of processes are currently based on batch and semi-batch processes, but besides their wide flexibility and versatility, they face considerable disadvantages. Examples are long reaction sequences, then occurrence of non-uniform conditions inside the vessels, limited potential to apply real time process monitoring, difficulties in automating the production, and so on. Furthermore, requirements to store intermediates and complicated cleaning procedures are activities that should be minimized in order to achieve the most profitable production of active pharmaceutical ingredients (APIs).

Continuous manufacturing of pharmaceuticals is one of the most suitable ways to avoid the typical problems associated with batch processing. These processes are eco-friendly and economical, including better usage of raw materials, smaller waste production, less energy consumption, higher throughput, improved yields, better temperature profiles inside chemical reactors, and so on. In addition, a high degree of automation and process controllability can be achieved and therefore the Process Analytical Technology (PAT) Initiative may be implemented in a very efficient way.

The main focus here is on one intermediate step in the production of Zuclopenthixol, a product of H. Lundbeck A/S. Dehydration of 9 Allyl 2 Chlorothioxanthen 9 Ol ("N714-Allylcarbinol") into cis/trans 9H thioxanthene,2 chloro 9 (2 propenylidene) (9CI) ("N746-Butadienes") is a manufacturing step that was traditionally performed in batch mode with quite long reaction times and plenty of difficulties hindering successful implementation of a PAT strategy. Performed in this way, the dehydration step required several supportive operations that were defined as non-value added activities (NVAs). More precisely, storage of substrates and products, as well as a solvent swap from tetrahydrofuran to toluene were neccesary in the traditional production of Zuclopenthixol.

Avoiding NVAs in such manufacturing step is here achieved by implementing a process intensification approach. More precisely, a switch from lab-scale batch processing to a high-pressure mesoscale laminar tubular reactor is performed. A pressure increase enables higher reaction temperature and therefore shorter reaction times. Furthermore, better mass and heat transfer in tubular reactors additionally allowed faster manufacturing of "N746-Butadienes". As a consequence, the chemical reaction was accelerated from 3 hours in batch mode to just 3 min in the tubular laminar reactor (considering total conversion of "N714-Allylcarbinol").

In addition to a new process design, successful implementation of real-time process monitoring was performed as well. The main focus is on in-line FT-NIR analysis that is validated with at-line FT-NIR and off-line HPLC analyses. Such monitoring is achieved by implementing a process chemometrics approach based on successful development of a Partial Least Squares (PLS) regression calibration model for the spectroscopic data. Very fast data acquisition is achieved, resulting in faster development of process understanding.

Furthermore, detailed analysis of substrates and products is performed in order to check the purity of these constituents. It was discovered that the presence of hydronium ions has a strong influence on the occurrence of side reactions. More precisely, high molar concentrations of Brøndsted acids caused the formation of undesired polymers in the reaction system caused by the opening of tetrahydrofuran rings. Chemical reactions between open rings were noticed and validated by using Size Exclusion Chromatography (SEC) and NMR spectroscopy.

Process intensification of the intermediate manufacturing step in the Zuclopenthixol production was successfully implemented by a transfer of the lab-scale batch process to a mesoscale tubular laminar reactor. Acceleration of the chemical reaction indeed allowed usage of mesoscale laminar reactors, combined with fast and accurate in-line process monitoring. The occurrence of side reactions was identified and changes in the overall Zuclopenthixol production were suggested. For instance, immediate neutralization of the hydronium ions after the dehydration reaction is considered to be necessary as a future work.

Isolation of Fatty Acids from Industrial Effluents

S. Lux, D. Painer, M. Siebenhofer

Motivation – Biobased Future

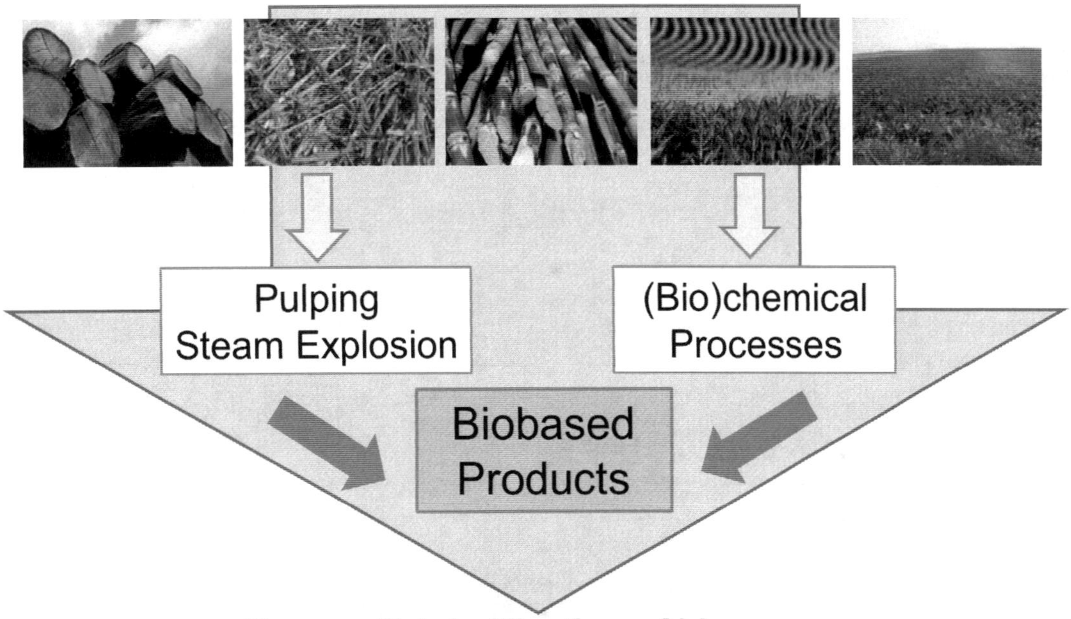

Focus: Total utilization of biomass

Isolation of Fatty Acids from Industrial Effluents

NAWI Graz TU Graz

Challenge…

multicomponent mixtures

aqueous effluents

uneconomic

Downstream Processing

wastewater treatment

azeotropes

low concentration

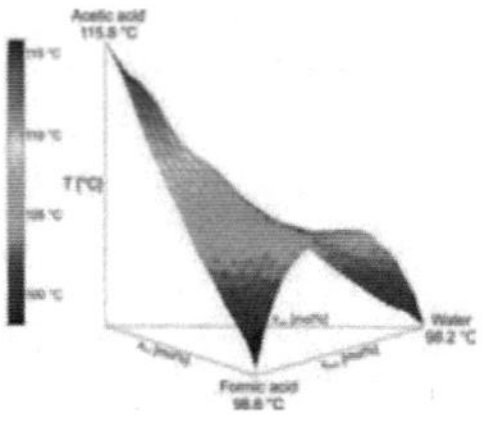

energy-intensive separation

… accepted

Isolation of Fatty Acids from Industrial Effluents

NAWI Graz TU Graz

e.g. Processing of Condensates

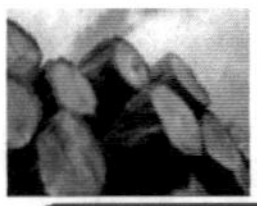 =  + condensate

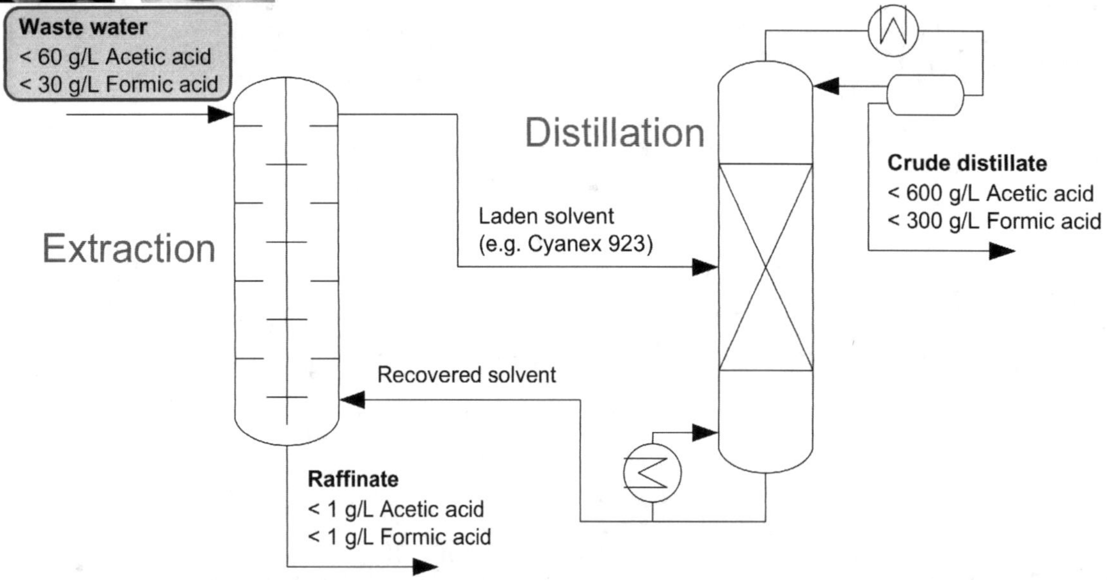

e.g. Processing of Condensates

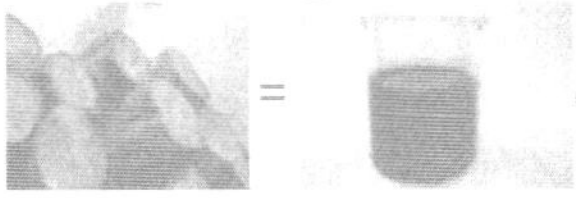

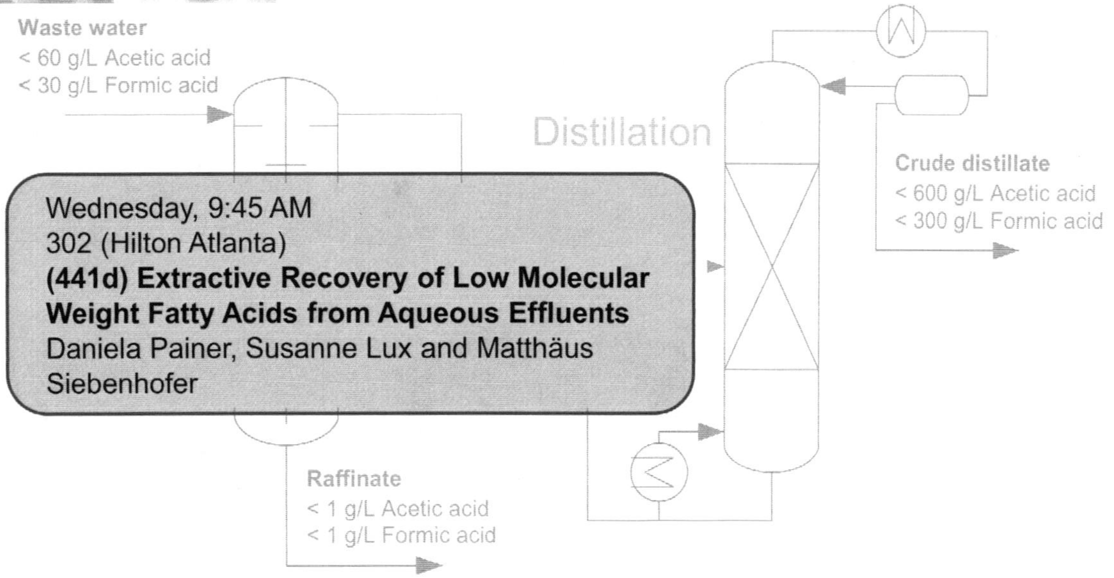

Wednesday, 9:45 AM
302 (Hilton Atlanta)
(441d) Extractive Recovery of Low Molecular Weight Fatty Acids from Aqueous Effluents
Daniela Painer, Susanne Lux and Matthäus Siebenhofer

e.g. Processing of Condensates

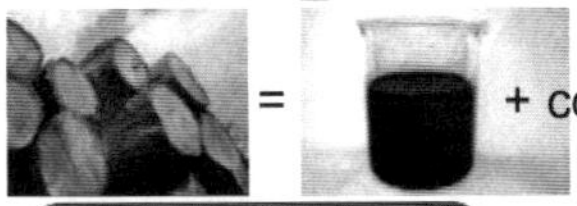

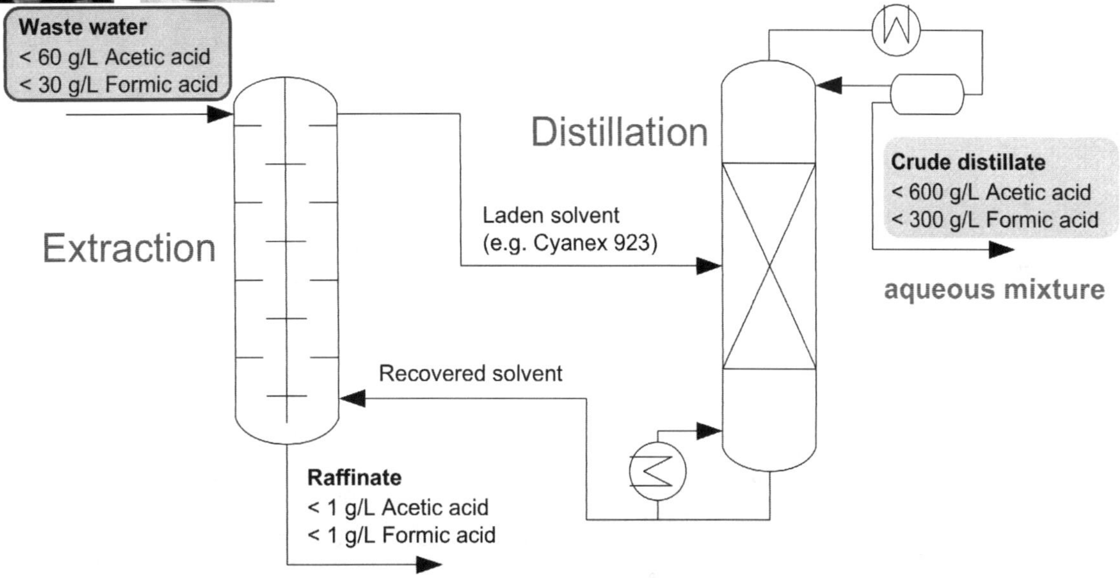

Formic Acid – Acetic acid – Water

Azeotropes: boiling points (950 mbar):

Binary high boiling azeotrope	105.6 °C
Ternary saddle point azeotrope	105.2 °C

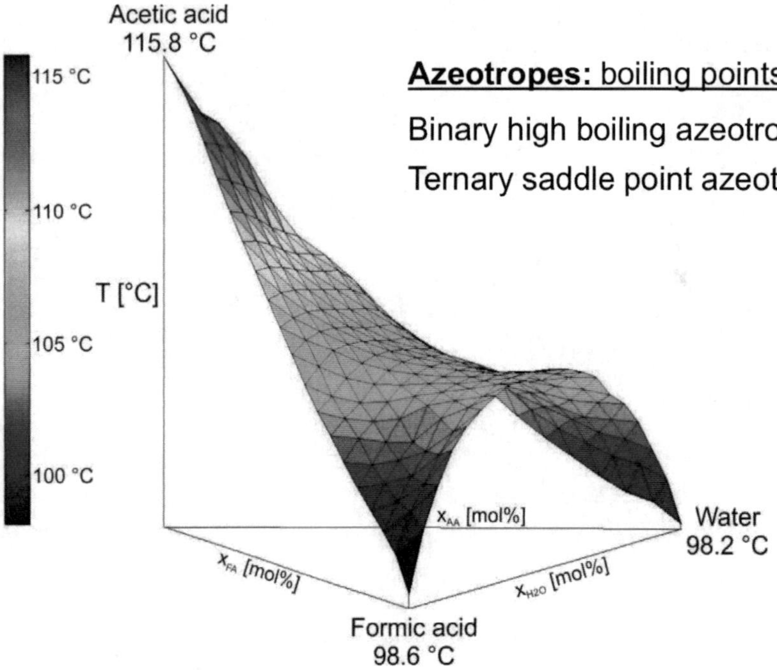

Effect of pressure

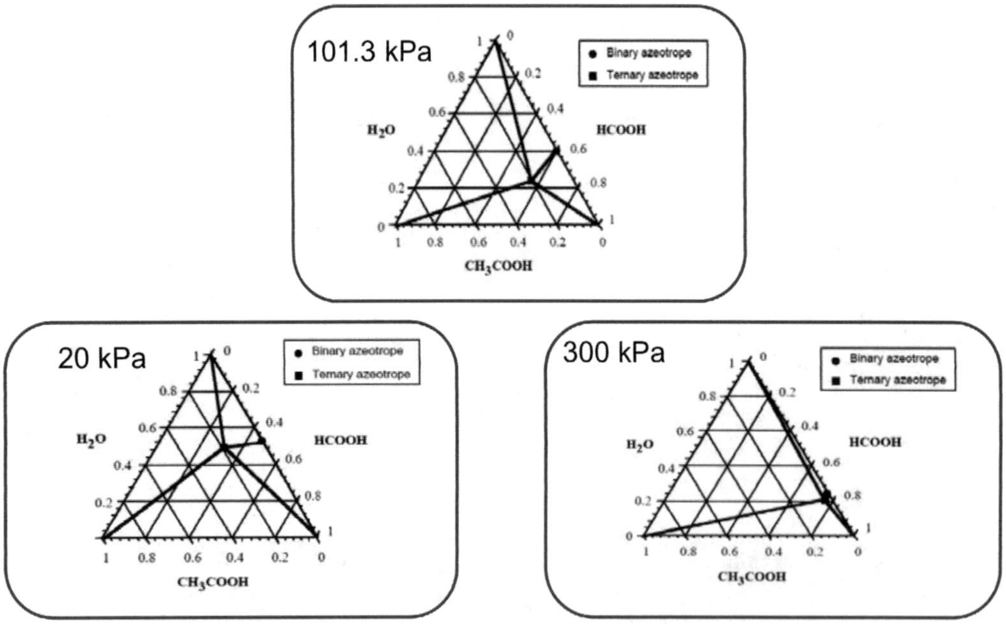

Muurinen, E.: Organosolv pulping. University of Oulu, PhD, 2000

Binary Systems

Dewatering (e.g. azeotropic distillation)?

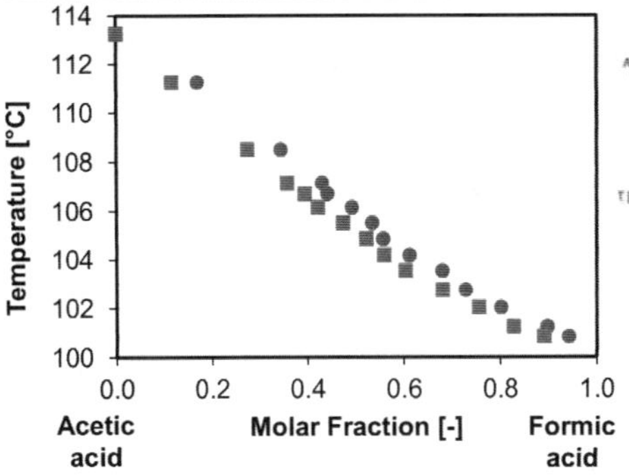

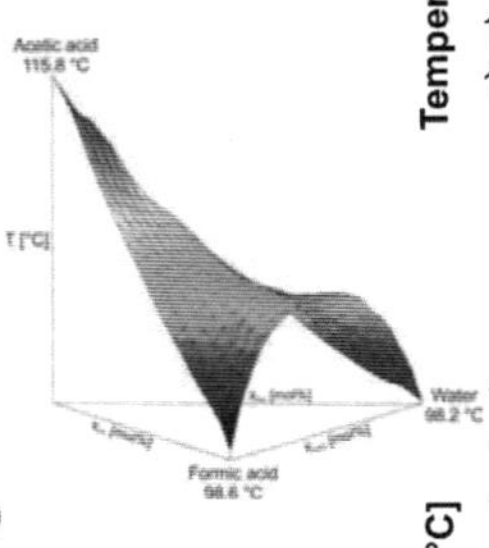

Decomposition of formic acid?

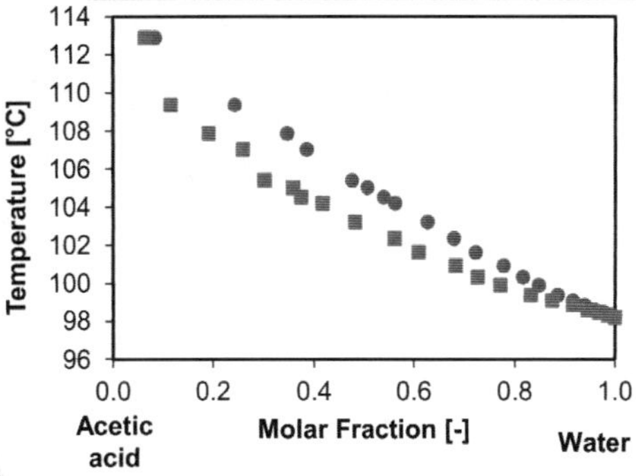

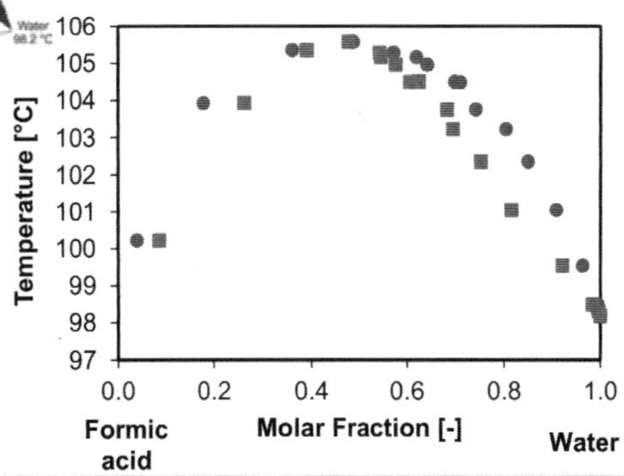

Solver Strategy

Change the properties!
Change the state!

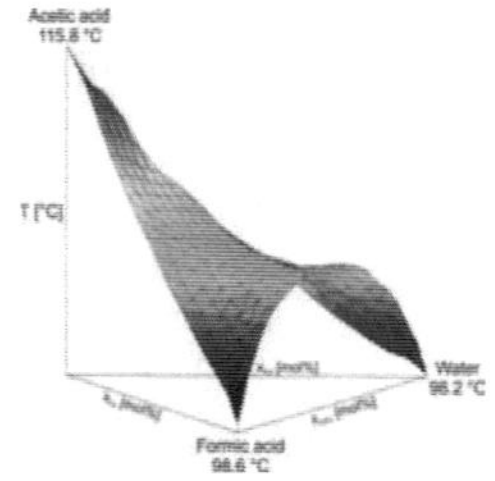

Building Blocks of PI

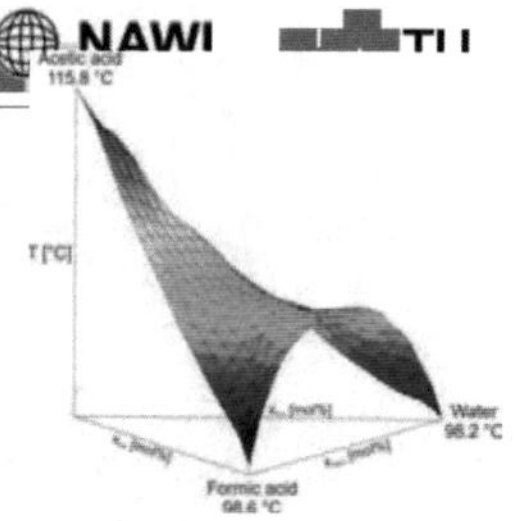

Chemical Reaction
e.g. Esterification

Membrane Process
e.g. Pervaporation

Distillation

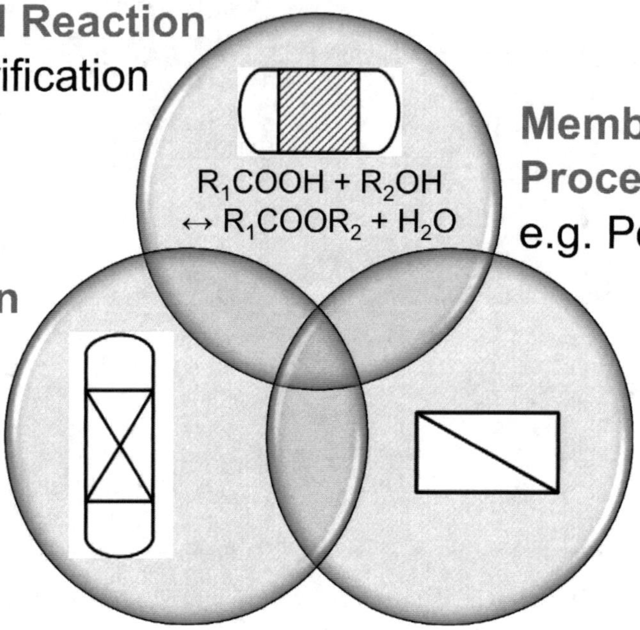

Reactive Distillation

Change the Properties

Esterification with Methanol

$$HCOOH + CH_3OH \overset{H^+}{\longleftrightarrow} HCOOCH_3 + H_2O$$

$$CH_3COOH + CH_3OH \overset{H^+}{\longleftrightarrow} CH_3COOCH_3 + H_2O$$

	Δhv [kJ/kg]	Boiling point [°C]
	Bp, 1.013 bar	1.013 bar
Water	2265	100
Formic acid	479	101
Acetic acid	398	117.9
Methanol	1097	64.6
Methyl formate	469	31.8
Methyl acetate	411	57

Continuous Reactive Distillation

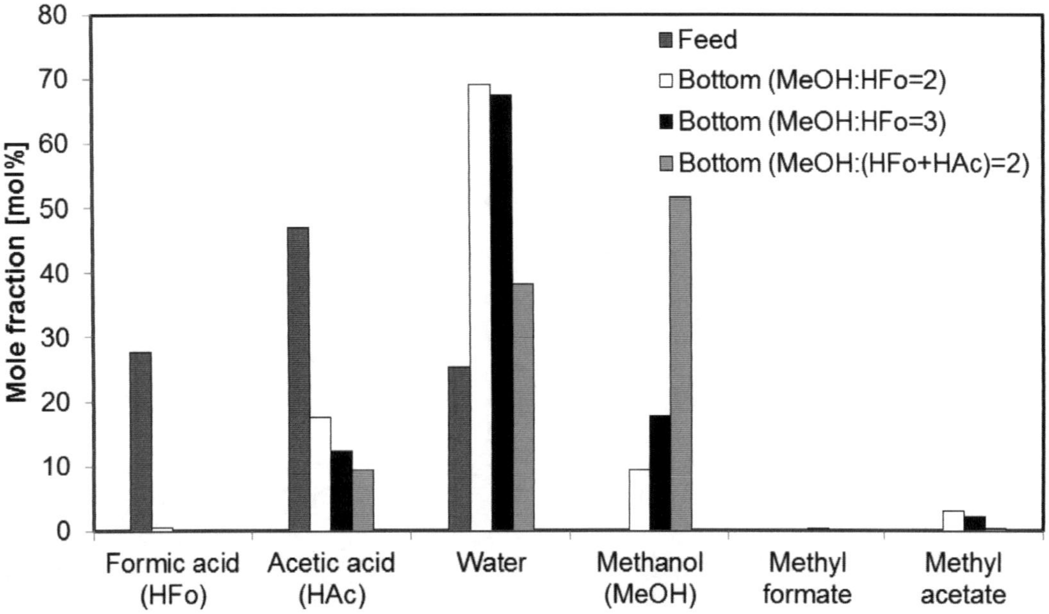

Continuous Reactive Distillation

	MeOH:HFo=2	MeOH:HFo=3	MeOH:(HFo+HAc)=2
Distillate	**mol%**	**mol%**	**mol%**
Methanol	20.4	24.4	46.0
Methyl formate	41.7	33.0	27.6
Methyl acetate	38.0	42.3	26.2

Conversion	**%**	**%**	**%**
Formic acid conversion	98.0	99.3	98.7
Acetic acid conversion	60.8	71.5	64.4

=> Complete removal of formic acid

Pervaporation

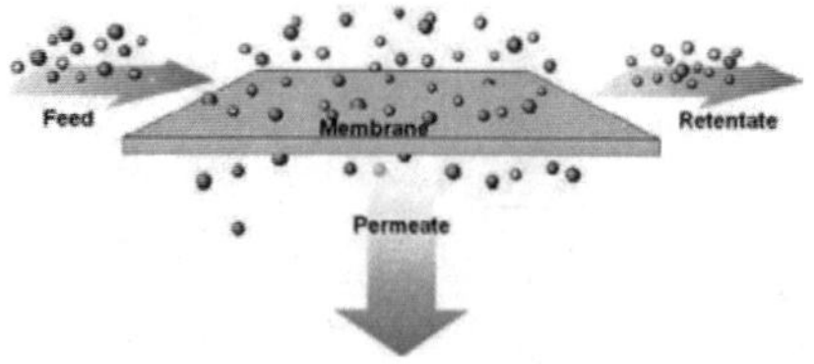

Processing of Condensates

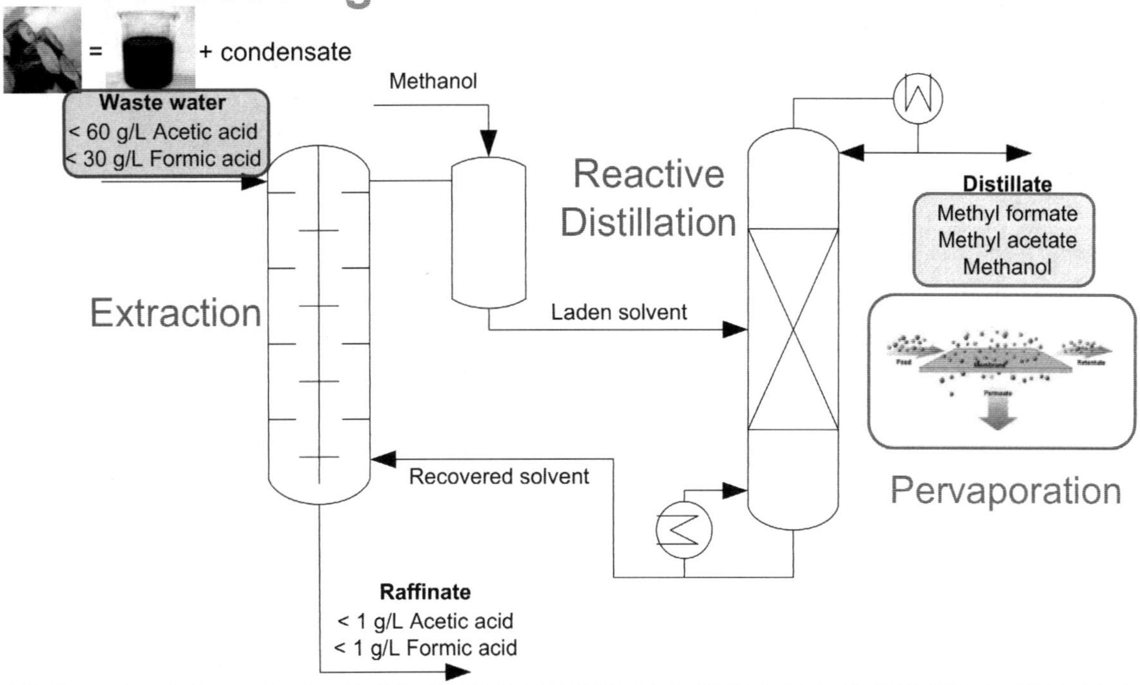

Pervaporation

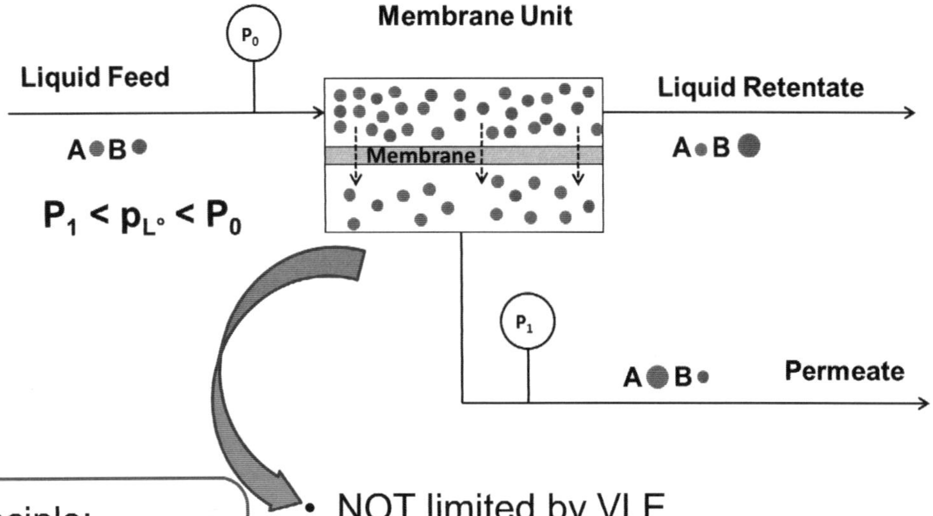

Principle:
Solution-diffusion properties

- NOT limited by VLE
- low operation temperature
- selective evaporation => energy savings

Membranes

permeable for e.g.
alcohol, H_2O

Composite Membranes

permeable for
ester

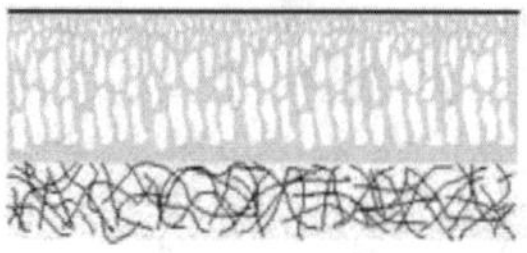

Hydrophilic surface properties

Separation Layer: **PVA**
Support Layer: **PAN**

PERVAP™ 2255-30 | **PERVAP™ 2255-70** | **PERVAP™ 2255-80**

Organophilic surface properties

PDMS: **PERVAP™ 4060** | **PDMS 04-075** | **PDMS 04-123**

POMS: **POMS 05-119**

Degree of cross-linking

Isolation of Fatty Acids from Industrial Effluents

Membranes

permeable for e.g.
alcohol, H_2O

Composite Membranes

permeable for
ester

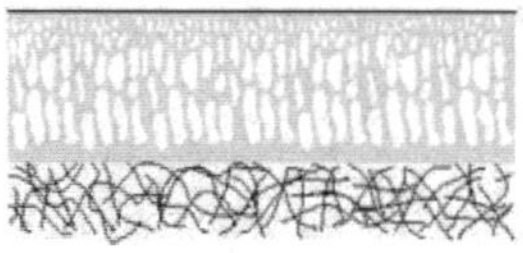

Hydrophilic surface properties

Separation Layer: **PVA**
Support Layer: **PAN**

PERVAP™ 2255-30 | **PERVAP™ 2255-70** | **PERVAP™ 2255-80**

Organophilic surface properties

PDMS: PERVAP™ 4060 | PDMS 04-075 | PDMS 04-123

POMS: POMS 05-119

Degree of cross-linking

Experimental Setup

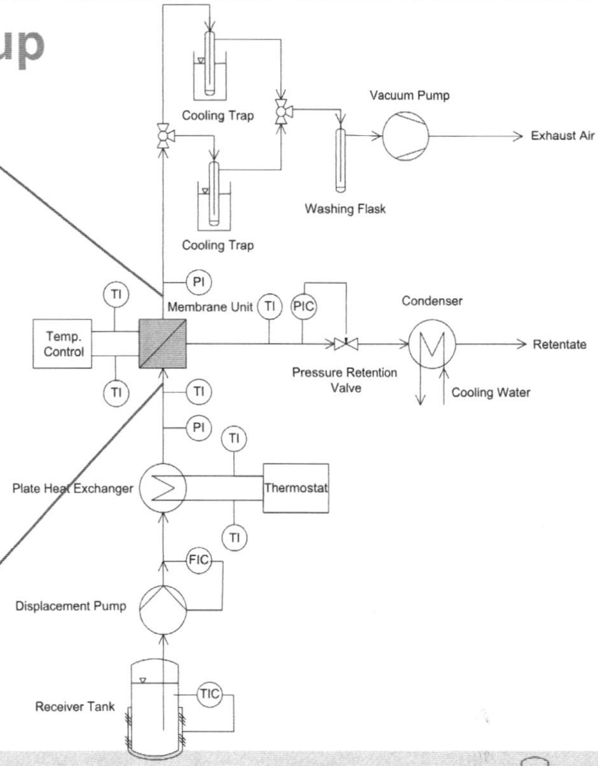

Membrane area: 27.5 cm^2

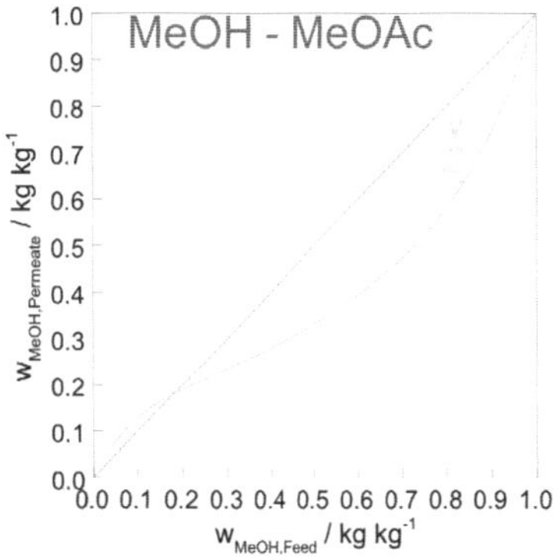

Pervaporation

Methanol – Methyl Acetate

Hydrophilic Membranes

Selectivity

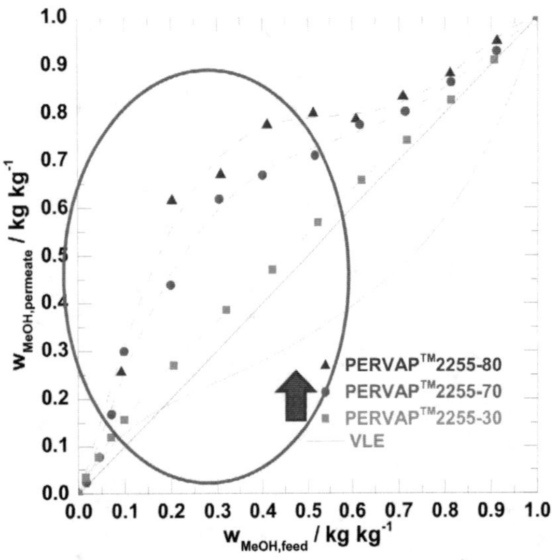

Permeate Flux

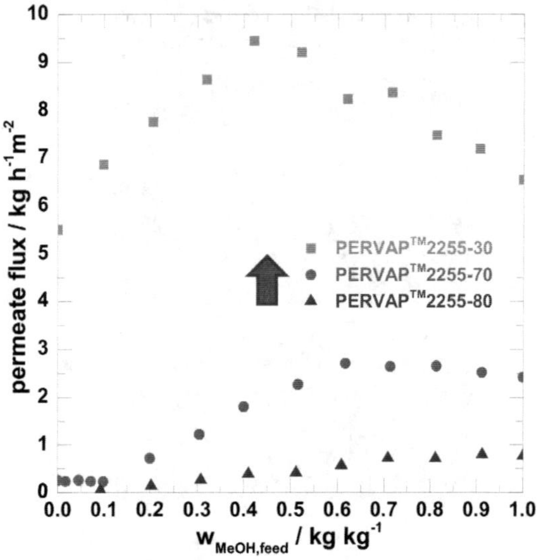

T = 50°C, p_{feed} = 1.5 bar, $p_{permeate}$ ... ar, $\dot{V}_{feed}$ = 15 ml min

High selectivity

Permeate Flux

MeOH + MeOAc
MeOH
MeOAc

PERVAP™2255-30

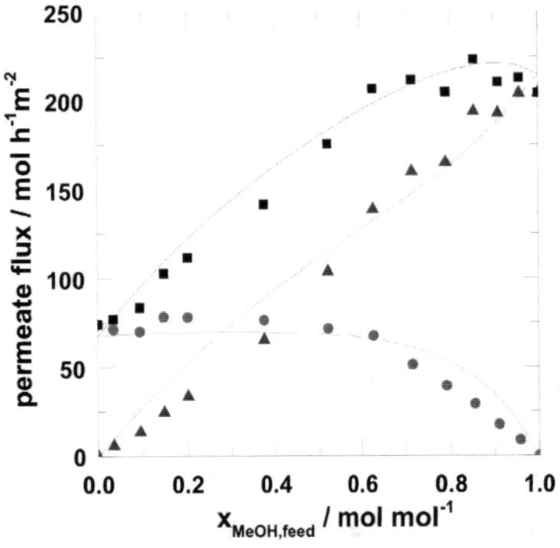

PERVAP™2255-80

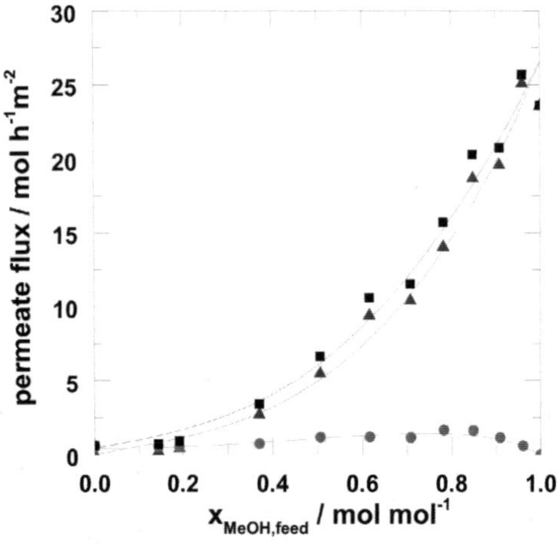

Solution-diffusion model: → swelling coefficient S

$$J_i = \overline{Q_i} \cdot \Delta p_i \cdot S$$

$$S = \frac{a + x_{SC,F}^{\,n}}{b - c \cdot x_{SC,F}^{\,n+1} + d \cdot x_{SC,F}^{\,n+2}}$$

Pervaporation

Methanol – Methyl Acetate – Methyl Formate

Integration of Reactive Separation & Pervaporation

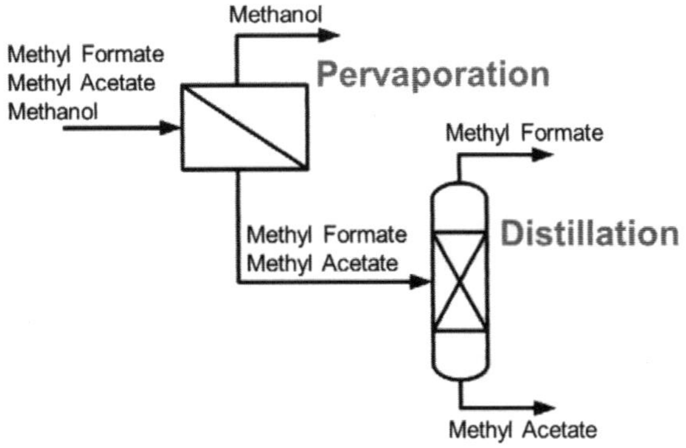

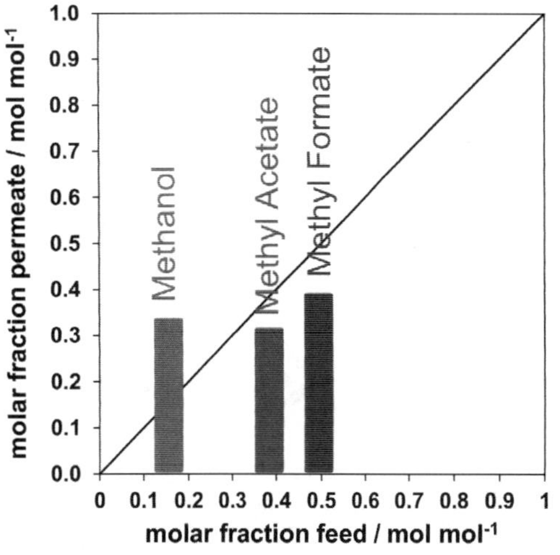

T = 40°C, p_{feed} = 1.5 bar,
$p_{permeate}$ = 50 mbar, $\dot{V}_{feed}$ = 15 ml·min^{-1}

Conclusion

- Integration of Reactive Separation and Pervaporation as an **efficient tool** (in e.g. Biorefinery)
- Recovery of **fatty acids** from aqueous effluents via **reactive distillation**
- **Valuable** bulk products **esters**
- Distillate processing via **pervaporation**

Isolation of Fatty Acids from Industrial Effluents

S. Lux, D. Painer, M. Siebenhofer

Acknowledgement: Sulzer Chemtech AG, Switzerland SULZER

376966 Effects of Ultrasonic Stimulation on Release of Internal Hydrophobic Dyes from Pluronic Micelles

Abstract:

In a general chemical industrial process, many reactors, mixture operations, and separation operations are needed, and the design and maintenance become complex. On the other hand, high performance drug carrier has been investigated in the field of drug delivery system. The use of Pluronic block copolymers in experimental medicine and pharmaceutical has been investigated long time. And, the application of ultrasound for drug delivery system has been investigated. In this study, the application of the drug carrier to a chemical reaction process was investigated from the view point of process intensification. Pluronic micelle and hydrophobic dye were used as a model reactant carrier and internal substances, respectively. The effects of ultrasonic irradiation on effluence of dye from Pluronic micelle were investigated. Especially, the relationships between ultrasonic frequency and type of Pluronic were investigated.

Pluronic was used as the micelle and NKX-1595, which is a hydrophobic organic dye, was used as the internal substance. A Pluronic micelle solution containing NKX-1595 was prepared by a dialysis method. The effects of ultrasonic stimulation on the release of NKX-1595 from the Pluronic micelle were investigated. Before ultrasonic irradiation, the Pluronic micelle solution (10 mL) containing NKX-1595 was charged into a reaction vessel. The sample was irradiated with ultrasound for 10 min, and the effects of ultrasonic frequency and power on the release of the dye were investigated. The ultrasonic power in the reactor was measured by calorimetry. To quantitatively evaluate the amount of dye released from the micelle, the absorbance of the sample solution before and after stimulation was measured at a wavelength of 480 nm using a UV–Vis spectrometer.

From the results of ultrasonic stimulation, the internal dye is readily released from the micelle at low ultrasonic frequencies. An ultrasonic physical effect is considered one of the main factors in the release of the hydrophobic dye from the Pluronic micelle. It also can be observed that degree of drug release (DDR) increases with the Pluronic micelle size. The release of the dye from the micelle by ultrasonic stimulation is not only affected by frequency but also by the molecular structure and micelle size. We focused on the inhibition effect when the dye moves to the corona from the core. Under strong ultrasonic physical effects at 22.8 kHz, DDR is not affected by the number of EO units per unit core surface area. However, under a weak physical effect at 490 kHz, DDR changes rapidly when the number of EO units per unit core surface area is approximately 0.4 nm^{-2}. This phenomenon is owing to the fact that when the internal substance moves to the corona from the core, the inhibition effect decreases with the number of EO units per unit core surface area.

Continuous Co-Crystallization As a Separation Technology: The Study of 1:2 Co-Crystals of Phenazine−Vanillin

Po-Yen Lin, Hsin-Rong Chen, Hong-Yu Lin, Hung-Lin Lee and Tu Lee*

Department of Chemical and Materials Engineering, National Central University, 300 Jhong-Da Road, Jhong-Li City 320, Taiwan R.O.C.

ABSTRACT

A potential separation technology in selective removal of vanillin by forming 1:2 co-crystals of phenazine-vanillin is demonstrated. From the initial solvent screening, 1:2 co-crystals of phenazine-vanillin can be formed by cooling co-crystallization in toluene. Form α phenazine crystals can be regenerated in acetone and vanillin is retained in the solution. Furthermore, the underlying mechanism of co-crystallization and regeneration is explained by solubility curves. Co-crystallization is successfully carried out in a continuous mode, and batch modes both with and without stirring giving 1:2 co-crystal of phenazine and vanillin. However, it is found that the efficient heat transfer due to the instant drop of bulk solution temperature to 25 °C in continuous co-crystallization at 90 mL/min gives many advantages due to rapid nucleation rate. The mixed suspension, mixed product removal (MSMPR) formalism for the population balance gives determining the nucleation rate, B_0, of $2.8 \times 10^4 \pm 4 \times 10^3$ no/l·s. The particle size distribution of 1:2 co-crystal is ranging from 75 to 500μm, a wider and flatter size distribution of 18 wt % for particle sizes of >75 to >250 and about 12 wt % of >300 to >500 μm by sieve analysis. Finally, the continuous operation gives a high yield of 51.2 ± 0.7 wt %, compared to the ones of 34.9 ± 0.4 wt % from batch operation with no agitation and

---

* Corresponding author. Telephone: +886-3-422-7151 ext. 34204. Fax: + 886-3-425-2296. E-mail: tulee@cc.ncu.edu.tw

57.7 ± 8.1 wt % from the batch mode with stirring. The wider size distribution of particle sizes exhibits a better filterability of the wet cake and flowability of the dried co-crystals.

A Framework For Combining Physics Based Approaches with Statistics and Fuzzy Input To Model Complex Systems

Narayan Ramesh, Shrikant Dhodapkar
The Dow Chemical Company, Freeport, TX

Complex systems with multiple sources of variability and subject to various mechanism that cause their performance to be affected have typically been modeled using statistical fits generated from design of experiments or detailed physical models that are not well suited to measure all the variability that the complex systems are subjected to. Some of the elements of variability may be controlled, others are noise variables and to complicate matters all the sources of variability are seldom known. In this talk we present a framework for combining inputs from actual system performance with fundamentals of underlying physics and statistical analysis to improve the fidelity and reliability of models. Two relevant examples are shared and future work is proposed.

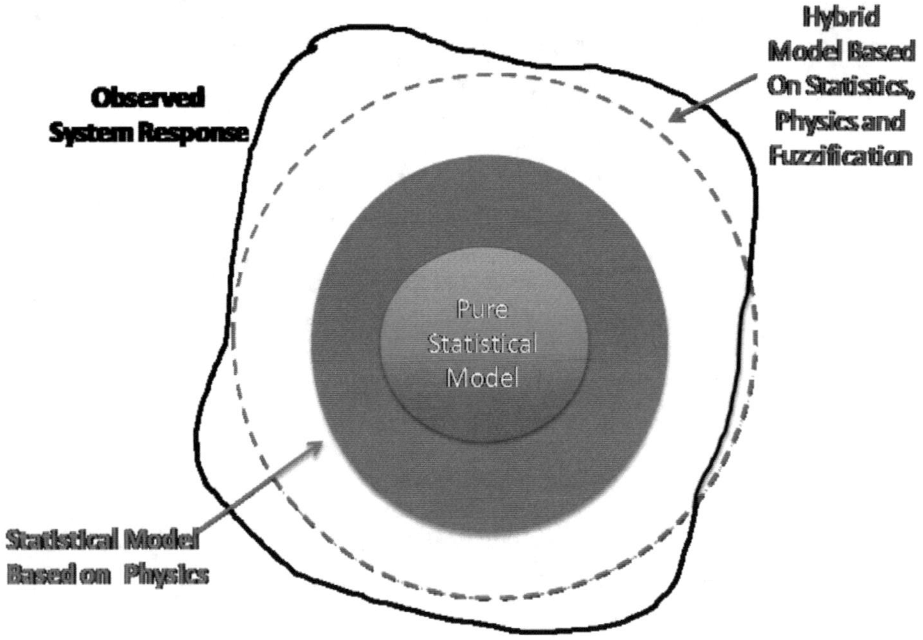

Early results for the improvement of algae production using partial emptying and filling

Michelle Low,[1*] David Ming,[1] Diane Hildebrandt,[2] David Glasser,[1,2] Tonderayi Matambo[2]

[1]University of the Witwatersrand, 1 Jan Smuts Avenue, Johannesburg, 2000,South Africa

[2]University of South Africa, Corner Pioneer and Christian De Wet Avenue, Florida,0003, South Africa

*Michelle.Low@wits.ac.za

Abstract

The objective of this research is to improve the batch production rate of algae by applying concepts from Ming et al. (2012). Improvements are achieved directly from experimentally measured data in the absence of a well-defined kinetic model. Specifically, this is achieved via a partial emptying and filling scheme. The basic methodology of the method is as follows: rather than emptying out the full volume after each cycle for a desired exit concentration, it is sometimes beneficial to retain a fraction of the current product volume for use in subsequent cycles. This is counterintuitive to the standard method of running a batch.

Improvements are estimated via a simple graphical approach – the batch growth profile is plotted in concentration-time space and inspected for concavities. If concavities in the batch profile are found it suggests potential opportunities for improving production rate via retaining a fraction of the batch. The benefit of this approach is that a theoretical model is not required in order to carry out the analysis and predictions may be obtained on experimental data alone. Furthermore, this approach also allows for a recommendation on the optimal fraction of retained volume, termed here as the retained fraction. It follows that this method may be applied on the experimental algae growth batch data to predict the production improvement at a certain retained fraction.

Production rate in concentration-time space may be conveniently represented as a straight line joining the initial and terminating conditions on the batch profile. The gradient of the line straight line, m, therefore provides a measure of batch production rate. This is shown in Figure 1. When a fraction of material is retained, we obtain an intermediate concentration C* from which a new production rate may be generated. Hence, if the intermediate gradient (m*) is smaller than the original gradient (m), then there is a concavity in the batch profile and an improvement in production rate may be achieved (Ming et al., 2012). Biological systems typically contain these gradients, such as in systems where Monod kinetics apply.

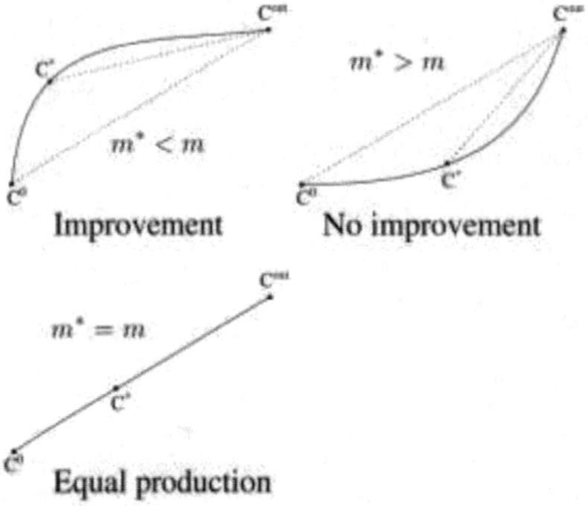

Figure 1: The following figure is obtained from Ming et al. (2012). The shape of the batch profile may be used to identify opportunities for improving production rate.

The experimental investigation of this work consisted of three parts:

1. To experimentally obtain the batch growth rate of an algae species and to determine if concavities existed.
2. To see if an improvement in the production rate could be obtained by the theoretical analysis described above.
3. Validation of the predicted values from chosen experimental results obtained at specific exit concentrations and retained fractions (0-80%). Comparisons with theory could then be carried out from point 2 above.

Early results indicate that the standard batch profile for the algae tested exhibits a concavity. This suggests that the growth rate of algae (production rate) may be improved via partial emptying and filling. When the method of partial emptying and filling is carried out, an improvement in batch production rate over the standard batch was observed. Additionally, the retained fraction obtained from experimental data matched the recommended retained fraction from theory for the desired exit concentration range.

The close correlation between experimental predictions and theory suggest that the method may be utilized with experimental data only. This is useful when a sophisticated kinetic model is not available. The method also aids in estimating a retained fraction in order to improve production rate. Further analysis of the method may be carried out which would allow for more reliable predictions on experimental data. This is currently an active area of investigation.

Reference:

D. Ming, D. Hildebrandt, D. Glasser. 2012. A graphical method of improving the production rate from batch reactors. Industrial & Engineering Chemistry Research, 51(42): 13562–13573.

380562 Dynamic Load Optimization for Ethylene Cracking Furnace System

Abstract:

Cracking furnace is the key equipment of an ethylene plant as it primarily determines the plant productivity. Because the furnace operation is semi-continuous, which involves periodical decoking operations, multiple furnaces have to be simultaneously employed to keep the plant production continuously. It should be noted that in reality the performance of each furnace is different when cracking different feedstock under different operating conditions. Thus, it is a very challenging task to optimize processing loads for each furnace under conditions that plant production recipe is changing or some furnace is experiencing the decoking operation.

In this paper, a general methodology framework for dynamic load optimization of ethylene cracking furnace systems has been developed to maximize the plant profitability. The methodology includes three stages of work. Firstly, predictive yield models for major cracking products such as C_2H_4, C_3H_6, H_2, C_4H_6 and C_6H_6 are developed based on a radical reaction mechanism. Secondly, the simplified heat transfer model and coking model are developed based on operating data and further validated according to the CO concentration in the furnace tail gas. These data-driven based models are used to predict the run length of each furnace batch operation. Finally, based on all the simulation models obtained from the above, an optimization model is developed and optimally solved to smartly allocate feedstock to different furnaces under various operating conditions, so as to maximize the total profitability of the ethylene plant furnace system. The modeling performance and different solving strategies will be extensively discussed in this study.

The University of Manchester

Monitoring Cleaning-In-Place (CIP) of Multiphase Formulated Products by Electrical Resistance Tomography (ERT)

Abstract ID: 382634

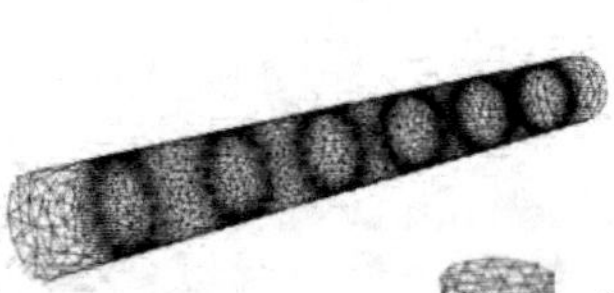

H. J. Uppal,[a] T. L. Rodgers,[a] M. Cooke,[a] L. Trinh,[a] A. J. Kowalski,[b] P. J. Martin[a]

[a]*Centre for Advances in Structured Liquid Engineering CASTLE, School of Chemical Engineering and Analytical Science CEAS, The University of Manchester, UK*
[b]*Process Science, Unilever R&D, Port Sunlight Laboratory, UK*

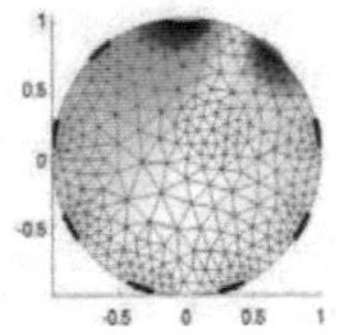

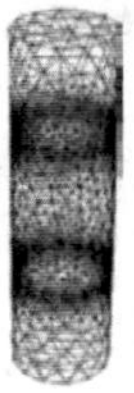

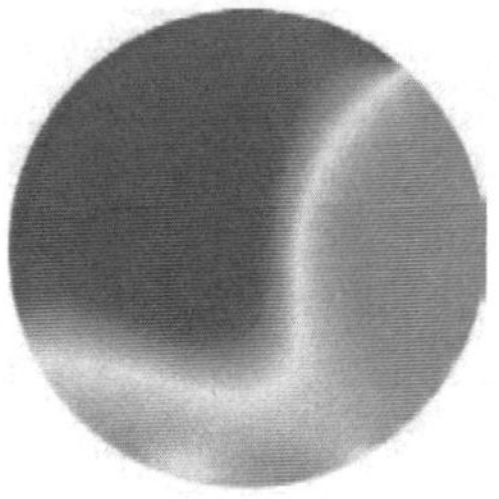

AIChE, Atlanta, GA, USA – Nov 2014

AIChE

1

The University of Manchester

Scope of Research

Investigating Cleaning-in-Place (CIP) strategies in non-Newtonian fluids

Research Objectives:

- Inline Electrical Resistance Tomography (ERT) experimental evaluation study for monitoring water–based Cleaning-in-Place (CIP),

- Implement scalable ERT sensors; a inline Perspex straight pipe section to image fluid displacement and a T-piece pipe section to mimic *'difficult to clean'*-type section,

2

AIChE

CONTENTS

- CIP Cleaning Protocols
- Electrical Resistance Tomography (ERT): complex geometries
- Non-Newtonian model-type fluids
- Process Parameters
- Perspex ERT pipe section; *'straight'* and *'difficult to clean'-type section*

- Results and Analysis

- CONCLUSIONS
- Acknowledgements

3

AIChE

[1] I. Palabiyik, et al., Chem. Eng. Res. and Design xxx (2014).

CIP Cleaning Protocols

- Non-Newtonian Cleaning-In-Place (CIP),
- Faster flexible cleaning protocols,
- A water–based cleaning; via continuous and semi–continuous process flow,
- 3D ERT visualization: radial/axial flow pattern, batch process mixing times quantification,

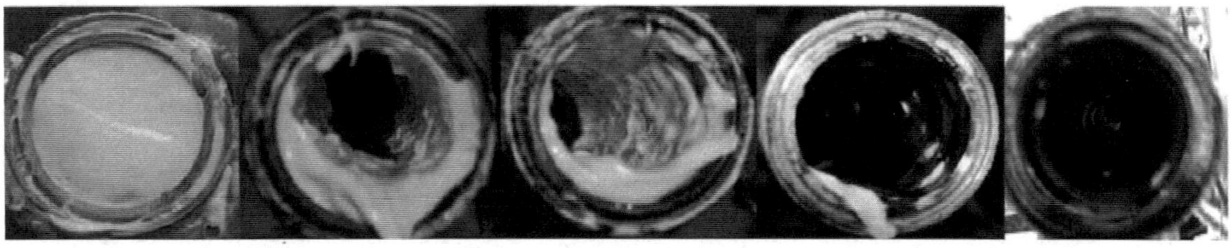

(i) (ii) (iii) (iv) (v)

Scope of Our Research

- To access ERT feasibility in monitoring *cleaning* of non-Newtonian fluids over industrially compliant flow rates (~ 1000–8000 kghr^{-1}, 0.270-2.463 m/s, 9152-83241 Re)

- *Radial CIP; thinlayer or thickness-type erosion,*

----Vs----

- *Axial CIP; core or pressure-type displacement,*

4

AIChE

ERT: Non-Invasive Imaging Technique
Circular Adjacent Conductivity Measurement Protocol

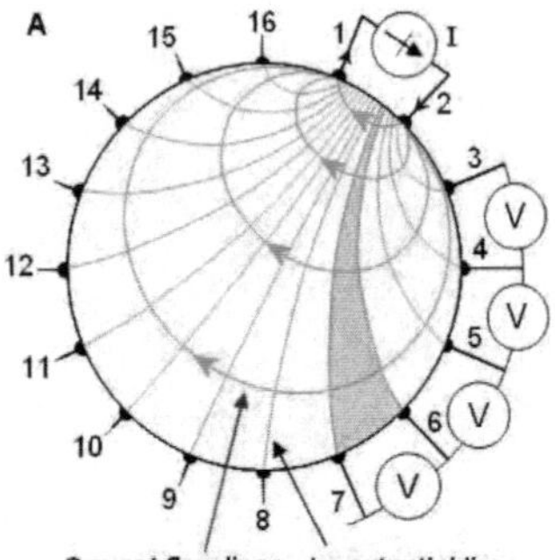

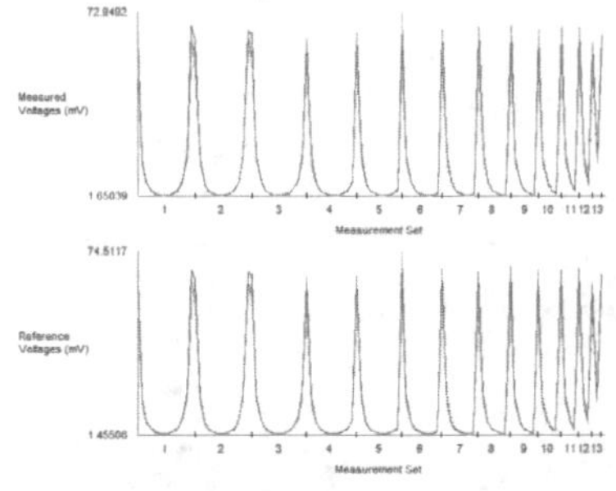

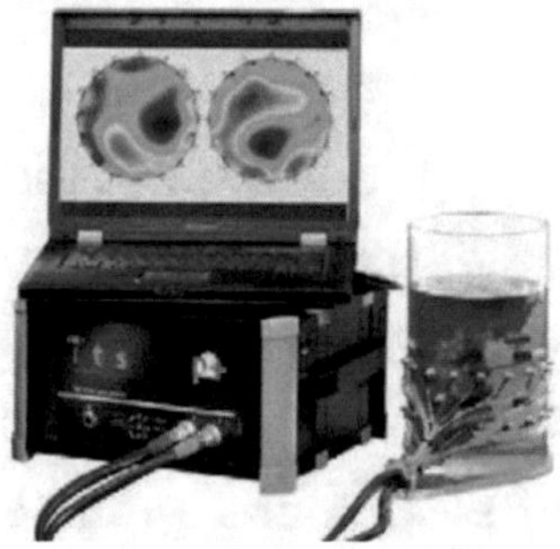

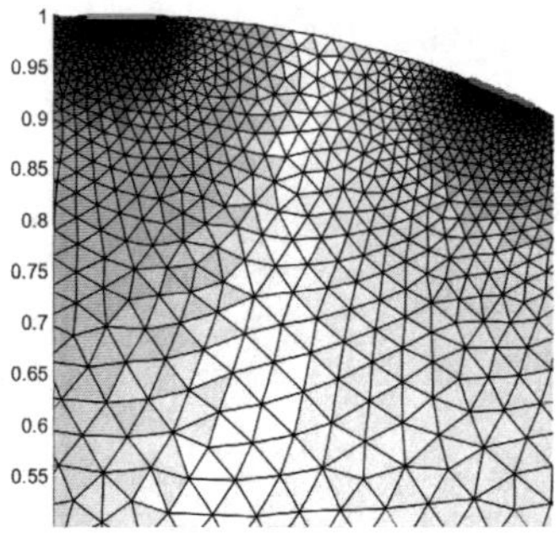

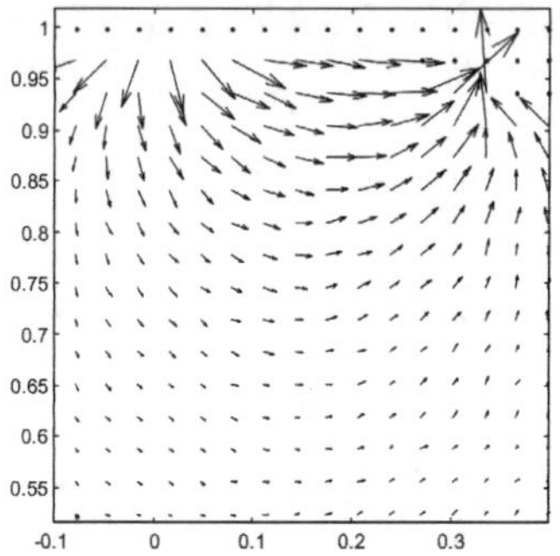

Voltage pattern /Current injection stimulation,

5

*EIDORS Emulator: NETGEN simulations of fwd/inverse solutions,
*Fraser Dickin and MI Wang, et al., Meas. Sci. Technol. 7 (1996) 247–260.

Experimental Setup: Process Parameters

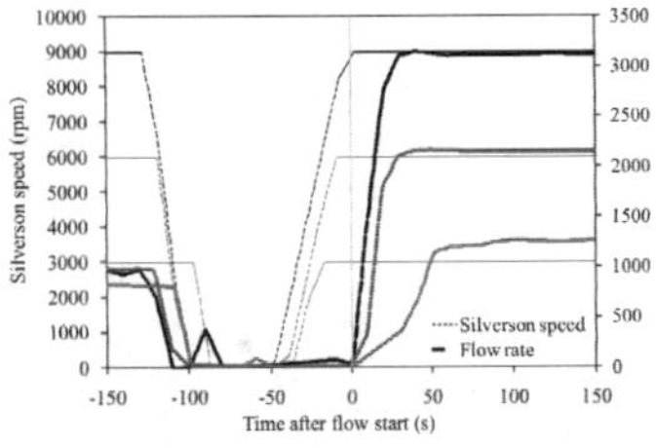

- Non-Newtonian Fluid (≈ 24 mS/cm @ 25°C) – as Fluid 1,
- water (ionic) (≈ 0.110 mS/cm) – as Fluid 2,
- Silverson: *Process Flow Rates @ 3000, 2000, 1000±25 Kg/Hr,*
- Control Valve V1; switching between Fluid 1 (loop) to Fluid 2 (open),

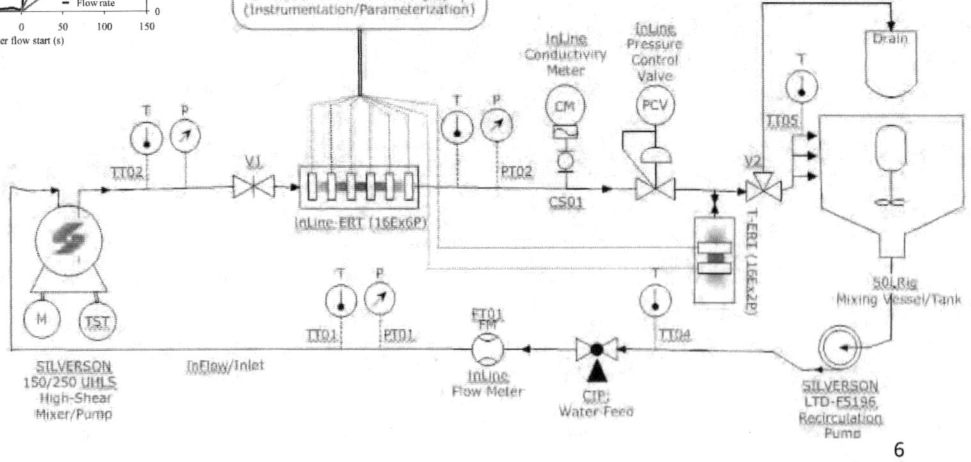

6

MANCHESTER 1824 | SILVERSON | The University of Manchester | MULTIPHASE PROCESSING | Unilever

Solid Geometry Reconstruction: FEM Meshing
PERSPEX InLine-ERT straight pipe and T-piece section

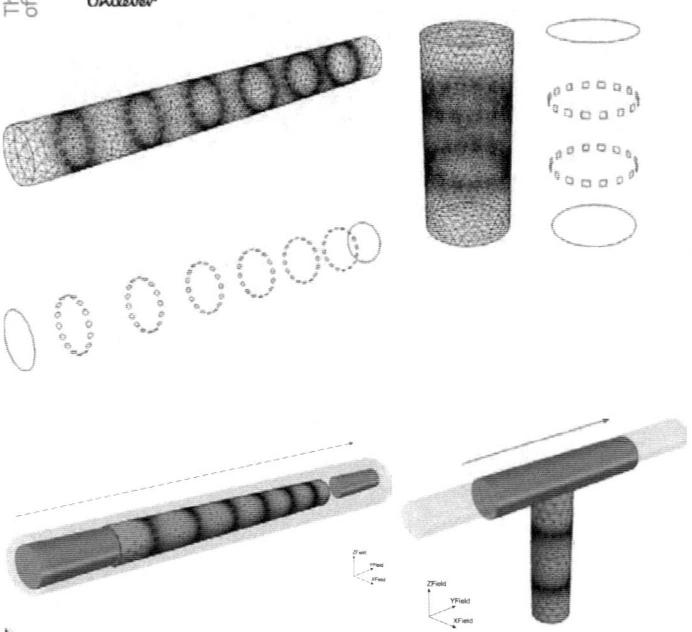

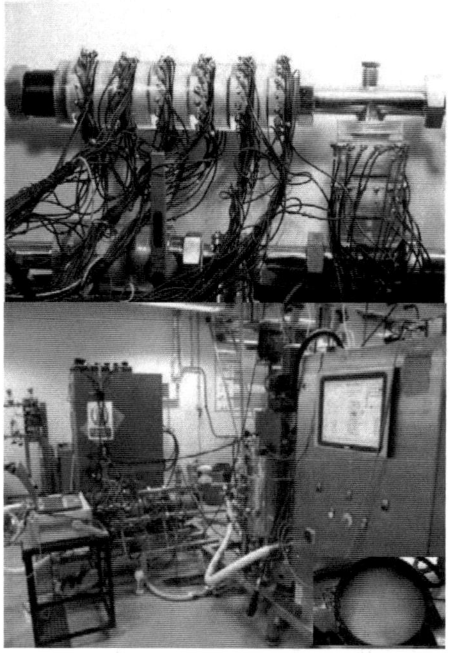

7

AIChE

MANCHESTER 1824 | SILVERSON | The University of Manchester | MULTIPHASE PROCESSING | Unilever

Non-Newtonian Viscoelastic Surfactant-type Micelle Fluid
Power-Law Index (Shear-Thinning)

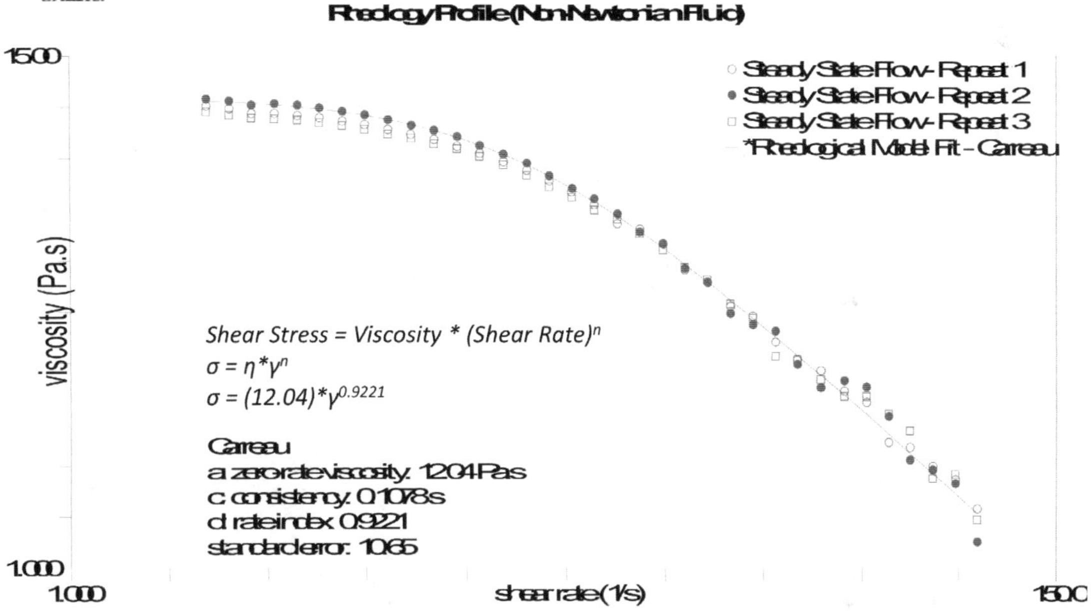

8

AIChE

Inline-ERT Cleaning Profile

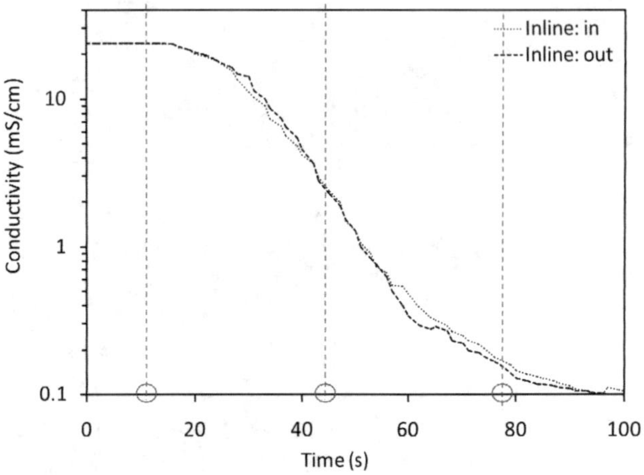

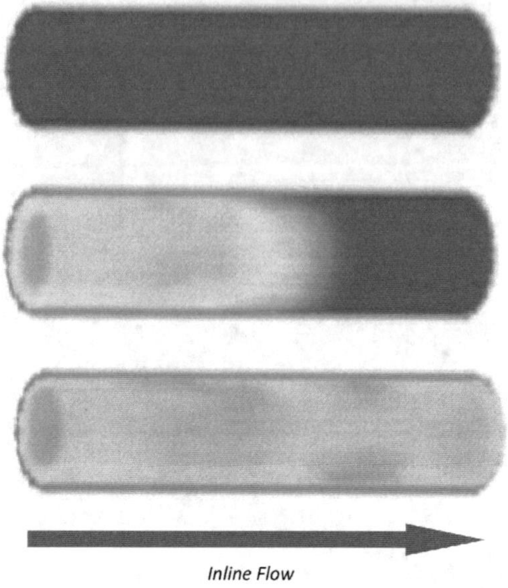

- CIP Sequence;
- *Phase-I; water PumpIN,*
- *Phase-II; water/product-interface FlushOUT,*
- *Phase-III; product core FlushOUT,*
- *Phase-IV; radial/axial post-displacement cleaning; core removal leaving thinlayer surface coating.*

9

AIChE

T-piece section Cleaning Profile

Tomography Profile (2D Spatial Conductivity) reconstructive visualization for InLine T-Pipework water-based CIP; MSBP image reconstruction algorithm; *for qualitative fast image processing.*

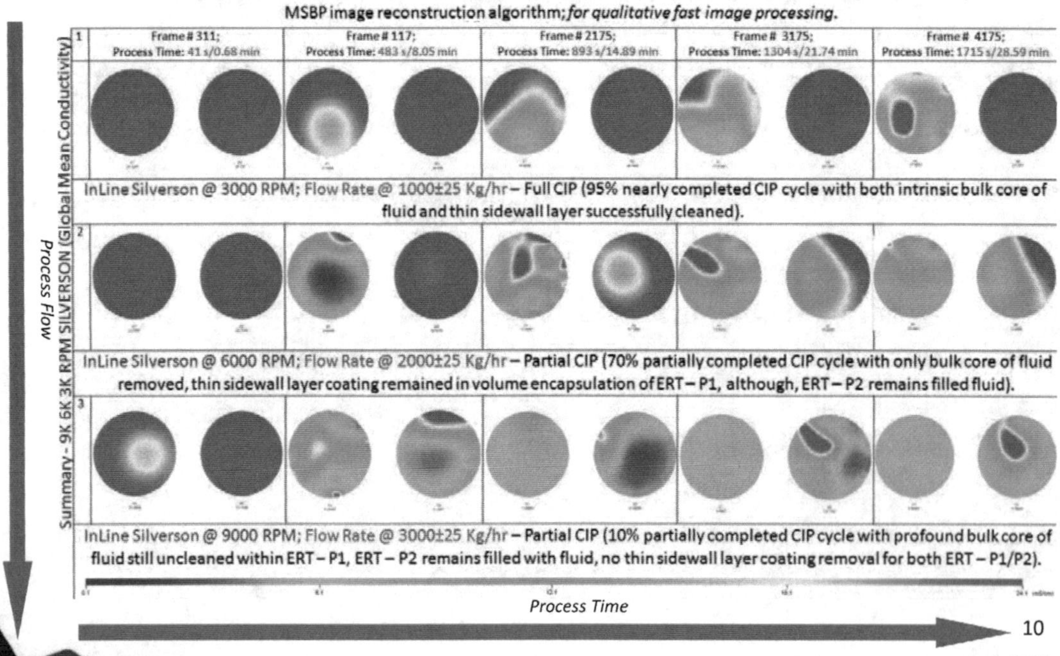

10

AIChE

IsoMetric Finite Element Modelling IsoFEM

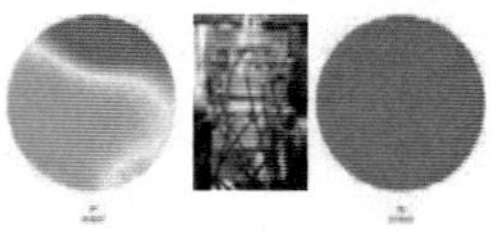

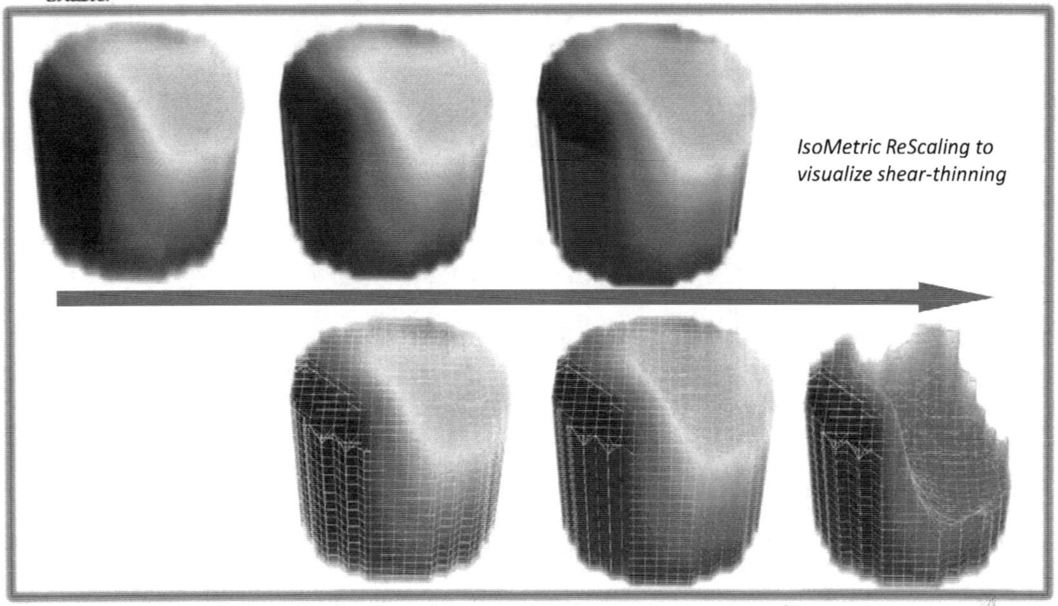

T-piece ERT sensor: IsoVisualization of Low- and High- Conductive Shear/Pressure Zones under:
Non-Newtonian Fluid to Water Cleaning; at InLine Process Flow Rate @ 2000±25 Kg/Hr. 11

Cleaning Profile Summary

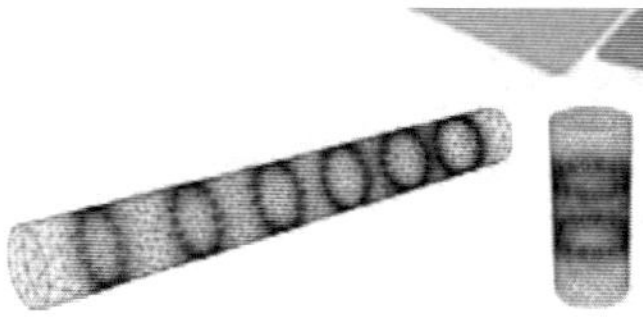

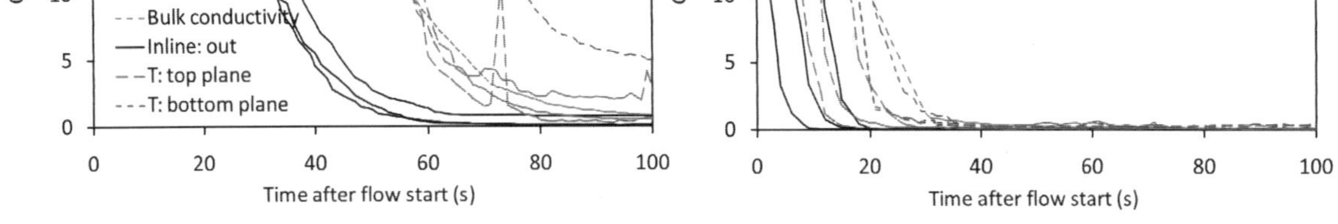

3 x Mean ERT plane conductivities against Fluid 2 flow time:
Non-Newtonian Fluid to Water Cleaning; Process Flow Rate (a) 1000±25 Kg/Hr, (b) 3000±25 Kg/Hr

12

Cleaning Times – Quantification, Method, Analysis
Conductivity RMS Method*

Conductivity-Concentration Maxwell Model:

$$\sigma_{ERT-ALGORITHM} = \left[\frac{(2 \propto (\sigma_1 - \sigma_2)) - (2\sigma_1 + \sigma_2)}{\left(2 + \frac{\sigma_2}{\sigma_1}\right) - \left(\propto (1 - \frac{\sigma_2}{\sigma_1})\right)}\right]$$

$$ln[\sigma_{RMS}] = \frac{1}{2} ln \left[\frac{1}{N}\sum_{i=1}^{N}\left(\left(\frac{\sigma_{t,i} - \sigma_{I,i}}{\sigma_{F,i} - \sigma_{I,i}}\right) - 1\right)^2\right]$$

σ_{ERT} = reconstructed conductivity (mS/cm), σ_1/σ_2 = conductivity of fluids,
$\propto$ = volume of fluids (m³ or cm³), σ_{RMS} = RMS conductivity,
N = finite conductivity elements i,
$\sigma_{(t,i)}$ = conductivity of element i, $\sigma_{(I/F,i)}$ = initial/final conductivity for element $i(n+1,t)$,

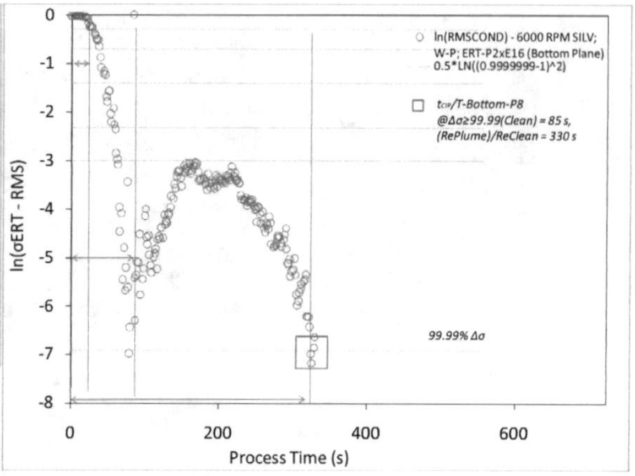

RMS Variance Method; conductivity sampling weight of *% Δσ* scaled towards the end elements *i(n,t)*.
*Δσ initial to end point value based on: for a *99.99% Δσ* variance, *0.5*LN((0.999-1)2) = -6.9077* threshold.
*CIP Times are quantified: when threshold is *> (-6.9077),* assuming homogeneous steady-state *99.99% Δσ*,

[29]*T.L. Rodgers, et al., Chem. Eng. Res. and Design 87 (2009) 615–626.*

CIP Cleaning Map; Summary

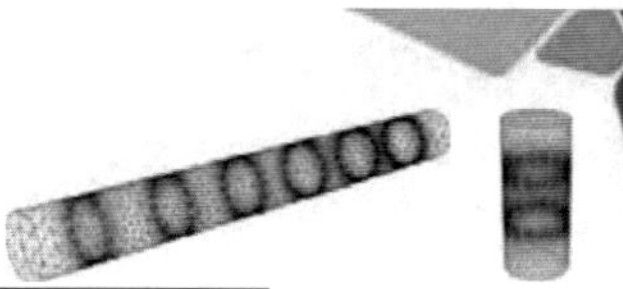

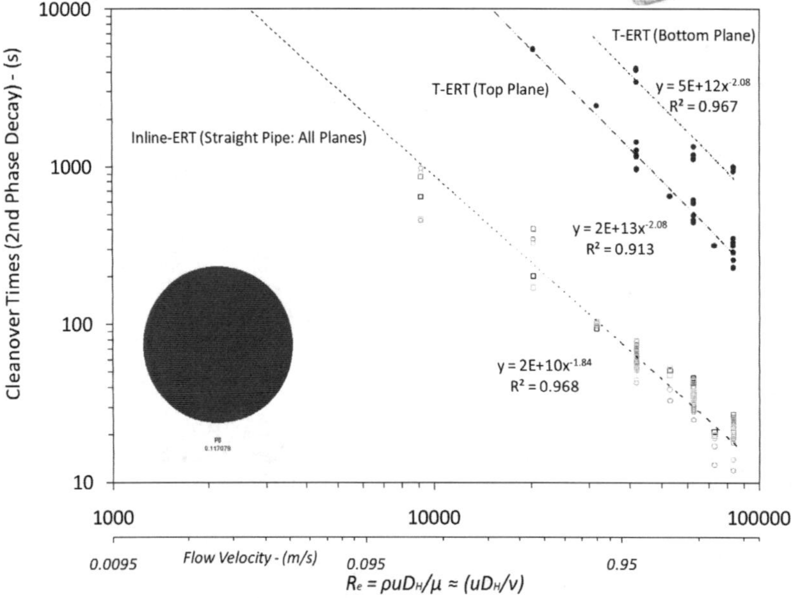

- InLine ERT section; *quicker product cleaning,*
- Top T-ERT plane: *similar trend to InLine cleaning profile,*
- Bottom T-ERT plane; *slower cleaning attributed to localized "difficult-to-clean" zones,*

MANCHESTER 1824 The University of Manchester

SILVERSON

MULTIPHASE PROCESSING

Unilever

CIP Cleaning Map; Summary

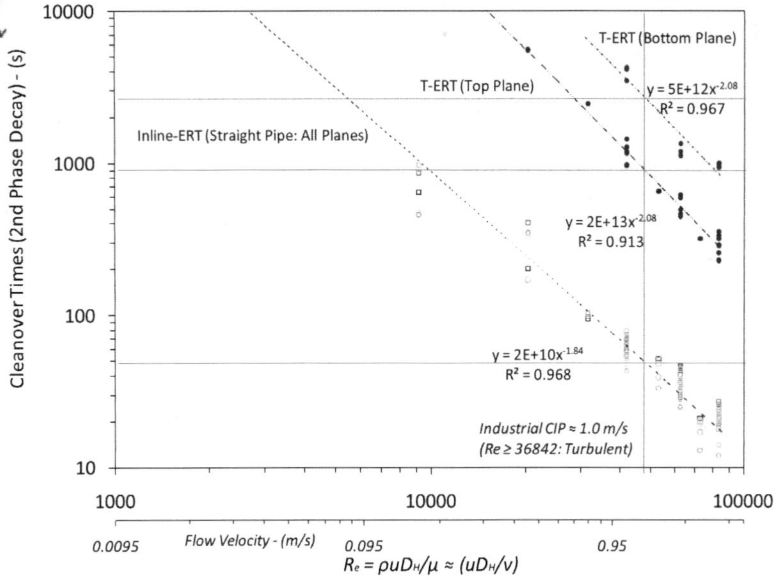

• Tomography-based CIP Index: *successful quantification of process-specific "difficult-to-clean" volumes of non-Newtonian "slower/quicker-to-displace" shear-thinning viscoelastic rheological-fluids*

15

AIChE

MANCHESTER 1824 The University of Manchester

SILVERSON

MULTIPHASE PROCESSING

Unilever

CONCLUSIONS

• Tomography - *a feasible technique;* different modelled geometries have been successfully demonstrated to monitor *different modes of CIP*;

• InLine-ERT and T-piece CIP Protocol:
• ***bulk core velocity/pressure-dependent cleaning;*** *faster CIP,*
• ***thinlayer sidewall thickness-dependent cleaning;*** *slower CIP,*

• Flexible and Agile Process Adaptation; a *"proof-of-concept"* monitoring tool where inline cleaning can be correlatively visualized with process normalized cleaning times ---- *a technique that can be extended to comprehensive product portfolios of non-Newtonian shear-thinning viscosity variant* $\Delta\eta_x$ *products.*

16

AIChE

Acknowledgments

The authors gratefully acknowledge the support of

- *Silverson Machines Ltd,*
- *Unilever UK,*
- *Project co-funded by; Innovate UK Technology Strategy Board and Engineering and Physical Sciences Research Council EPSRC through the Boosting Innovation in Manufacturing Competencies programme 'Project No 101271'*

Innovate UK
Technology Strategy Board

EPSRC
Engineering and Physical Sciences Research Council

SILVERSON
HIGH SHEAR MIXERS/EMULSIFIERS

MANCHESTER 1824
The University of Manchester

Unilever

17

The University of Manchester

Experimental Setup: Process Parameters

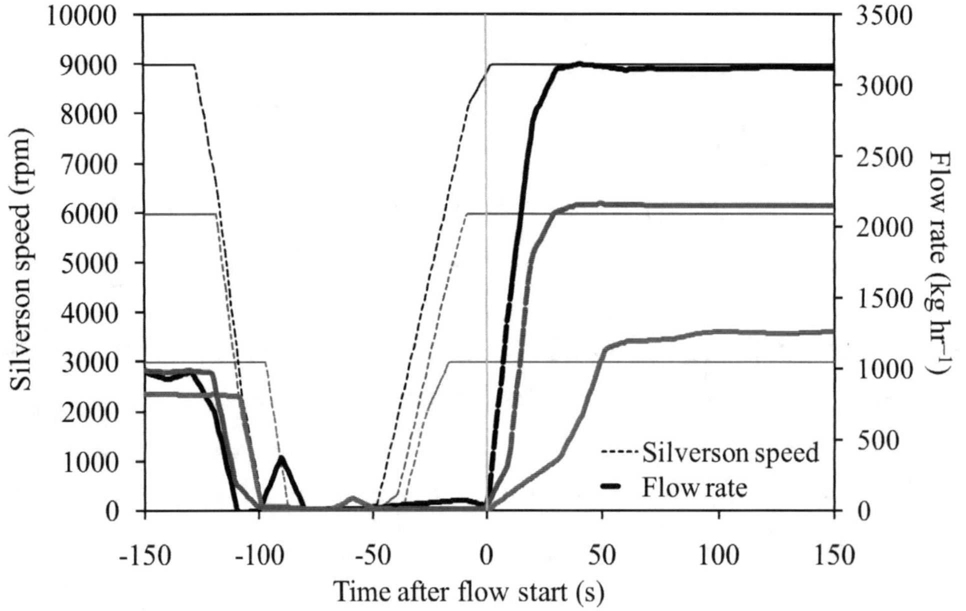

18

AIChE

Strategy to avoid catalyst deactivation in telescoped Miyaura borylation/Suzuki cross-coupling reaction

Marimuthu Andiappan, Jeremy Merritt, Gregory Brown, Jennifer Kuehne-Willmore
Small Molecules Design and Development (SMDD)
Eli Lilly and Company, Indianapolis, USA
AICHE Annual Meeting 2014

Outline

- Introduction: Telescoped Miyaura borylation/Suzuki cross-coupling reaction
- Problem: Variable Suzuki reaction completion time
- Hypothesis: Palladium (Pd) blackout at the end of borylation (when aryl halide is fully consumed)
- Strategy to avoid Pd blackout: Change the LR to borylation reagent instead of aryl halide
- Experimental results
- Summary & Conclusions

2

Telescoped Miyaura borylation/Suzuki cross-coupling reaction

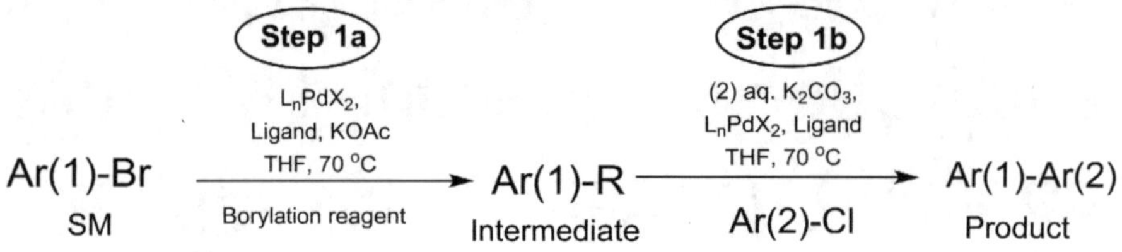

- Step 1a – Borylation
  - 1 equiv. Aryl bromide (LR), 1.4 equiv. borylation reagent - BisPin, 0.03 equiv. $L_nPd^{II}X_2$ catalyst and 0.03 equiv. ligand
- Step 1b – Suzuki reaction
  - 1.5 equiv. Aryl chloride, 0.02 equiv. $L_nPd^{II}X_2$ catalyst and 0.02 equiv. ligand

3

Problem: Variable Suzuki reaction completion time

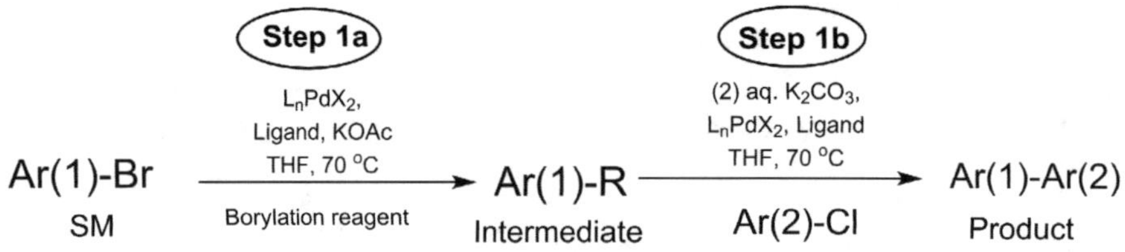

- Step 1a – Borylation
  - Reaction completion time in the lab: ~5.5 hrs
- Step 1b – Suzuki reaction
  - Reaction completion time in the lab: ~12-24 hrs
- <u>Problem</u>: Variable Suzuki reaction completion time
- <u>Desired</u>: Consistent Suzuki reaction completion time
  - To design robust, consistent and predictable process (if possible)
  - To get consistent model(able) data – to aid development, QbD, simulation of PARs, predict risk of reaction stalling, risk of impurity formation

4

Problem: Variable Suzuki reaction completion time in the plant (DME process)

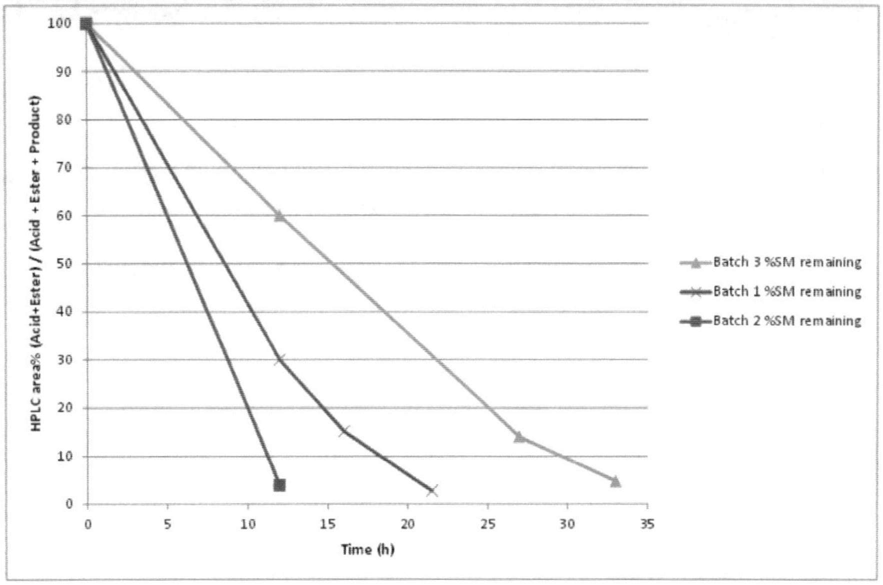

- Suzuki reaction completion time in the plant: ~12-34 hrs
- Reaction completion time in the lab (THF process): ~12-24 hrs

5

Step 1a (Borylation) Pictures

During borylation (0 to ~97% completion)

Near end borylation (~97-99% completion)

Borylation completed mixture (~100% Ar-Br conversion)

6

Hypothesis: Macroscopic view

During borylation (0 to ~97% completion)

Near end borylation (~97% completion)

Borylation completed mixture (~100% Ar-Br conversion)

- Hypothesis: Brown/Black color is clear indication of Pd blackout (Pd blackout happens at the end of complete consumption of aryl bromide, LR)

7

Hypothesis: Mechanistic view

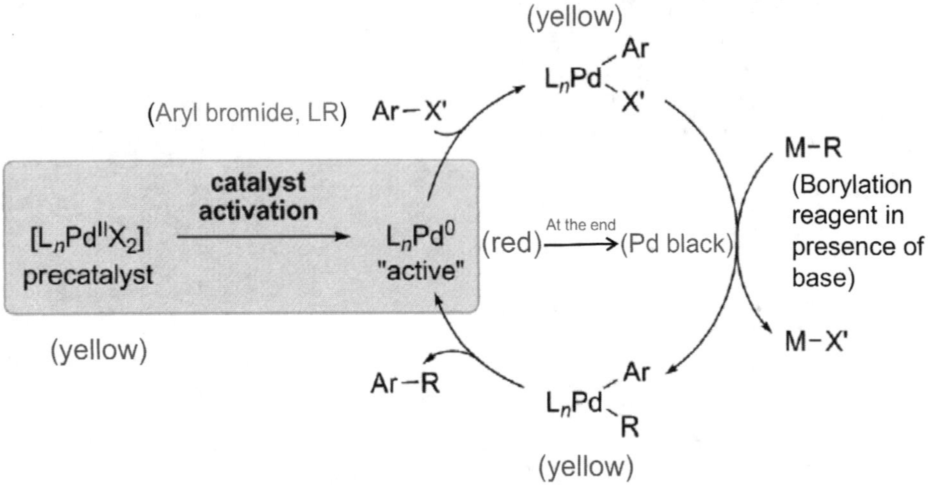

- Transmetallation step (M-R → M-X') is the rate limiting step: L_nPd-Ar-X` might be the resting state giving the yellow color to the borylation mixture
- Hypothesis: Brown/Black color is clear indication of Pd blackout (Pd blackout happens at the end of complete consumption of aryl halide, LR)

Reference for the mechanism: Angew. Chem. Int. Ed. 2013, 52, 5822

8

Strategy to avoid Pd blackout

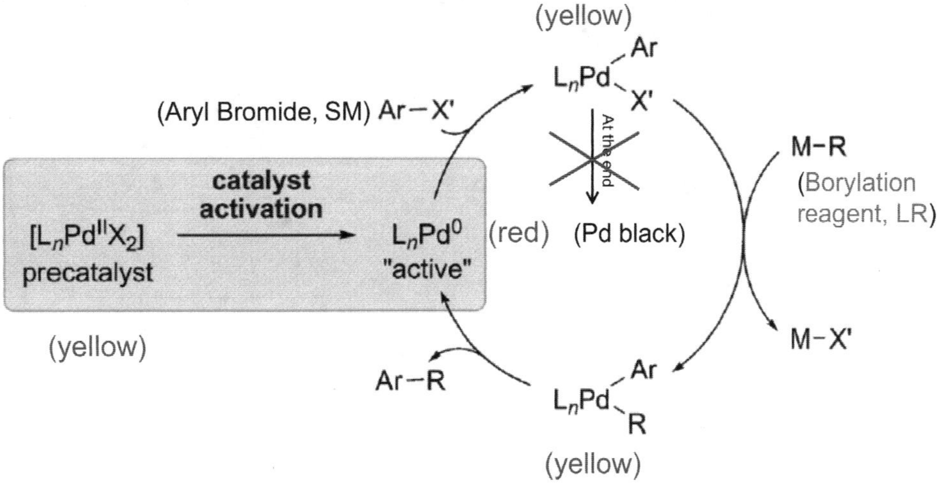

- Change the LR to borylation reagent instead of aryl halide
  - Say only ~97% aryl halide conversion (~0.97 equiv. borylation reagent)
  - Yellow color throughout the reaction could be expected

9

Strategy to avoid Pd blackout

- Hypothesis: Brown/Black color at the end of borylation (Reaction 1a, ~100% Ar-Br conversion) is clear indication of Pd blackout – Blackout conditions

- How to avoid Pd blackout: ~97% Ar-Br conversion (0.97 equiv. borylation reagent, BisPin): Non-blackout conditions

- What could be expected:
  - I. No color change under non-blackout conditions (yellow color throughout)
  - II. Faster Suzuki (reaction 1b) under non-blackout conditions compared to blackout conditions

10

Suzuki kinetic data –Blackout condition (continued from ~100% Ar-Br conversion)

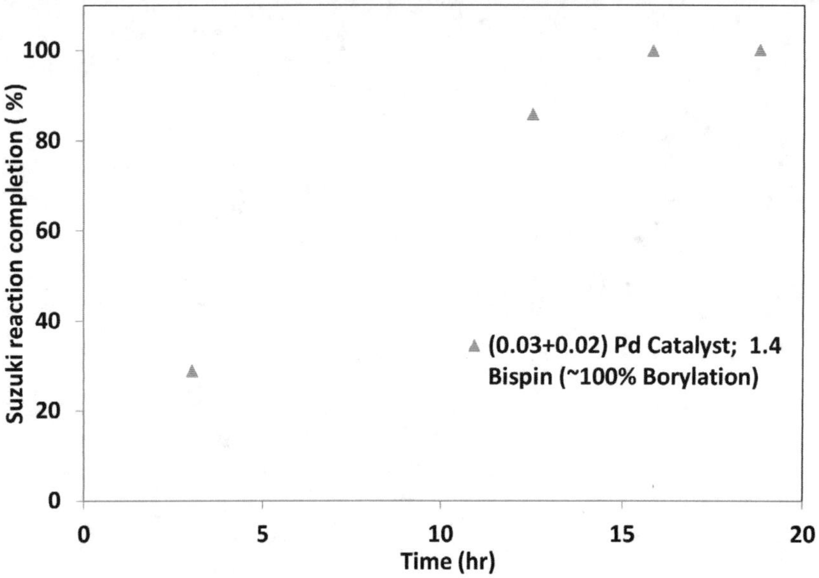

- Suzuki reaction (after Pd blackout in Borylation) completes generally in ~12-24 hrs (~16 hrs in this experiment)

11

Step 1a (Borylation) and 1b (Suzuki) Pictures – Nonblackout conditions (~0.96-0.97 equiv. borylation reagent)

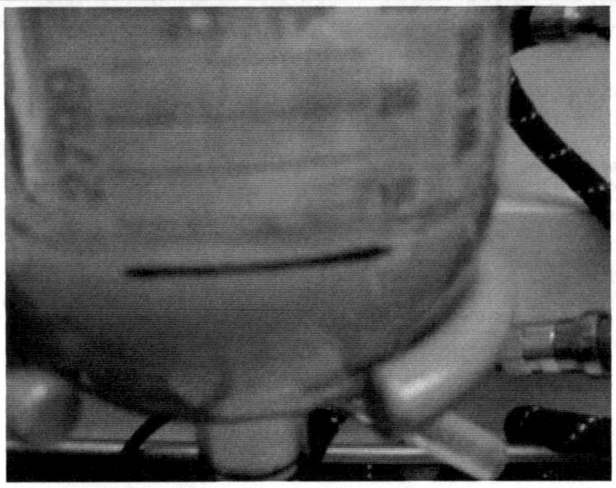

Borylation (98.95 % conversion)

During Suzuki reaction

Diluted after Suzuki reaction

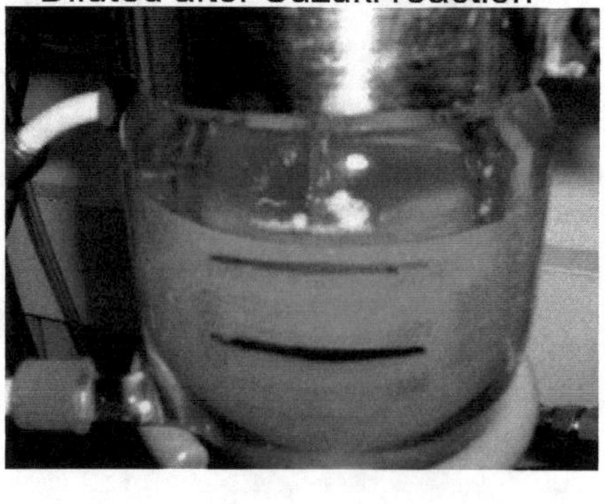

- Color is yellow throughout the reaction

Suzuki kinetic data – Blackout condition vs. Nonblackout condition

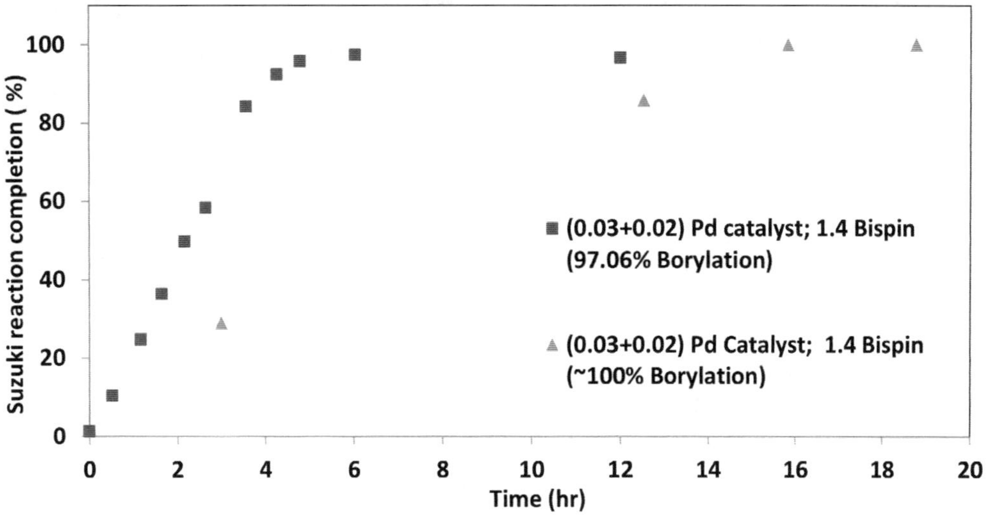

- Suzuki kinetic data - Blackout condition (continued from ~100% Ar-Br conversion completion) vs. Nonblackout condition (continued from ~97% Ar-Br conversion)
  - Suzuki reaction (under blackout conditions) completes generally in ~12-24 hrs
  - Suzuki reaction is done in ~5 hrs when blackout is prevented (non-blackout condition)

13

Suzuki kinetic data – Blackout condition vs. Nonblackout conditions

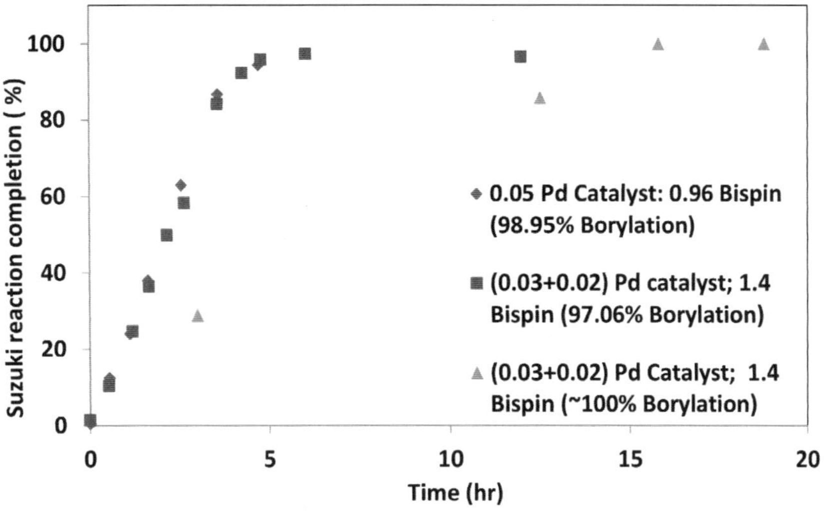

- Suzuki kinetic data - Blackout condition (continued from ~100% Ar-Br conversion) vs. Nonblackout condition (continued from ~97% Ar-Br conversion in Borylation)
  - No significant loss of catalytic activity under non-blackout conditions (0.03+0.02=0.05 equiv. active catalyst)
  - Under blackout conditions, the catalytic activity is mainly from the kicker charge (0.02 equiv. of catalyst)

14

Other things to consider under non-blackout conditions

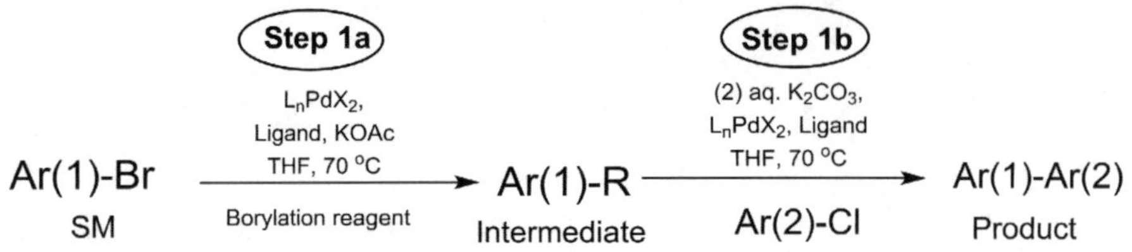

- Non-blackout conditions: ~97% Ar(1)-Br conversion (~0.97 equiv. Borylation reagent)
  - Ar(2)-Cl needs to be charged first before aqueous K_2CO_3 (or)
  - Ar(2)-Cl and aqueous K_2CO_3 need to be charged simultaneously (co-addition)
  - Charging aqueous K_2CO_3 first leads to reaction between remaining SM (~3%) and intermediate – Eventually Pd blackout

15

Summary – Reason for variable Suzuki reaction completion time

- Reason for variable Suzuki reaction completion time
  - Pd blackout at the end of borylation (step 1a)
  - Interaction between blackedout Pd and fresh Pd catalyst used in the kicker charge – Blackedout Pd could quench the portion of fresh Pd catalyst (0.02 equiv. of kicker charge)

- How to avoid Pd blackout, variability in Suzuki reaction rate
  - Developed reaction conditions (non-blackout conditions) to achieve consistent Suzuki end of reaction time

16

Alternate strategy to avoid Pd blackout

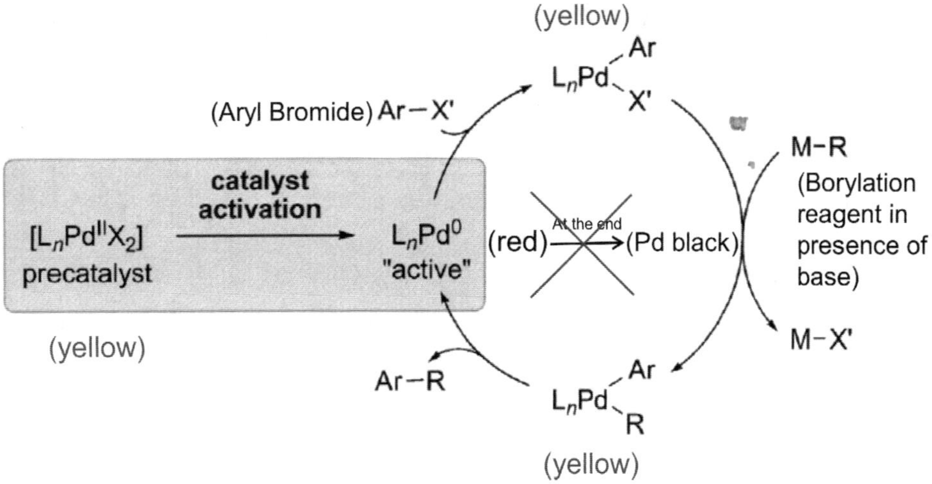

- Alternate catalyst/ligand – Stable Pd(0) state

17

Alternate strategy to avoid Pd blackout

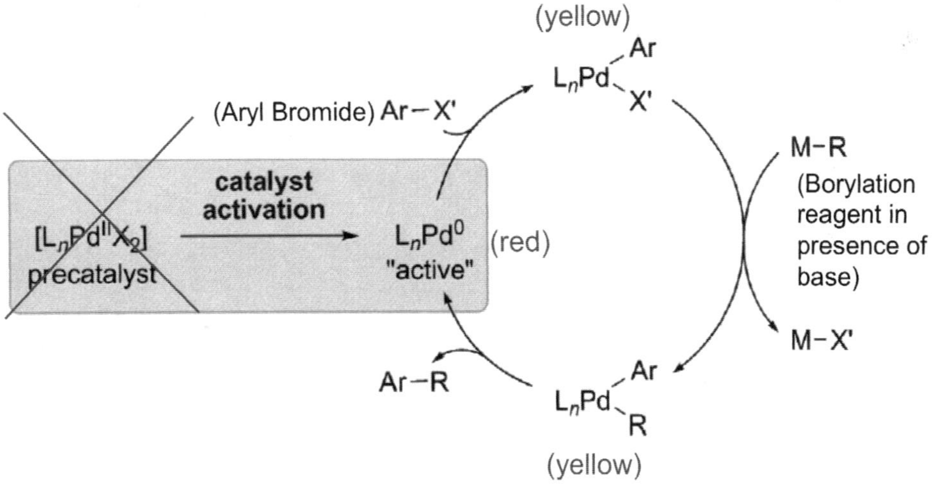

- Alternate catalyst/ligand – Stable Pd(0) state

18

Summary & Conclusions

- Strategy to avoid Pd blackout
  - Change the LR to borylation reagent instead of aryl halide (non-blackout conditions)
    - Consistent and fast Suzuki reaction completion time
  - Alternate catalyst/ligand – Stable Pd(0) state
- Suzuki reaction under non-blackout conditions
  - Helped in understanding the variations observed in the lab and plant
  - Led to better process understanding and improved process
  - Helped to get kinetic data for the model – Aided the development, QbD, simulation of PARs, risk of reaction stalling and risk of impurity formation

19

Acknowledgement

- Step 1 team - Jeremy Merritt, Kevin Sullivan, Rachel Richey, Mark Pietz, Greg Brown, Alison Campbell, Brad Held, Nathan Furr, Jen Kuehne-Willmore, Nick Magnus
- Stan Kolis
- Doug Kjell
- Utpal Singh
- Eric Moher
- Kevin Seibert
- Marv Hansen
- Candice Wong, Carla Luciani and Chris Burcham

20

390774 Optimization of Formaldehyde Production from Methanol By Utilizing Silver Catalyst Process

Abstract:

Formaldehyde, the target product of the present work, is an organic compound representing the most elementary configuration of the aldehydes. It behaves as a synthesis baseline for many other chemical compounds, including phenol formaldehyde, urea formaldehyde and melamine resin. The aim of the present work is to reach a 98% conversion of methanol by means of taking out the distillation column and replacing it with a higher duty cooler to bring the product to 37 wt. % of formaldehyde. Simulation was carried out for the formaldehyde production process from the methanol using a silver catalyst. Detailed calculations were performed in this report for all equipment and streams in the plant, taking into account the required process conditions to achieve a production capacity of 60000 tons/year of formalin. Plant's units and operations were simulated by using the chemical simulation software Aspen plus to gain an optimized position of the operation conditions. With a conversion of 98%, the alternative design produces 5481 kilo/hr. Simulation results predict that a much higher net volume for the reactor in order to reach the specified conversion (98%).

392019 Aligning Innovation, Patent Strategies, and Business Models

Abstract:

An iterative approach to balancing research efforts, patent strategies, and business models can identify new opportunities as well as provide an attractive package for management or investment support. An overview of the beneficial use of intellectual property is provided and specific examples are shown.

364257 Building an Excellent Entrepreneurial Team

Abstract:

An entrepreneurial team is built on excellence, confidence and passion. All team members must be comfortable with chaos. The cube farm mentality has to be destroyed. The passion to change the world and make a difference has to be unleashed. The recruiting relationship is built on the abnormal. There is no room for conformity on an entrepreneurial team. However, the same traits that make it great also tends to blow it apart.

Entrepreneurial engineering leaders are either superstars or oddballs. Both traits are needed. The team leader must find the mix of innovative off the wall creativity with the ability to drive fast-paced targets and get things done. Decisions must be made on the fly without all the information, but the aspects of safety and ethics can never be compromised. Innovation is unleashed by having mundane tasks taken out of the equation.

An entrepreneurial team is a paradox. Like the grand unified force theory, the moment you think you have handle on how it works, a new opportunity emerges and bang ! pivot to a new direction.

There is nothing more fulfilling, exciting or satisfying.

392020 How to Generate Revenue from Your Technology: Strategies for Licensing Your Intellectual Property

Abstract:

Licensing of intellectual property can be a significant source of revenue for companies. This session will provide practical lessons for generating and increasing value from IP through licensing. Learn how to identify and protect your IP while avoiding common mistakes when monetizing it.

consulting engineers and scientists

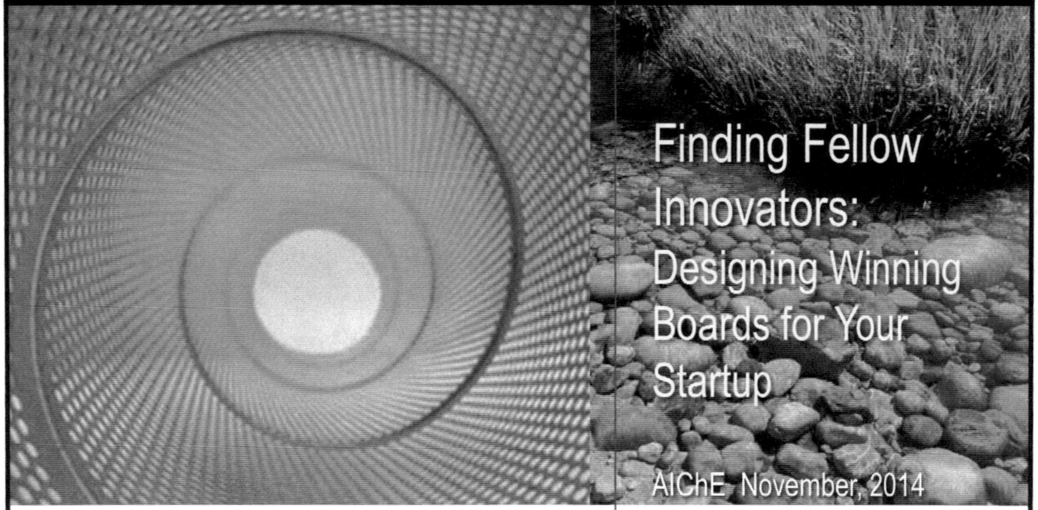

GEI Consultants

John T. Finn, P.E.
GEI Consultants, Inc.

Dedication

In Memory of My Father

Robert K. Finn, PhD.
Professor of Chemical Engineering
Cornell University
Member AIChE 1946 - 2012

GEI Consultants

Create Winning Boards for Your Startup

You Want to Be Here

You Are Here

GEI Consultants

Purpose ?

- What do you want from outside help?
- Depends on:

1. Unique Identity and Circumstances
2. Stage: Idea → Startup ➡ Growth

GEI Consultants

Directors, Advisors, and Supporters

- Firm reports to the Directors

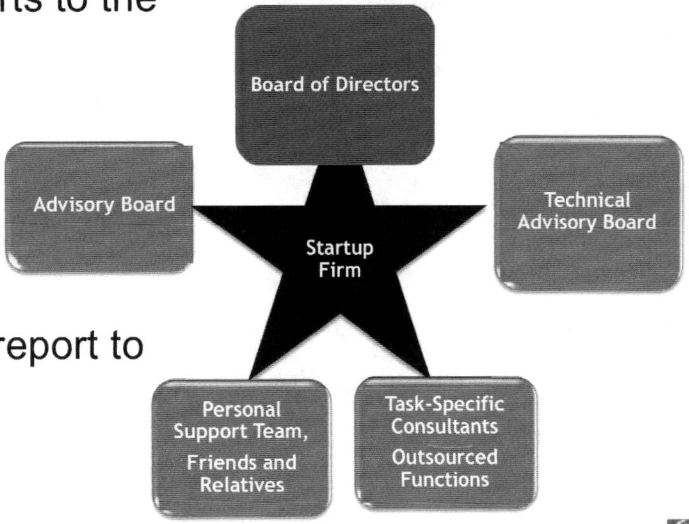

- Advisors report to the Firm

GEI Consultants

Three Types of Boards

Board	Legal Status	Primary Role
Directors	Statutory	Makes Big Decisions, Goals, Selects CEO
Advisory (General)	Informal	Advises on vision, goals, strategy, the market, legal, staffing, financial, governance, etc.
Technical Advisory	Informal	Advises on technical and scientific aspects of the process and product, R&D, external related technologies, etc.

GEI Consultants

Directors

- Usually called "The Board"
- Usually after Idea and Initial Startup
- Founders, VC Investors and Outside Members
- How Many: 7 to 9 to fulfill legally mandated activities
- Commitment: 250 hours/year Chair or active startup 450 hours/ year

GEI Consultants

General Advisors

- Provide critical business-related help, introductions to other key supporters
- Focus on your path to the next stage
- 3 to 5 persons, with a range of backgrounds
- For the number of advisors, consider your own bandwidth. 1 hr each per month, min.

GEI Consultants

Technical Advisory Board

- Keep it small and focus on your challenges
- Academic and Industry backgrounds
- Include your technical team members
- Assign members 10 minute presentations on an aspect of the field
- Encourage constructive dialogue
- Stress confidentiality of your data

GEI Consultants

TAB Special Considerations

- Certainty vs Risk for engineers.
- Audacity (arrogance ?) for Chem Es!
- Humility is the greatest virtue for the TAB

GEI Consultants

Finding Fellow Innovators

- Meet a lot of people. Reach outside your circle, referrals from college profs, grad students, conferences, accelerator meetings, AIChE, LinkedIn

...let people know you are looking.

- Board membership benefits the members: Pitch
- Be picky! Proven track record
- Passion about your startup
- Bandwidth to be able to help
  - test this by giving a small assignment
- Good chemistry with you, and board members
- Collaborative, yet challenge assumptions

GEI Consultants

Getting the Most from Advisors

- Meet both individually and in groups
- Run the meetings efficiently, with pre-distributed agenda, minutes.
- Give them homework, specific tasks
- Set the next meeting date
- Use calls, email and chat effectively
- Consider a secure site Wiki with company updates, financials, tech developments, marketing tools, Suggestion Box, etc.

GEI Consultants

Compensation

Board	Vested Equity	Cash	Meeting Expenses, Lunch
Directors	0.5% to 2% if not a VC or Founder	Not usually	Yes
Advisory (General)	0 to 1%	Small	Usually
Technical Advisory	0 to 1%	Small	Usually

5% Total equity pool set aside for advisors is typical

GEI Consultants

Standard Agreement

- Founder Institute

Founder/Advisor Standard Template "FAST"

- Easily defines

scope, time commitments, compensation

- Proportional to stage: Idea, Start, or Growth
- Term limit, typically 1 year

GEI Consultants

Advisors Compensation

- Wide Range: Lunches at first, to Equity

FAST Scope	Idea Stage	Startup Stage	Growth Stage
Standard: Monthly Meetings	0.25%	0.20%	0.15%
Strategic: Add Recruiting	0.50%	0.40%	0.30%
Expert: Add Contacts & Projects	1.00%	0.80%	0.60%

Source: The Founder Institute

GEI Consultants

Supporters and Consultants

- Asking the Deeper Questions
- Improve Structure
- Project Management

Systems

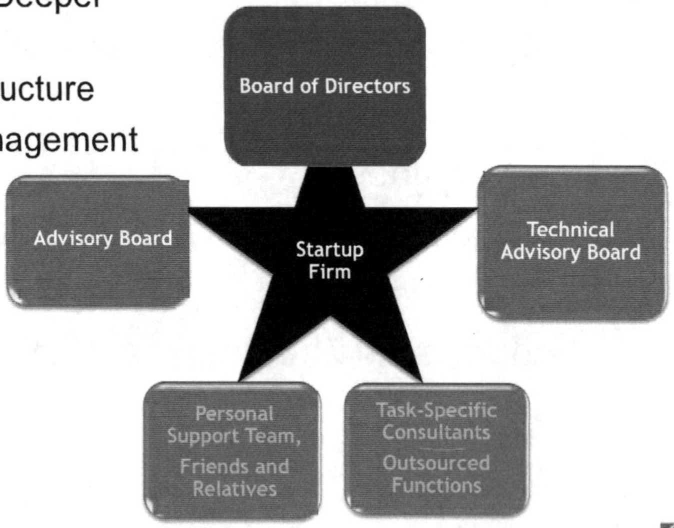

GEI Consultants

Critical Points

- Advisory Boards are essential for most Startups to thrive and progress
- Invest time and attention in setting up and maintaining the board
- Consider general business advisors as well as expert and diverse technical advisors
- Tailor board makeup, scope, compensation to your stage
- Use consultants for specific outsourced tasks
- Friends and family provide personal support

GEI Consultants

Thank You

Questions and Discussion ?

GEI Consultants

Resources

- Entrepreneur article: www.entrepreneur.com/article/71684
- Ventureneer blog: bloghttp://ventureneer.com/vblog/how-advisory-boards-can-help-your-startup
- RockHealth Video: www.youtube.com/watch?v=sTcpUCmUu4A
- FAST document:

http://fi.co/contents/fast#

- John Finn jfinn@geiconsultants.com

GEI Consultants

392021 The I-Corps Program at NSF (and NIH)

Abstract:

There have been profound recent developments that have changed the way universities and the nation's funding agencies foster the creation of startups. Come learn about this new approach, how the NSF and NIH are implementing it in their new I-Corps program, and what this means to you as you seek to commercialize your innovation

381194 Study of the Effect of Alkali Distillation Liquid on Simultaneous Saccharification and Co Fermentation (SSCF) of Alkaline Pretreated Solid State Fermented Sweet Sorghum Bagasse with Zymomonas Mobilis

Abstract:

Sweet sorghum is a promising energy crop to produce bioethanol. An integrated process is necessary to reduce the product cost and increase the utility efficient of feedstock. An alkaline pretratment combined distillation based on the advanced solid-state fermentation technology (ASSF) has been designed to improve the bioconversion of lignocellulose into fermentable sugar. After alkaline distillation step, most of lignin moved into aqueous state from lignocellulosic substrates. Effects of these alkali liquid on the performance of enzymatic saccharification and sugars fermentation of SS with *Zymomonasmobilis*is required to be investigated because it is important to decide the following process.

In this study, effects of lignin content, ionic strength, pH and washing operation on enzymatic saccharification and fermentation were studied. The results showed that: 1) The free lignin content has no obvious negative effect on enzymatic hydrolysis. 2) Ionic strength has no effects on enzymatic hydrolysis, but high ionic strength inhibits the fermentation of *Zymomonasmobilis*. 3) Final ethanol yield was highly dependent on the monomer sugar yields during prehydrolysis step in the present of various alkali distillation liquid concentration. 4) Highest ethanol yield was achieved when the solid from solid-liquid separation was washed once.

2014 AIChE Annual Meeting
T4 International Congress on Energy (ICE)
T4A07 Biomass Conversion: Processing of Solids
November 16-21, 2014
Atlanta, GA

Multi-Scale Investigation for the Effect of Mechanical Refining for Pretreated Lignocellulosic Biomass on the Subsequent Enzymatic Hydrolysis

Junyeong Park [1,*], Wenhui Geng [1,2], Kabindra Kafle [3], Brandon Jones [1], Seong H. Kim [3], Sunkyu Park [1]

[1] Department of Forest Biomaterials, North Carolina State University, Raleigh, NC 27695
[2] Jiangsu Provincial Key Lab of Pulp and Paper Science and Technology, Nanjing Forestry University, Nanjing, China
[3] Department of Chemical Engineering, Pennsylvania State University, University Park, PA 16802
* Presenting author: jpark7@ncsu.edu

Mechanical refining, an established technique widely used in the pulp and paper industry, has recently received significant attention from industries due to its potentials as a promising process for pretreated lignocellulosic biomass. Several studies have been reported about the effects of mechanical refining on the subsequent enzymatic hydrolysis, such as increased yield of carbohydrate conversion, reduced dosage of cellulolytic enzymes to get target yields, and reduced pretreatment severity that make fermentation or catalytic upgrading more effective. In addition, it is commercially proven technology at process flows of ~1,500 dry tons/day of biomass. The details about how refining affects the enzymatic hydrolysis, however, have not been fully elucidated so far. In this study, the changes after mechanical refining of pretreated biomass were investigated by using various tools with different observation scales. Water retention value (WRV) of unrefined and refined samples, which can represent the changes in surface area and pore volume of the biomass, showed close correlations with increased yields of carbohydrate conversion. Optical microscopy and fiber quality analysis (FQA) can also support this observation. TEM provided the evidence of structural changes in nano scale, including delamination of cell wall and external fibrillation. It was also expected that the refining process will affect the molecular structure of pretreated biomass, such as crystallinity of celluloses. XRD solid-state NMR, and SFG spectroscopies were carried out to observe alterations in molecular-scale level, but no change in cellulose crystallinity was observed by those tools. Based on the results, the effect of refining on enzymatic hydrolysis seems to be mainly from the structural modification from the nano to the micron level. More details about this observation with varying factors in conversion yields, like type of pretreatment, refining severity, enzymatic hydrolysis condition, will be discussed.

376760 Effect of Temperature on Pelletizing SPORL Pretreated Forest Biomass and Subsequent Enzymatic Saccharification and Fermentation

Abstract:

Pelletizing of pretreated lignocelluloses to facilitate biomass supply chain logistics was simulated using hot-pressing. The effects of fiber hornification induced by hot-pressing on enzymatic digestibilities of lodgepople pine and poplar NE222 wood chips pretreated by Sulfite pretreatment to overcome recalcitrance of lignocelluloses (SPROL) were examined. The pretreated wood chips were pressed at 25ºC, 70ºC, 90ºC, 110ºC, and 177ºC. The cellulose accessibilities of the pressed and unpressed substrates were evaluated using water retention value and direct cellulase adsorption measurements. Hot-pressing below 110ºC produced a degree of hornification (DH) below 0.26 and had limited effect on cellulose accessibility and enzymatic digestibility. Hot-pressing at 177 ºC produced a DH of 0.86 that significantly hornified the fibers and resulted near zero saccharification. The saccharification results were consistent with cellulose accessibility data. Ethanol fermentation studies at 18% solids suggest that a pressing below 110ºC is preferred to reduce its effect on biofuel yield.

382102 Optimization of Operating Conditions of Aerobic Digester for Effective Conversion of Biomass to Bio-Gas

Abstract:

Pakistan is facing energy crises from last 20 years and share of the bio-fuel in the energy mix is also negligible. According to vision 2020 plan of Pakistan, share of alternative resources/ renewable will be 2,000MW out of 20,000 MW. Various biomasses are available but Conversion of biomass to bio-fuel at low cost is a challenge.

This paper investigates the yield of bio-gas from biomass (like molasses, fruit (banana, apple, water melon)/vegetable wastes, Fruit/vegetable peels, lignocelluloses material and dairy waste etc). Anaerobic digester (LS-26 102) is used to measure the bio-gas yield and concentration at various applied conditions like temperature, residence (time) and behavior of pH value and conductivity is recorded. Enhanced yield of bio-gas is recorded at 40 ^{0}C than 22 ^{0}C at similar other applied condition. An experimental investigation shows that conversion of all biomass at 40 ^{0}C for bio-gas formation is much higher than at 22 ^{0}C. Biogas yield for fruit and vegetable wastes is about 30-35% higher than peels of same fruits and vegetables. Increasing trend of conductivity is observed over the time which also confirms the change in Ph value (decreasing trend) and is according to the literature [1, 2]. Moreover yield of bio-gas at 40 ^{0}C, from molasses, fruit /vegetable waste, and peels of fruit / vegetable is compared. Results reveal that molasses produce higher volume of bio-gas than wastes of fruits/vegetable and peels etc under similar applied condition.

Regional potential of province of Punjab Pakistan is calculated for bio-gas from biomasses and 1500MW can be generated to achieve the vision 2020-25.

[1] Sagagi,B. S., B. Garba and N. S. Usman Bayero, Journal of Pure and Applied Sciences, 2(1): 115 – 118 Received: October, 2008 Accepted: February, 2009, STUDIES ON BIOGAS PRODUCTION FROM FRUITS AND VEGETABLE WASTE

[2]http://www.unigraphinternational.com/en/pdf_infotech/pH_and_conductivity.pd

Abstract:

Bio-based chemicals constitute a promising alternative platform for industrial chemical production, allowing for feedstock diversification, utilization of renewable resources, and potentially reduced environmental burdens relative to petrochemicals. Bio-based chemicals also present new market opportunities in a range of end-use applications, from coatings to polymers. Global production of chemicals and polymers from petroleum is about 330 million tons, and it mainly consists of basic platform chemical that are basis for production of fine chemical and polymers. Studies have shown that most of these building blocks can be produced from biomass resources.[1] Current bio-based chemical production, from both woody and herbaceous resources, is estimated to be around 50 million tons.[1] Among all available sources, lignin with annual production of 200 million metric tons/yr. is the second most abundant natural polymer on earth after cellulose.[2, 3] Lignin accounts for approximately 25–35 % of the organic matrix of wood. The monomer structure has been reported to consist of a phenyl propanoid skeleton.[4] Transformations of lignin have been developed with the goal of preserving the phenolic structure for more ready production of complex aromatic chemicals , but these routes have been difficult and relatively energy consumptive.[4, 5] There are several transformation methods such as pyrolysis and liquefaction techniques, acidic or basic hydrolysis, and more selective depolymerization through catalytic processes,[5, 6] but there is little modeling work on the economic and environmental feasibility of large-scale bio-based chemical production via these routes.

Here we consider bio-based production of catechols from catalytic depolymerization of lignin from candlenut nutshells. These nutshells have 12% (wt. %) lignin in their structure[5] and are typically burned or piled as waste. Target product, catechol, is an organic compound with molecular formula $C_6H_4(OH)_2$ consumed in the production of pesticides, the remainder being used as a precursor to fine chemicals such as perfumes and pharmaceuticals. Integrated process consists of nutshell preparation (rinsing, drying and milling) and lignin extraction with methanol followed by lignin depolymerization in presence of Cu-doped porous metal oxide (Cu-PMO) catalyst. Our study is mainly focused on catalyst synthesis and lignin depolymerization.

A life cycle assessment (LCA) was conducted in order to model predicted energy use and environmental impacts associated with this bio-based production scheme at industrial scales and to compare catechol production from both lignin and petroleum sources. All major upstream processes were considered, including candlenut cultivation, harvesting, and transport, for the bio-based route and oil&gas extraction, transport, and petrochemical refining for the petrochemical route. Lignin extraction and modification steps were modeled by scaling up bench-scale process and estimating energy consumption. Conde-Mejía, C., et al. [7] was modified based on solvent to wood ratio and used. For catalyst synthesis and lignin depolymerization, commercial-scale process simulation was performed using ASPEN plus v.8.4. Input chemicals from original research [5, 8] and data from the NREL biomass ASPEN data bank [9] were used to model the reaction components and kinetics. Simulations were based on input lignin flow rate of 75 tons/day (matching current US industrial scales for lignin processing).

Catalyst degradation (5000 hr lifetime [10]) and solvent recycling rates of 97% were assumed. Figures 1 and 2 show the ASPEN process flow sheets for these steps.

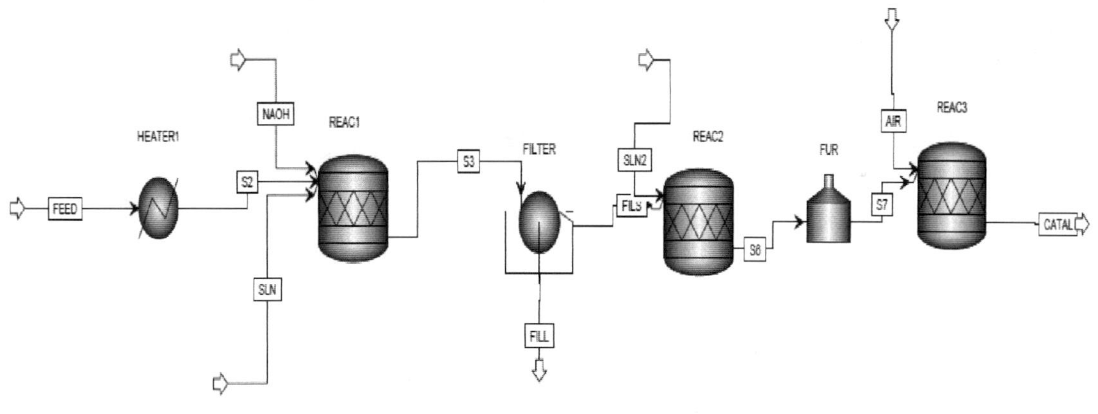

Figure 1. Flow diagram of catalyst synthesis

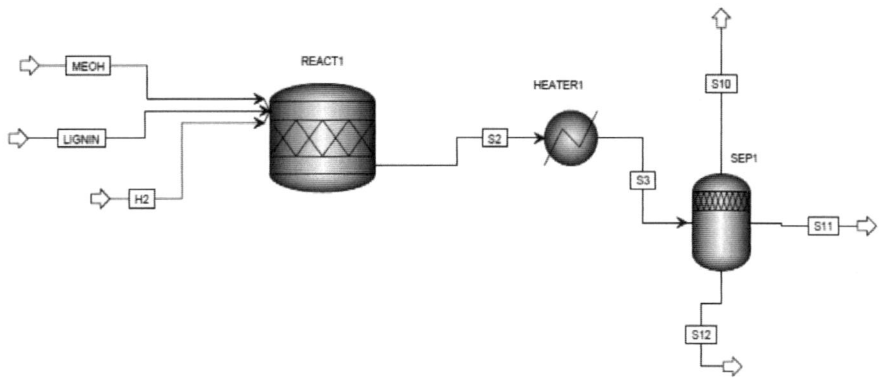

Figure 2. Flow diagram of lignin depolymerization

Energy use for catalyst synthesis is estimated as 5,920 kWh/kg of Cu-PMO catalyst, while depolymerization requires 2.46 kWh/kg of catechol. Phenol and ethanol are co-products of depolymerization process with daily production of 10 and 27 tons. Ethanol will be recycled back to extraction and used as solvent and phenol can be sold to the market.

Resource inputs (energy and material) were modeled in SimaPro 8.01 LCA software using ecoinvent 2.2 life cycle inventory unit processes adjusted for the US energy system. The U.S. EPA's Tool for the Reduction of Chemical and Other Environmental Impacts (TRACI) 2.1 fate-effect model was used for estimation of environmental impacts. Associated impacts were normalized based on 1 kg catechol in output stream, considering burden allocation for catalyst synthesis over its lifetime and two types of credits; one from carbon sequestration during candlenut tree growth prior to harvesting and the second from phenol as a co-product of depolymerization.

Figure 3 demonstrates the relative process contribution, normalized in each category, for the field-to-gate production of catechols from candlenut lignin.

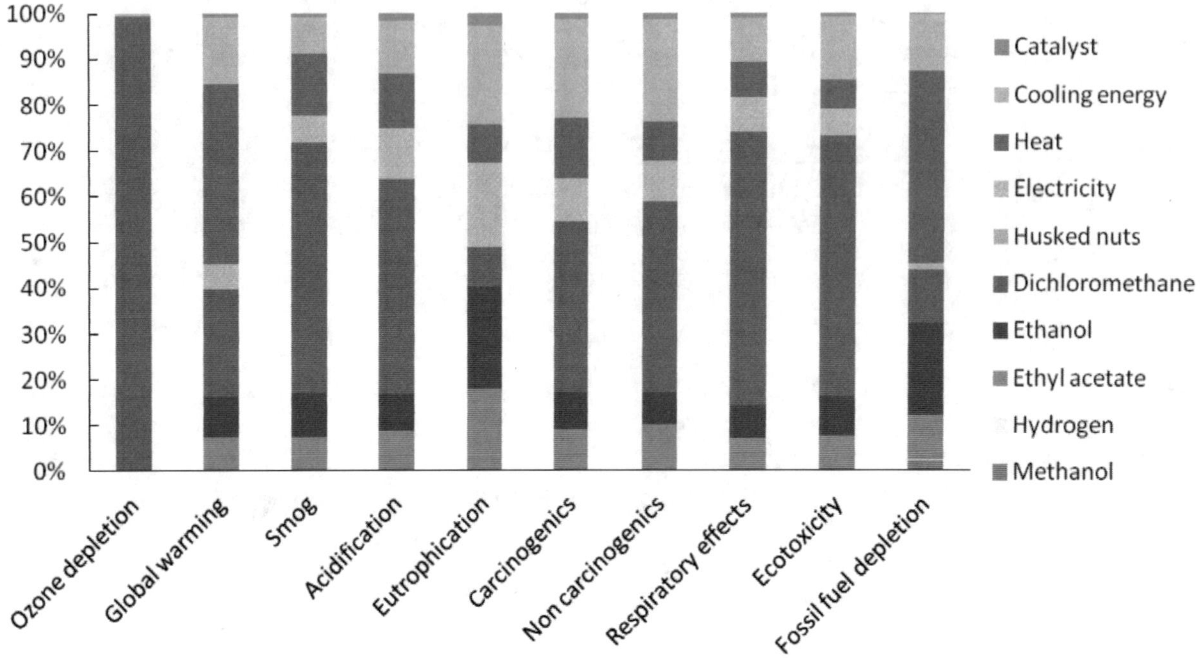

Figure 3. Process contribution of catechol production from lignin, considering extraction, depolymerization and catalyst synthesis

From the chart, it can be seen that almost in all categories, dichloromethane and heat used in extraction process drive life cycle impacts. Dichloromethane is produced from methyl chloride at high temperatures (400-500 °C) and it is highly volatile, which makes it an inhalation and atmospheric hazard. It is a potent ozone-depleting substance and its contribution is 97% in ozone depletion and 62% in human respiratory effects.

Catalyst contribution is not that significant compared to other processes. The production of 1 kg catechol requires 0.76 kg of catalyst. Results demonstrated that maintaining operational conditions for high retention time of catalyst synthesis process leads to excessive electricity consumption. Electricity production is the dominant contributor in all categories with about >99% of total burdens, while the actual catalyst materials themselves are relatively insignificant.

Final step is comparing environmental burden of proposed catechol production with convenient catechol from petrochemical phenols. Phenol reacts with hydrogen peroxide to form catechol based on equation 1.[11]

$$C_6H_5OH + H_2O_2 \longrightarrow C_6H_4(OH)_2 + H_2O\ ,\ \Delta H_{reaction} = -200.4\ (kJ/mol) \quad \textbf{eq. 1}$$

The process was simulated in SimaPro and associated burden was estimated. Its process contribution is shown in figure 4.

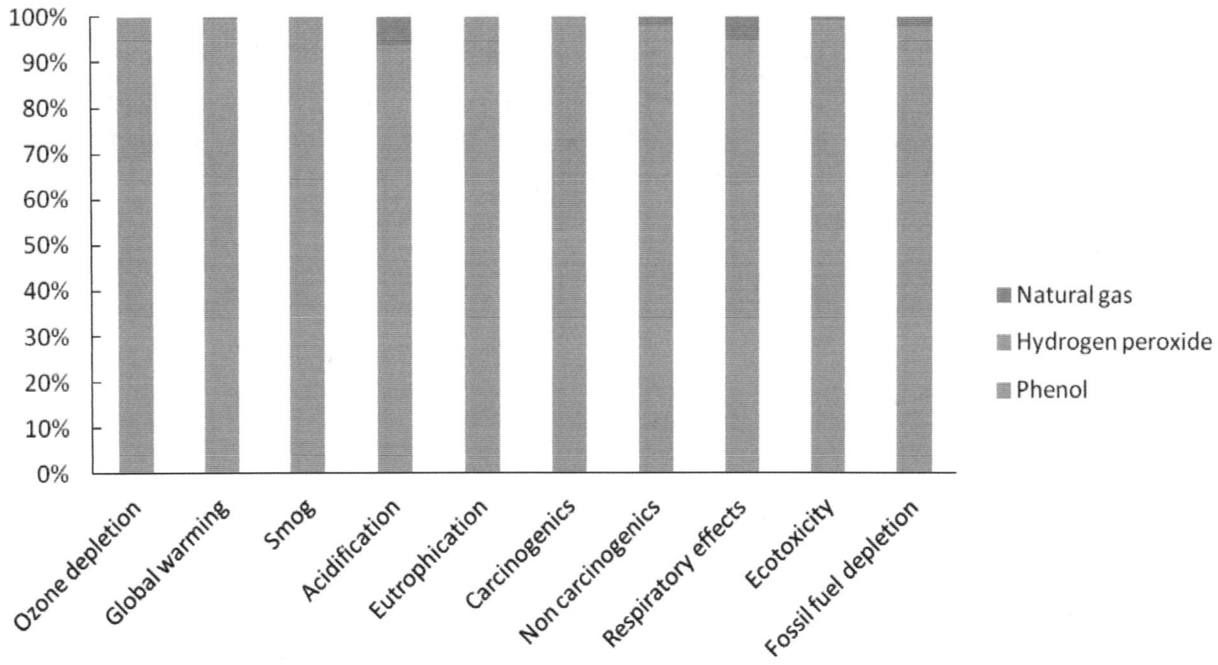

Figure 4. Process contribution in catechol production from petroleum based phenol

Based on above figure, phenol production at first place is the most dominant contributor almost in all environmental impact categories.

Comparing petroleum-based and lignin-based catechol illustrates interesting results. Table 1 shows total impacts for each of the production routs.

Impact category	Unit	Total (lignin-based catechol)	Total (petroleum-based catechol)
Ozone depletion	kg CFC-11 eq	1.6E-04	1.5E-07
Global warming	kg CO2 eq	3.0E+01	4.0E+00
Smog	kg O3 eq	1.5E+00	1.7E-01
Acidification	kg SO2 eq	1.1E-01	1.6E-02
Eutrophication	kg N eq	3.6E-02	8.8E-03
Carcinogenics	CTUh	1.0E-06	2.4E-07
Non carcinogenics	CTUh	2.5E-06	3.0E-07
Respiratory effects	kg PM2.5 eq	1.0E-02	1.2E-03
Ecotoxicity	CTUe	2.7E+01	8.5E+00
Fossil fuel depletion	MJ surplus	6.9E+01	1.3E+01

Table 1. Total environmental burden of lignin-based and petroleum-based catechol

Petroleum-based catechol seems to have lower environmental impacts compared to the bio-based catechol and that might be because of high complexity and resistance in lignin structure that needs more effort for conversion to simpler aromatic compounds. Utilization of dichloromethane as a solvent for modification of crude lignin is a driver of negative environmental impacts and its substitution should be a target of further research.

Finally, the results points that proposed pathway for bio-based catechol is not as promising as expected. More development and process optimizations are needed for industrial scale production of such chemicals. Choosing appropriate solvents and reaction modifications can be the first step while using catalyzed reactions to mitigate energy consumption of the whole process can be another consideration.

References:

1. de Jong, E., *Bio-based Chemicals Value Added Products from Biorefineries.* IEA Bioenergy, Task42 Biorefinery, 2012.
2. Ragauskas, A.J., et al., *The path forward for biofuels and biomaterials.* Science, 2006. **311**(5760): p. 484-489.
3. Percival Zhang, Y.-H., M.E. Himmel, and J.R. Mielenz, *Outlook for cellulase improvement: screening and selection strategies.* Biotechnology advances, 2006. **24**(5): p. 452-481.
4. Kleinert, M. and T. Barth, *Phenols from lignin.* Chemical engineering & technology, 2008. **31**(5): p. 736-745.
5. Barta, K., et al., *Depolymerization of organosolv lignin to aromatic compounds over Cu-doped porous metal oxides.* Green Chemistry, 2014. **16**(1): p. 191-196.
6. Huber, G.W., S. Iborra, and A. Corma, *Synthesis of transportation fuels from biomass: chemistry, catalysts, and engineering.* Chemical reviews, 2006. **106**(9): p. 4044-4098.
7. Conde-Mejía, C., A. Jiménez-Gutiérrez, and M. El-Halwagi, *A comparison of pretreatment methods for bioethanol production from lignocellulosic materials.* Process safety and environmental protection, 2012. **90**(3): p. 189-202.
8. Hansen, T.S., et al., *One-pot reduction of 5-hydroxymethylfurfural via hydrogen transfer from supercritical methanol.* Green Chemistry, 2012. **14**(9): p. 2457-2461.
9. Wooley, R.J. and V. Putsche, *Development of an ASPEN PLUS physical property database for biofuels components*. 1996: National Renewable Energy Laboratory Golden, CO.
10. Griffiths, O.G., et al., *Using life cycle assessment to measure the environmental performance of catalysts and directing research in the conversion of CO 2 into commodity chemicals: a look at the potential for fuels from 'thin-air'.* RSC Advances, 2013. **3**(30): p. 12244-12254.
11. Sienel, G., R. Rieth, and K. Rowbottom, *Ullmann's encyclopedia of industrial chemistry.* Wiley-VCH, Weinheim, 2000.

ID# 369173

TECHNICAL, ECONOMIC AND ENVIRONMENTAL VIABILITY OF OFFSHORE CO_2 REUSE FROM NATURAL GAS BY DRY REFORMING

Bruna C. de S. Lima, Federal University of Rio de Janeiro, Rio de Janeiro, RJ, Brazil
Ofélia Q. F. de Araújo, Federal University of Rio de Janeiro, Rio de Janeiro, RJ, Brazil
José L. de Medeiros, Federal University of Rio de Janeiro, Rio de Janeiro, RJ, Brazil
Cláudia R. V. Morgado, Federal University of Rio de Janeiro, Rio de Janeiro, RJ, Brazil

ABSTRACT

The recent discoveries of Pre-Salt layers in Brazil require process developments for enhanced sustainability as these reservoirs have oil with associated natural gas exhibiting an expressive amount of CO_2 (FORMIGLI, 2007; GAFFNEY, 2010). The challenge is thus to expand the supply of this fossil raw material to produce energy and chemicals in a carbon-constrained economy. In this scenario, CO_2 capture and reuse techniques are important routes for a moderate transition from the present fossil-based economy to a long-term sustainable future. This work evaluates two process alternatives for a floating unit consisting of physical absorption of CO_2 from natural gas, with propylene carbonate, producing a natural gas stream poor in CO_2 and a stream consisting of a mixture of CO_2 and hydrocarbons. Two alternatives are evaluated for processing the CO_2 rich stream to yield methanol. Technical, economic and environmental analyses of the process flowsheets are performed for comparison purposes. Alternative 1 combines dry and steam reforming in one reactor (one pot reactor, Bi-Reforming), while Alternative 2 segregates the two reactions: dry reforming (Dry-Reforming) occurs in one reactor and water gas-shift reaction takes place in a subsequent reactor. The two Alternatives are evaluated with ASPEN HYSYS® (ASPENTECH Inc) simulator for calculation of energy and mass balances necessary to assess economic and environment performances. The economic evaluation employed the software Capital Cost Estimation (CAPCOST) for calculations of CAPEX and OPEX. The software Waste Reduction Algorithm (WAR, EPA) was used to evaluate the potential environmental impacts, by PEI (Potential Environmental Impact). Results indicate that the performance of Alternative 1 was better than Alternative 2: (a) methanol production (17905 kg/h) 4 times higher, (b) CAPEX (U$$75.497.127), 21,7% lower, (c) Sales Revenue (U$$144,890,686) 181% greater, and (d) the index for environment impact (868 PEI/hr) 47,7% lower.

INTRODUCTION

The recent discoveries of Pre-Salt layers in Brazil require process developments for enhanced sustainability as these reservoirs have oil with associated natural gas exhibiting an expressive amount of CO_2 (FORMIGLI, 2007; GAFFNEY, 2010). Carbon dioxide is one of the greenhouse gases and its emission should hence be avoided or reduced. The challenge is thus to expand energy supply in a carbon-constrained economy. In this scenario, CO_2 capture and reuse techniques are important routes for a moderate transition from the present fossil-based economy to a long-term sustainable future (ARAUJO et al., 2014).

In this sense, sustainable principles for designing natural gas conditioning and utilization processes must be inserted in the synthesis and operation phases. The sustainability concept states that a sustainable progress is only possible when considered, simultaneously, economics, environmental and social aspects (Johnston et al., 2007). With this vision, this work presents a study of natural gas production, concomitantly with CO_2 capture and reutilization in methanol synthesis as well as in enhanced oil recovery (CO_2-EOR).

Two offshore process alternatives are herein evaluated where CO_2 is captured from natural gas, by physical absorption with propylene carbonate (PC), producing a natural gas stream poor in CO_2 and a stream consisting of a mixture of CO_2 and hydrocarbons, which is processed by Bi-Reforming (Alternative 1) and Dry-Reforming (Alternative 2), to yield methanol. This solvent has a high affinity for CO_2 and its use has advantages for the environment, since PC is a nontoxic and biodegradable chemical (GADELHA, 2010), and its production route involves the utilization of CO_2 as a reactant along with propylene oxide (DAI et al., 2010), a relevant aspect in a life cycle approach.

The selection of PC as solvent has an important advantage: its regeneration does not require heat input, and involves an expansion valve and a flash vessel. Also relevant in the scope of the downstream processing of the CO_2 rich stream is that PC has also affinity for hydrocarbons, yet lower than its affinity for CO_2. This aspect is often viewed as a disadvantage. However, this will be a benefit for the proposed alternative processes since the hydrocarbon which "contaminates" CO_2 will react with CH_4 from natural gas in dry reforming reaction.

Dry reforming to yield synthesis gas (syngas) is one alternative to reduce CO_2 emissions, adding value to it (ARAUJO et al., 2014). The interest in this reforming route is due to the production of a syngas with a H_2/CO molar ratio suitable for a variety of products, and because the reaction consumes two types of greenhouse gases, CO_2 and CH_4 (ZHANG et al., 2003, GANGADHARAN et al., 2012). This reaction can be represented as

$$CH_4 + CO_2 \rightarrow 2CO + 2H_2 \qquad \Delta H = +247.3 \text{ kJ/mol} \qquad (1)$$

The biggest limitation to the dry reforming process is the availability of a suitable catalyst. This reaction requires high temperature, what contributes to carbon deposition. In CO_2 reforming, coke deposition on the catalyst is reported to be very fast. However, when CO_2 reforming occurs simultaneously with steam reforming, coke deposition is drastically reduced. (GANGADHARAN et al., 2012). This combination, named bi-reforming (Eqs. 2, 3 and 4), can generates syngas with a H_2/CO ratio of 2, ideal for the synthesis of methanol (OLAH et al., 2009).

$$\text{Steam reforming: } CH_4 + H_2O \rightarrow CO + 3H_2 \qquad \Delta H = +206.3 \text{ kJ/molCO}_2 \qquad (2)$$
$$\text{Dry reforming: } CH_4 + CO_2 \rightarrow 2CO + 2H_2 \qquad \Delta H = +247.3 \text{ kJ/molCO}_2 \qquad (3)$$
$$\text{Bi-reforming: } 3CH_4 + 2H_2O + CO_2 \rightarrow 4CO + 8H_2 \qquad (4)$$

Although dry reforming generates syngas H_2/CO ratio suboptimal of for methanol synthesis, the water gas-shift (WGS, Eq. 5) reaction is responsible for adjusting the syngas ratio (GANGADHARAN et al., 2012). The reversibility of Eq 5 (WGS and RWGS) is important, once that H_2/CO ratio can be adjust according process needs just changing the reaction conditions.

$$H_2 + CO_2 \leftrightarrow CO + H_2O \qquad \Delta H = +51 \text{ kJ/mol} \qquad (5)$$

On the other hand, syngas from bi-reforming is appropriate for methanol production (Eq. 6).

$$CO + 2H_2 \rightarrow CH_3OH \qquad \Delta H = -90.6 \text{ kJ/mol} \qquad (6)$$

This work presents the synthesis of two processes for processing the CO_2 rich stream, Alternative 1 and Alternative 2, to yield methanol. Technical, economic and environmental analyses of the process flowsheets are performed for comparison purposes. Alternative 1 combines dry and steam reforming in one reactor (*one pot reactor*), while Alternative 2 segregates the two reactions: dry reforming occurs in one reactor and water gas-shift reaction takes place in a subsequent reactor.

The two Alternatives are evaluated with Aspen HYSYS® simulator for calculation of energy and mass balances necessary to assess economic and environment performances, employing Peng-Robinson EOS. The economic evaluation employed Capital Cost Estimation Software (CAPCOST) for calculations of CAPEX and OPEX, and the Waste Reduction Algorithm Software (WAR) was used to evaluate the potential environmental impacts. It is worth noting that besides the environmental motivation, the alternatives were driven by the fact that methanol is one of the most important commodity chemicals, used as raw material in several processes, what would bring economic advantages, and it is a route of gas to liquid(s).

This work was conducted at CE-GN – Center of Excellence in Natural Gas, a joint venture between UFRJ and PETROBRAS/CENPES.

PROCESS SIMULATION

The two processes - Bi-Reforming (Alternative 1) and Dry Reforming(Alternative 2) - were evaluated with Aspen HYSYS® simulator for calculation of energy and mass balances necessary to assess economic and environment performances. The first process steps relates to CO_2 capture and are common to both alternatives, which is fed with raw a natural gas stream with composition shown in Table 1.

Table 1. Natural Gas Composition

Component	% mol
CH_4	60
CO_2	20
C_2H_6	12
C_3H_8	06

The study assumes the following process premises:

- The natural gas, previously dehydrated, enters at 70 bar and 40 °C with a volumetric flow rate of 1MMNm^3/d.;
- Treated gas meets selling specifications (ANP, 2008), i.e., CO_2 poor gas leaves PC absorption columns exhibiting less than 3% in CO_2.
- Propylene Carbonate mass flow was set up according to the molar ratio of CH_4 to CO_2 necessary for reforming reactions, resulting in a value of 435ton/h.

Figure 1 shows the Aspen HYSYS® process flow diagram (PFD) of the CO_2 PC absorption process, common for Dry Reforming and Bi Reforming processes.

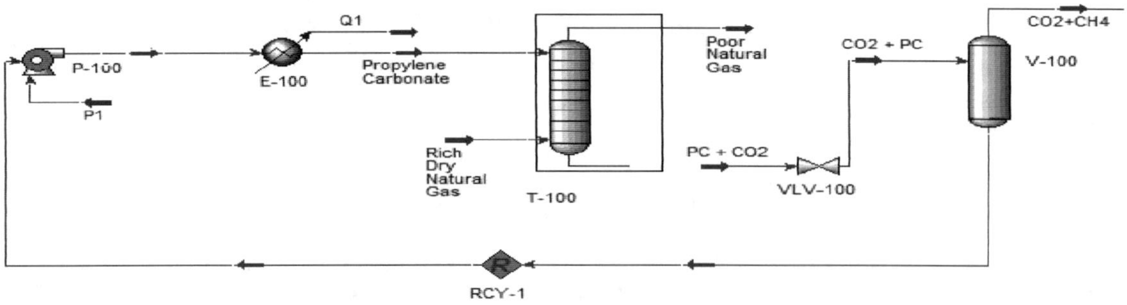

Figure 1. Process flow diagram for CO_2 absorption with PC - Dry Reforming and Bi Reforming processes.

The natural gas stream enters at the bottom of absorber column (T-100), countercurrent to Propylene Carbonate (absorbent). The stream with PC and CO_2 is recovered at the bottom of T-100 and goes to a flash vessel (V-100) for PC regeneration. Natural Gas correctly specified is the top product of absorber. The PC rich stream that leaves the bottom of T-100 is depressurized (at VLV-100) and enters in V-100 for PC regeneration. Then, the gas that leaves V-100, predominant in CO_2 and CH_4, but with residual hydrocarbons components, goes to reaction and the liquid product (PC) is recycled and returns to T-100.

Bi-Reforming Process

The stream 'CO_2 + CH_4' is divided into two parts: one goes to a compression train, for CO_2 reinjection, and the other goes to reaction. In the downstream section, this stream goes to a reactor where both dry and steam reforming reactions occur, shown in Figure 2.

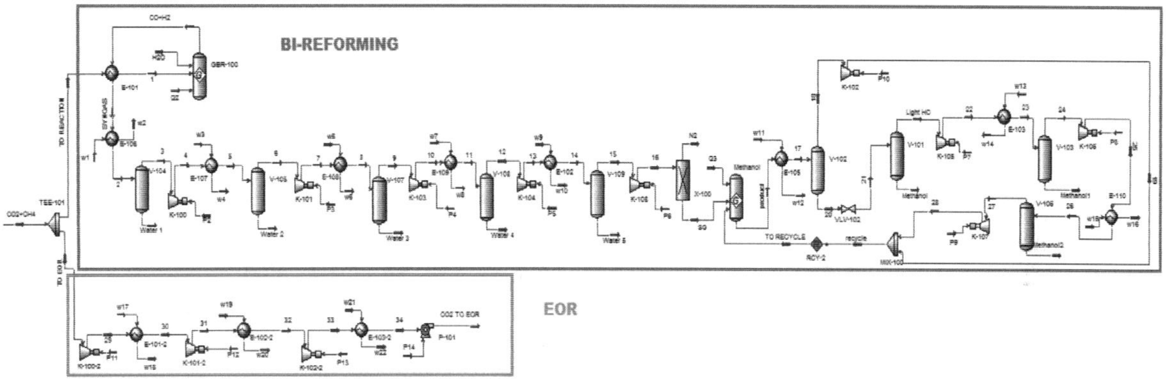

Figure 2. Process flow diagram for EOR and Bi-Reforming process – Alternative 1.

The stream 'TO EOR' is pressurized to 400 bar to meet requirements of injection for the CO_2-EOR. Table 2 shows the operational conditions for compressors: compression ratio, suction pressure, discharge pressures and discharge temperatures. After the compression train, pump P-101 finally pressurizes the stream that goes to EOR. The criteria utilized for compression was that temperature discharge is limited to 170 °C.

Table 2. Operational conditions of compressors for EOR

Compressor	Suction Pressure (bar)	Discharge Pressure (bar)	Discharge Temperature (°C)	Compression ratio
K-100-2	2.0	9.5	167.2	4.75
K-101-2	9.0	47	166.6	5.20
K-102-2	46.5	240	169.3	5.16

On the other hand, the stream 'To Reaction' proceeds to reactor GBR-100, passing through E-101 for energy integration. The reactor model used for simulation was the Gibbs Reactor, which works by finding the equilibrium state with the lowest Gibbs free energy. At the feed ratio of CO_2/CH_4 of 1:1, the optimum temperature for dry reforming reaction is between 870 and 1040 °C to prevent carbon deposition at pressure of 1 atm (WANG, 1996). However, this process combines dry and steam reforming. For steam reforming, the reaction temperature is between 700 °C and 980 °C, and pressure around 20 bar (Pacífico, 2004). In this case, GBR-100 operates at 900°C and 1 bar, to favor dry reforming. Water vapor is added in reactor at 10.81 ton/h. Table 3 shows the gas composition (% mol) before and after reaction.

Table 3. Gas composition (% mol) before and after Bi-Reforming

Before Reaction		After Reaction	
CH_4	35.10	CH_4	0.040
CO_2	35.10	CO_2	3.890
C_2H_6	19.20	C_2H_6	-
C_3H_8	10.40	C_3H_8	-
N_2	0.200	N_2	0.040
CO	-	CO	27.71
H_2	-	H_2	57.86

After reaction, the stream 'SYNGAS' passes through a sequence of compressors, heat exchangers and vessels, in order to eliminate water and pressurize gas to 70 bar for methanol synthesis. These heat exchangers utilize seawater for refrigeration. It is important to add that before reaction, nitrogen is removed with a splitter component, to simulate molecular sieve adsorption process (CAVENATI et al, 2006).

For methanol production from synthesis gas, the operational conditions for reaction are: temperature around 260 °C and pressure between 50 and 100 bar (AQUINO, 2008, KIM et al., 1998). In this work, Methanol reactor operates at 260 °C and 70 bar. The stream leaving the Methanol reactor goes to V-102, a flash vessel at high pressure. The stream that leaves V-102 at top returns to the Methanol reactor, while the liquid stream (bottom product of V-102) is depressurized to 5 bar for liquid methanol recovery at V-101. Light hydrocarbons and residual methanol are pressurized at K-105, refrigerated at E-103, and methanol is recovered at V-103. The same sequence occurs with K-106, E-100 and V-106. Finally, compressor K-107 pressurizes the remaining hydrocarbons that return to Methanol reactor.

Dry Reforming Process

Similarly to the Bi-Reforming process, the stream 'CO_2 + CH_4' is divided into two parts and the stream "TO EOR" has the same destiny on both processes. On the other hand, the stream 'To Reaction' proceeds to the GBR-100 reactor, where occurs exclusively the dry reforming. This stream passes through E-101 for energy integration. In this case, GBR-100 operates at 870°C and 1 bar. After reaction, stream 'SYNGAS' passes through a compression train, with four compressors and five intercoolers, which utilize seawater for refrigeration. This stream is pressurized to 70.5 bar before entering the Adjust reactor. Water vapor is added at a rate of 2 ton/h, and the water shift reaction occurs at 90 °C. The product stream for this reactor goes to a split component operation, used to simulate N_2 removal by molecular sieve adsorption. In the sequence, stream "feed" enters in the Methanol reactor, which operates at 260 °C and 70 bar. The stream that leaves the Methanol reactor has a high quantity of unreacted methane, carbon monoxide and hydrogen. Hence, after methanol recovery, it will be necessary to recycle this stream to GBR-100. For recovering methanol, stream "product" is refrigerated by E-105, E-112 and E-104 before enter in V-101. The top product from this vessel goes to recycle, and the bottom product is depressurized and enters in V-102 for light hydrocarbon removal from methanol. Methanol is recovered at V-102, and those hydrocarbons are recycled. Stream "Recycle 1" works as a cooler for E-105 and is heated before entering a turbine (K-105) to reach GBR-100 condition. After K-105, this stream passes through E-112 also working as a cooler to reach a higher temperature to enter in GBR-100. This stream is divided and part of it goes to CRV-100 and is combusted. "Recycle 2" stream goes directly to GBR-100, since it already has the same pressure of this reactor. Figure 3 shows the PFD for this process.

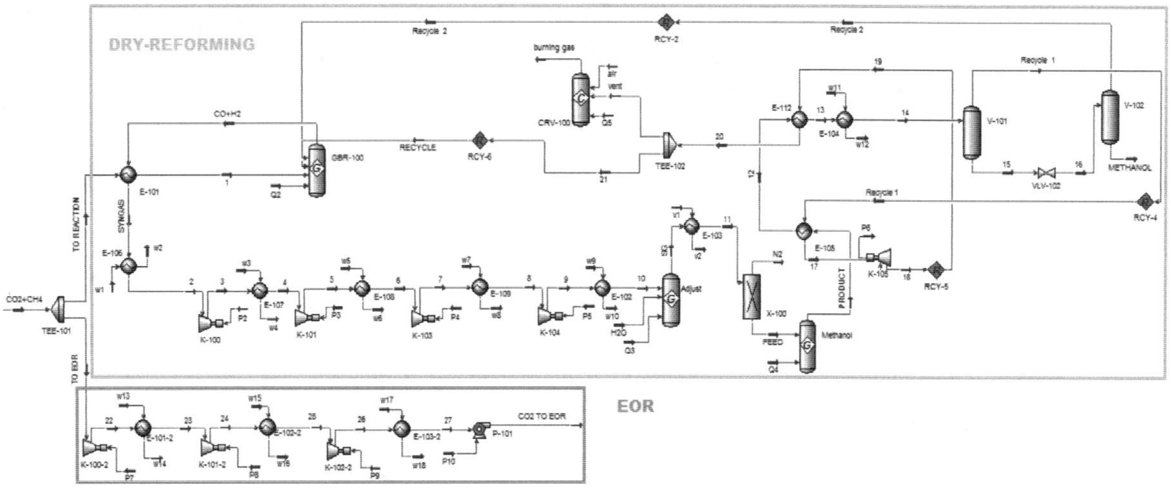

Figure 3. Process flow diagram for EOR and Dry-Reforming process – Alternative 2

ECONOMIC ANALYSIS

The two process alternatives are evaluated economically, through CAPEX (Capital Expenditure) and OPEX (Operational Expenditure) calculations. A prior step to the CAPEX calculation is sizing of process main equipment necessary for cost estimation through correlations (TURTON et al., 2009). Raw materials costs, costs of utilities, operating labor (direct costs) and others general expenses were equally estimated based on Turton et al. (2009). Equipment sizing was performed according to procedures described in the literature (e.g., CAMPBELL, 2004; PERRY, 1997; TOURTON et al., 2009). Economic premises used in the evaluation of the economic performance of the proposed alternatives are presented in Table 4.

Table 4. Economic Premises Adopted

Economic Variable	Value Adopted
CEPCI (Chemical Engineering Plant Cost Index)	572.6
Operating labor cost	U$$136.363/year
Conditioned natural gas price	U$$ 16/GJ
Raw natural gas price	U$$ 8/GJ
Propylene carbonate price	U$$ 1.19/kg
Methanol price	U$$ 0.81/kg
CO_2 for EOR	1ton – 2barrel: U$$118.94
Offshore electricity price	U$$ 0.154
13X Molecular Sieve	U$$ 1.60/kg
Catalyst	U$$ 10/kg
Desalinated water	U$$ 0.25/m^3
Water for refrigeration	none

For the economic analysis, the MS Excel spreadsheet CAPCOST, developed by Turton et al (2009), was employed, allowing the estimation of Capital Expenditure and Operational Expenditure.

Table 5 presents the main results from Bi-Reforming (Alternative 1) and Dry-Reforming (Alternative 2) Processes, indicating the Total Module Cost per piece of equipment and the resulting CAPEX for each one. Compressors, turbines and pumps are responsible for the major share of CAPEX. The CAPEX of Dry-Reforming process is higher when compared to Bi-Reforming, mainly because Dry-Reforming requires more compression steps. On the other hand, the OPEX of Bi-Reforming is higher than the OPEX of the Dry-Reforming. Raw material and operating labor costs are similar. The utilities costs are responsible for the major share of OPEX for both processes. However, Bi-Reforming requires a large amount of electric energy due to the GBR-100 reactor. Table 6 shows these results.

Table 5. Equipment Cost and CAPEX

Total Module Cost (U$$)	Bi-Reforming	Dry-Reforming
Compressors, turbines and pumps	16,941,000	22,208,000
Drives	3,519,800	3,485,000
Heat Exchanger	3,000,465	3,275,000
Vessels	665,400	486,900
Reactors	280,100	450,061
Absorber column	317,000	317,000
Adsorbers	441,944	1,923,945
CAPEX (U$$)	**25,165,709**	**32,145,906**

Table 6. Operational Cost (OPEX)

Cost (U$$/year)	Bi-Reforming	Dry-Reforming
Raw Material	15,933,520	15,917,834
Utilities	79,130,000	60,200,000
Operating Labor	409,089	681,815
OPEX (U$$/year)	**122,575,216**	**101,189,415**

The revenue considers the sum of methanol and natural gas sales, and also takes in account that for each ton of CO_2 reinjected, 2 barrels of oil is produced and sold. The revenue is calculated based on the price market for each product and their rates of production. The results are presented at Table 7, which shows that Bi-Reforming process (Alternative 1) has the highest revenue. Methanol production for this process (17960 kg/h) was 4 times higher than for the Dry-Reforming Process (Alternative 2).

Table 7.Revenue Sales

Revenue (U$$/year)	Bi-Reforming	Dry-Reforming
Natural Gas	7,906,748	7,925,725
CO_2 to EOR	14,107,236	13,995,744
Methanol	122,876,702	29,542,288
TOTAL (U$$/year)	**144,890,686**	**51,463,757**

ENVIRONMENTAL ANALYSIS

The WAR Algorithm (CABEZZAS et al.,1999) was used to quantify the environmental performance of the alternative processes, based on eight categories of environmental impacts, equally weighted:

- Human Toxicity Potential by Ingestion (HTPI)
- Human Toxicity Potential by Exposure (HTPE)

- Aquatic Toxicity Potential (ATP)
- Terrestrial Toxicity Potential (TTP)
- Global Warming Potential (GWP)
- Ozone Depletion Potential (ODP)
- Smog Formation Potential (PCOP)
- Acidification Potential (AP)

WAR measures performance as output rate of PEI (Potential of Environmental Impact). PEI can be described through rates of inlet (Îin), outlet (Îout) and generation (Îgen) of impact in a given system. At steady state, PEI input and output fluxes are constant, and their difference equals the PEI generation (Eq.7). Then, if Îgen is negative, it can be conclude that the process reduces the environmental impact.

$$0 = \hat{I}in - \hat{I}out + \hat{I}gen \qquad (7)$$

The mass flow and composition obtained from ASPEN HYSYS simulations were used in the WAR algorithm, with calculations presented in Figure 4. For both processes, PCOP is the category that most contributes to PEI, followed by HTPI and TTP. That is because methanol has high scores for these categories. Comparing the total impact for these processes, Bi-Reforming has a lower impact (868 PEI/hr) than Dry-Reforming (1661 PEI/hr). Another conclusion that arises is that Îgen is negative for Bi-Reforming and positive for Dry-Reforming. Consequently, the Bi-Reforming process shows superior environmental performance when compared to the Dry-Reforming alternative.

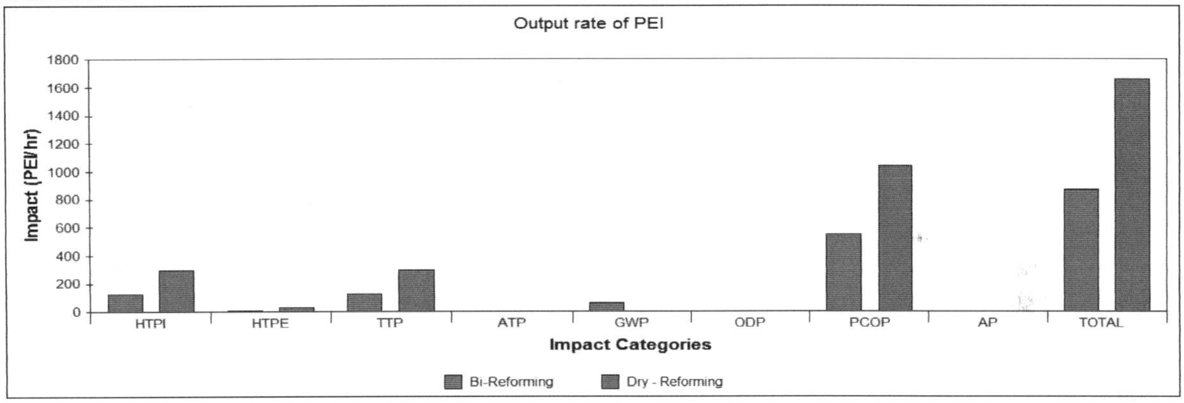

Figure 4. Potential Environmental Impact for Bi and Dry-Reforming Processes

DISCUSSION AND CONCLUSIONS

This work combines CO_2 capture to CO_2 utilization, producing methanol, natural gas and CO_2 for EOR, in an offshore installation. Two alternatives for converting CO_2 and CH_4 into methanol were evaluated - Bi-Reforming and Dry-Reforming, on economic and environmental grounds. It is worth noting that the work evaluates a hybrid alternative encompassing gas-to-liquid technology (bi-reforming and dry-reforming), that allows natural gas deployment to remote locations, gas conditioning to pipeline transportation, which would service main large-scale consumption points, and CO_2-EOR. Furthermore, both proposed processes can be implemented in a methanol floating production system, an MFPSO, where the regeneration of PC would present a competitive advantage over chemical absorption in aqueous solutions of ethanolamines or membranes, since weight and space are the key considerations for MFPSO plants. The methodology adopted employed ASPEN HYSYS®, for calculation of mass and energy balances, necessary for equipment sizing needed in economic evaluation (using CAPCOST) and measuring environmental performance (with WAR Algorithm). The main results for Bi-Reforming and Dry-Reforming Processes are displayed in Table 8 and Table 9, respectively.

Table 8. Main Results for Bi-Reforming Process

	Poor Natural Gas	CO_2 TO EOR	Methanol	Methanol 1	Methanol 2
Volumetric flow (m^3/h)	248.7	27.38	24.05	0.1624	0.1208
Mass flow (kg/h)	13190	14230	19130	133.3	107.7
Composition (% mol)					
CH_4	90.19	35.08	-	0.01	0.10
CO_2	1.67	35.12	3.45	13.63	-
C_2H_6	3.31	19.19	-	-	-
C_3H_8	0.63	10.43	-	-	71.84
N_2	4.20	0.18	-	-	-
CO	-	-	-	0.02	0.27
H_2	-	-	-	-	0.05
H_2O	-	-	4.13	0.73	0.05
CH_3OH	-	-	92.42	85.61	27.69

Table 9. Main Results for Dry-Reforming Process

	Poor Natural Gas	CO_2 TO EOR	Methanol
Volumetric flow (m^3/h)	249.0	27.05	5.52
Mass flow (kg/h)	13210	14230	4341
Composition (% mol)			
CH_4	90.14	35.04	0.05
CO_2	1.69	35.16	0.28
C_2H_6	3.33	19.18	-
C_3H_8	0.64	10.44	-
N_2	4.20	0.18	-
CO	-	-	-
H_2	-	-	-
H_2O	-	-	0.6
CH_3OH	-	-	99.07

The main difference of the proposed processes is the fact that at Bi-Reforming just one reactor is utilized for both dry and steam reforming. On the other hand, for Dry-Reforming two reactors are conceived: one for dry reforming and other to shift reaction. This difference poses a large discrepancy of methanol productivity, as methanol production by Bi-Reforming was 17905 kg/h, 4 times higher compared to Dry-Reforming. It was also observed that dry reforming occurs not only for methane, but also for ethane and propane, and that these last reactions are more favored than the first one. In consequence, methane is not consumed, but produced in the Dry-Reforming alternative, and requires recycle to the reformer reactor. This conclusion explains the low production of methanol for this process.

The economic analysis shows that the Bi-Reforming process has advantages due to the larger methanol production. Although its OPEX is higher than the OPEX for Dry-Reforming, methanol production capacity is sufficiently high to compensate the increased operational expenditure. For By-Reforming, the CAPEX was U$$25,165,709, considering an Onshore Plant, at United States. It was assumed that the offshore cost is the double of onshore cost, and that the "Brazilian Cost Factor" (cost for a plant installed at Brazil) is about 1.5. Then, the real CAPEX for this process would be U$$75,497,127. For OPEX, this process has a cost of U$$122,575,216, and the Revenue Sales U$$144,890,686, what turns Bi-Reforming technically and economically feasible.

The potential environmental impact of both processes were calculated (in PEI/hour) as 868 PEI/h and 1661 PEI/h, respectively for Bi-Reforming and Dry-Reforming. This result shows that Bi-Reforming is environmentally more favorable than Dry-Reforming. Another important point is that the potential impact generation rate is negative for Bi-Reforming, what strengthens the conclusion of environmental feasibility.

Therefore, the preliminary calculations indicate that Bi-Reforming is technically, economically and environmentally attractive. Future studies are required such as (among others): (i) sensitivity analysis to design and economic premises (e.g., cost of offshore desalinated water, price of raw natural gas), (ii) flowsheet optimization, (iii) study impact of production scale production, (iv) analysis of EOR process with pure CO_2; (v) detailed analysis of N_2 removal.

ACKNOWLEDGMENT

The authors gratefully acknowledge the financial support from the National Petroleum Agency (ANP) – PRH41, for scholarship of Bruna C.S. Lima.

REFERENCES

Aquino, A. S. **Analysis of alternative synthesis routes for chemical sequestration of co2: methanol, syn gas and acetic acid production**. 2008. 87f. Monografia - Curso de Engenharia Química, Departamento da Escola de Química, Universidade Federal do Rio de Janeiro, Rio de Janeiro, 2008.

ARAUJO, O. Q. F.; MEDEIROS, J. L. de; ALVES, R. M. B. CO_2 utilization: a process systems engineering vision. In: MORGADO, Claudia do Rosário Vaz ; ESTEVES, Victor Paulo Peçanha (Ed.). **CO_2 sequestration and valorization.** InTech, 2014. cap.2. ISBN 978-953-51-1225-9.

CABEZAS, H.; BARE, J. C.; MALLICK, S. K. Pollution prevention with chemical process simulators: the generalized waste reduction (WAR) algorithm - full version. **Computers and Chemical Engineering**, v. 23, n. 4–5, p. 623–634, 1999.

CAMPBELL, J. M. **Gas conditioning and processing: the equipment modules**. 8. ed. Norman: John M Campbell & Company, 2004. 2 v.

CAVENATI, S.; GRANDE, C. A.; RODRIGUES, A. E. Separation of CH4/CO2/N2 mixtures by layered pressure swing adsorption for upgrade of natural gas. **Chemical Engineering Science**, Elsevier, v. 61, p. 3893-3906, 2006.

DAI, W. L., et al. Synthesis of propylene carbonate from carbon dioxide and propylene oxide using Zn-Mg-Al composite oxide as high-efficiency catalyst. **Catalysis Letters**, v. 136, n. 1-2, p. 35-44, 2010.

FORMIGLI, J. **Pre-salt reservoirs offshore Brazil: perspectives and challenges**. Miami. nov. 2007.

GAFFNEY, CLINE & ASSOCIATES. **Exame e avaliação de dez descobertas e prospectos selecionadas no play do pré-sal em águas profundas na Bacia de Santos, Brasil.** CG/JW/RLG/C1820.00/gcaba.1914. ANP, 2010.

GANGADHARAN, P.; KANCHI, K.C.; LOU, H. H. Evaluation of the economic and environmental impact of combining dry reforming with steam reforming of methane. **Chem. Eng. Res. Des.**, v. 90, n. 11, p. 1956–1968, 2012.

JOHNSTON, P. et al. Reclaiming the definition of sustainability. **Env. Sci. Pollut. Res**. v. 14, n. 1, p. 60–66, 2007.

KIM, W., et al. Methanol synthesis from syngas over supported palladium catalysts prepared using water-oil-microemulsion. Applied Catalysis A: General, v. 169, p. 157-164, 1998.

OLAH, G.A. et al. Bi-reforming of methane from any source with steam and carbon dioxide exclusively to metgas (CO–2H2) for methanol and hydrocarbon synthesis. **J. Am. Chem. Soc.**, v. 135, n. 2, p. 648–650, jan. 2013.

PACÍFICO, J. A. **Cinética do processo de reforma catalítica do metano com dióxido de carbono. Aplicação à modelagem e simulação da operação em reator de leito fluidizado.** 2004. 159f. Dissertação (Mestrado em Engenharia Química) – Departamento de Engenharia Química, Universidade Federal de Pernambuco, Recife, 2014.

PERRY, R. H.; GREEN, D. W.. **Perry's Chemical Engineers's Handbook**. 7th ed.. McGraw-Hill, 1997.

TURTON, R. et al. **Analysis, Synthesis, and Design of Chemical Processes**. 3. ed. New Jersey: Prentice Hall, 2009.

WANG, S. B.; LU, G. Q. M.; Millar, G. J. Carbon dioxide reforming of methane to produce synthesis gas over metal-supported catalysts: state of the art. **Energy Fuels**, v. 10, n. 4, p. 896–904, 1996.

ZHANG, X., et al. Carbon dioxide reforming of methane with Pt catalysts using microwave dielectric heating. Catalysis Letters, v. 88, n. 3–4, p. 129-139, 2003.

371846 Improved CO2 Capture Process : Rich Vapor Recompression with Split Flow

Abstract:

CO_2 capture process using MEA scrubbing is the most mature technology for reducing the CO_2 emission into the atmosphere. However the MEA scrubbing process requires a lot of heat energy for solvent regeneration in the stripper reboiler. The reboiler heat requirement is a function of preheating duty and stripper reflux ratio. The reboiler heat requirement decreases with preheating duty increases in the lean/rich heat exchanger. When the stripper reflux ratio decreases, the reboiler heat duty is reduced also. In order to reduce the reboiler heat requirement, large number of advanced configurations have been suggested already. The split flow configuration, well known advanced configuration, reduces the stripper reflux ratio by lowering the stripper top temperature. As a part of cold rich solvent is fed to the stripper top without passing through the lean/rich heat exchanger, the cold rich solvent is directly cooling off the stripper top. Although this split flow configuration reduces the stripper reflux ratio, the heat exchanger preheating duty is decreased due to the cold side flow rate reduction. It is hard to reduce the stripper reflux ratio while maintaining the preheating duty. In order to solve this problem, this paper suggests an innovative configuration of MEA scrubbing process; rich vapor recompression with split flow. The rich vapor recompression is a new configuration which is vaporizing the cold rich solvent at the lean/rich heat exchanger. It is similar concept with lean vapor recompression which is vaporizing the hot lean solvent at the flash drum. However the energy reduction effect is different when split flow is combined. In the suggested configuration, rich vapor recompression with split flow process, a part of cold rich solvent is fed to the stripper top without passing through the lean/rich heat exchanger. The other part of cold rich solvent is vaporized at the lean/rich heat exchanger. The generated vapor is fed to the stripper bottom as steam energy source. The remaining liquid is fed to the stripper middle. Although the heat exchanger inlet feed flow rate is reduced, the cold side heat capacity increases through the solvent vaporization. Consequently, the stripper reflux ratio is decreased with maintaining the preheating duty in the lean/rich heat exchanger. As a result, the reboiler heat requirement was 2.91 GJ/tonCO_2, 3.07 GJ/tonCO_2, and 3.38 GJ/tonCO_2 in the rich vapor recompression with split flow, lean vapor recompression with split flow and conventional MEA process, respectively.

378844 Development of CO2 Capture and Utilization Technologies for Alkaline Wastes for Building Waste-to-Resource Supply Chain: Theoretical Consideration

Abstract:

In this study, accelerated carbonation of alkaline solid wastes using a rotating packed bed (RPB) for "CO_2 capture" and their utilization as "green construction materials" have been developed for economic feasibility and environmental sustainability. The CO_2 capture capacity and overall energy consumption of the developed carbonation processes were critically assessed from the engineering and environmental aspects. The challenges of CO_2 capture by carbonation of alkaline waste were to accelerate the reaction and exploit the heat of reaction to minimize energy and material losses. Therefore, the objectives of this research work were to evaluate the performance of RPB process by introducing life cycle assessment (LCA) and response surface methodology (RSM) for minimizing the environmental impacts and maximizing the overall CO_2 capture capacity.

In addition, a prediction model for carbonation within an RPB was developed by combining the process chemistry, reaction kinetics, and mass transfer. The enhancement factors for mass transfer coefficient in case of carbonation of different types of wastes within an RPB were determined accordingly. The important factors include particle size of solid waste, reaction temperature, rotation speed, and liquid-to-solid (L/S) ratio. According to the observations of TGA, XRD and SEM-EDX, the alkaline wastes were found to be successfully carbonated with CO_2 in an RPB, where calcite ($CaCO_3$) was identified as the main product.

The LCA was utilized to determine the net performance of carbon offsetting by considering (1) the energy consumption and carbon emission associated with RPB operation; (2) benefits of offering alternative materials for cement or concrete; (3) the CO_2 caused by raw material pre-processing & conditioning of carbonate product for final products. The functional unit is set to be capture of 1 kg CO_2 by carbonation processes within the system boundary while carrying out the LCA. Energy consumption of the overall processes including grinding, sieving, pressuring, heating, pumping, and rotating processes was measured. A balance for material and energy flows is drawn up for calculating and analyzing the investigated system, which consists of all input (e.g., land, energy, raw materials, auxiliary agents, etc.) and output (e.g., products, emissions, waste, etc.)

Furthermore, the developed RPB process were designed and analyzed by RSM using Design Expert statistical software to visualize the effects of operating parameters on the carbonation conversion of solid wastes for overall maximization. Two types of response parameters including the carbonation conversion of alkaline waste and environmental impact were systematically evaluated in this study. The factors affecting the above two types of responses were coded with low and high levels using central composite design (CCD) method. The carbonation conversion and environmental impact could be maximized and minimized, respectively, by observation of the location of the saddle point on the 3D-surface imaging from different sets of response surfaces. With the results of the prediction model, LCA and RSM, the best achievable technology (BAT) for carbonation using an RPB was proposed for system optimization. It was thus concluded that accelerated carbonation of alkaline wastes using an RPB is an effective and efficient method for CO_2 capture and utilization due to its higher mass transfer rate and carbonation conversion.

380371 CO2 Fixation through Carbonation of Waste Cement and Concrete

Abstract:

CO_2 capturing and fixation processes with waste concrete via mineral carbonation reactions have been developed. The waste concrete is either waste concrete generated from demolished concrete buildings (waste concrete, hereafter) or concrete sludge. Waste concrete is composed of aggregates and waste (hydrated) cement, and waste cement powder of which the diameter is under about 100 μm would be separated in the recycling process of aggregates. The concrete sludge is waste fresh concrete in the form of slurry generated at the concrete using sites. Both concrete wastes are alkaline containing calcium that could be used as a capturing and fixation material for CO_2 via carbonation reaction. Several options are considered as the CO_2 fixation process with these waste concrete, and process feasibility has been evaluated based on the laboratory experimental studies. The followings are the options examined in this study.

(1) Direct fixation of CO_2 with concrete sludge

This option is a direct fixation process of CO_2 with waste concrete sludge. Flue gas is introduced to the slurry of concrete sludge, and CO_2 will react with hydrating cement components, mainly calcium hydroxide, to form calcium carbonate. The effect of hydration reaction time, gas flow rate, water content, CO_2 concentrations on the CO_2 fixation performances were examined in a small-scale laboratory experiments.

(2) Indirect multistep fixation of CO_2 with waste cement powder

This option is four-step process composed of (i) dissolution of calcium ions into an acid, (ii) CO_2 capture with an alkaline solution, (iii) formation of calcium carbonate by mixing the above two solutions, and (iv) regeneration of the acid and the alkaline solutions via electrodialysis. Parameters affecting the process performances were examined for each step in laboratory experimental studies.

Based on the experimental studies, the feasibilities of these processes have been evaluated in views of the process power consumption as well as the cost.

14AIChE
Annual Meeting, Atlanta, GA

Technological Trends in CO_2 Capture, Transport and Utilization

Rita M. Brito Alves
Fernanda M. Fontes
José L. Medeiros
Ofélia Q. F. Araújo

USP

H2CIN

*Motivation

*The development of economies results in increased energy consumptions.

*In the coming decades, the supply of such expanding demand will still remain based on fossil fuels technologies.

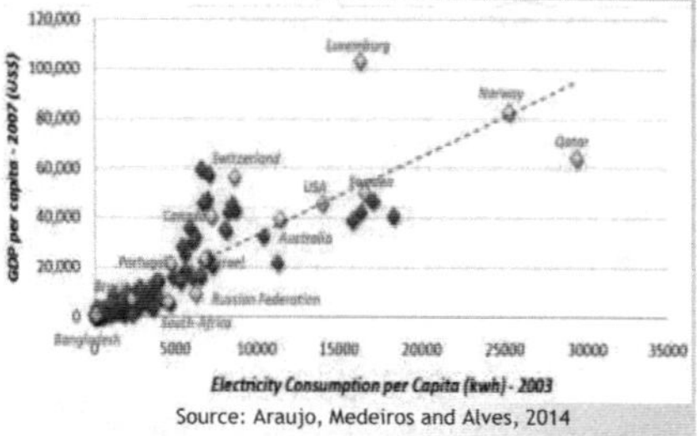

Source: Araujo, Medeiros and Alves, 2014

*Fossil fuel based energy is recognized as carbon emitter with serious consequences for the environment.

*Measures have been taken not only to reduce but also to try to reverse the current situation of intensification of global warming due to the greenhouse effect.

*These measures include reduction of the dependence on fossil fuels by modern society, and methods to control the emission of greenhouse gases, especially CO_2.

*The challenge is thus to expand energy supply in a carbon-constrained economy.

*Motivation

* According to the World Economic Forum (King, 2010), no truly low-carbon technology will be able to penetrate the mass market in the short term.
* In the context of energy and climate policy, governments place much emphasis on carbon capture and geological storage (CCS) as an option that can reduce substantial amounts of greenhouse gas emissions.
* However, there is also the possibility of CO_2 utilization in building material production, for fuels or in the chemical industry, which could contribute to a green economy.
* On the other hand, it is reasonable to expect that CO_2 conversion and biomass-based processing alternatives will coexist for a while with fossil-based technologies.
    * So, the use of installed fossil processing infrastructure, with co-processing of biomass and fossil feedstock and capturing and utilization of emitted CO_2 is the *"escape route"* for a moderate transition from the present to a long-term sustainable future.

*Objectives

* The main objective of this work is to to build technological maps in order to support decision-makers aligned to initiatives for reduction of CO_2 emission.
* In this framework, this work presents the CO_2 Capture Cycle as well as promising alternatives of its reutilization.
    * The work is oriented to cover CO_2 as feedstock, i.e., use of CO_2 as raw material to chemicals and fuels with economic application.

*CCS

*Carbon capture and storage (CCS) is most commonly defined as the capture of CO2 from an industrial or power-sector point source combined with its transport and its storage in geological formations (IEA, 2009).

*CCS in some sectors provides cost-effective emission reductions, but has significant shortcomings:

  *Costs of CCS are considered to be highest for the capture component, but storage and especially transport costs can also be significant.

  *the potential storage capacity has uncertainties

  *public resistance to CCS has been increasing

  *and it costs energy.

CCS

*According to the IPCC (2005) the components of CCS are classified as being in different stages of technological maturity.

*Some options, such as capture from a limited number of industrial processes, are considered mature as they are implemented routinely, for instance in hydrogen production and gas processing plants.

*Other components are seen as in or just beyond the demonstration phase – significant-scale demonstrations are operative but the technology is not yet proven at full scale or in commercial conditions in the power sector.

CCS - CO_2 Capture

* Current state of the art post-combustion CCS uses a variety of capture agents; however these are mainly based on amines.
  * Corrosivity/volatility/hazardous emission
  * Capex
* There is therefore considerable effort required to produce new capture agents with better environmental credentials.
  * Recent studies have shown that solid capture agents such as functionalized silicas, poly(ionic liquid)s, activated carbons and calcium oxide can be used to not only enhance capture but also reduce the capital expenditure required on plant construction.

CCS - CO_2 Storage

* Geological storage

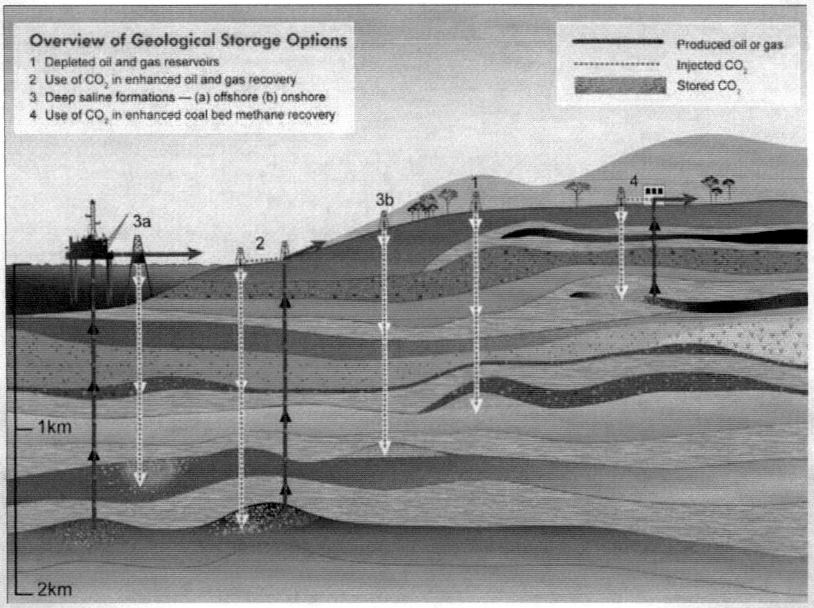

Methods for storing CO_2 in deep underground geological formations. Two methods may be combined with the recovery of hydrocarbons: EOR (2) and ECBM (4)

*CCS - CO_2 Storage

*Geological Storage

*Geological storage of CO_2 is accomplished by injecting it in dense form into a rock formation below the earth's surface.

*The option of storing CO_2 in coal beds and enhancing methane production is still in the demonstration phase

*Geological storage of CO_2 is ongoing in three industrial-scale projects (~1 $MtCO_2$/yr): the Sleipner project in the North Sea, the Weyburn project in Canada and the In Salah project in Algeria.

*About 3–4 $MtCO_2$ that would otherwise be released to the atmosphere is captured and stored annually in geological formations.

*CCS - CO_2 Storage

*CO_2-EOR

The US has been commercially operating EOR programs for over 40 years.

It currently provides nearly 300,000 barrels of oil per day in North America from 125 distinct EOR projects

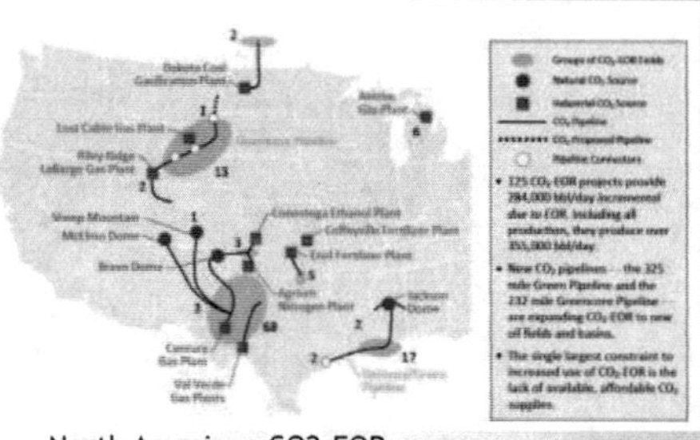

North American CO2-EOR sources

Source: Oil&Gas Journal, 2012

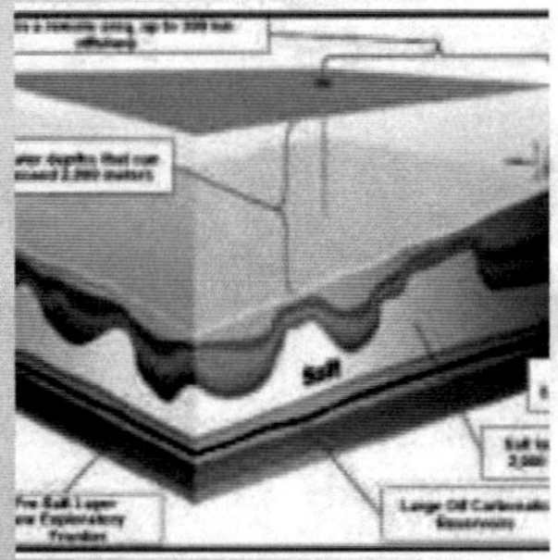

Brazil basin pre-salt environment.

Outside North America, only a few CO_2-EOR projects are underway in Brazil, Turkey, and Trinidad.

However, the development of universally agreeable predictive models and comprehensive monitoring techniques must still be established.

*CO_2 Storage

* EOR and CGS constitute, under present conditions, the only CO_2 processing chain that has some steps ready to be put into operation at high scales.
* These encompass some separation technologies for CO_2 post-combustion capture, CO_2 compression and CO_2 transportation via long pipelines.
* EOR and CGS can be reinforced if oxy-fuel technologies finally attain industrial maturity, which may occur within the next 5 decades.
* Pre-combustion technologies are also potential powerful contributors to EOR-CGS systems, which should be fully developed within the next 5 decades.
* Pre-combustion relies on some separation technologies also present in post- combustion alternatives.

*Transport of CO_2

* Pipelines today operate as a mature market technology and are the most common method for transporting CO2.
* In some situations or locations, transport of CO_2, as a liquid, by ship may be economically more attractive, particularly when the CO_2 has to be moved over large distances or overseas.
* Road and rail tankers also are technically feasible options.

* Alternative routes for CO_2 valorization

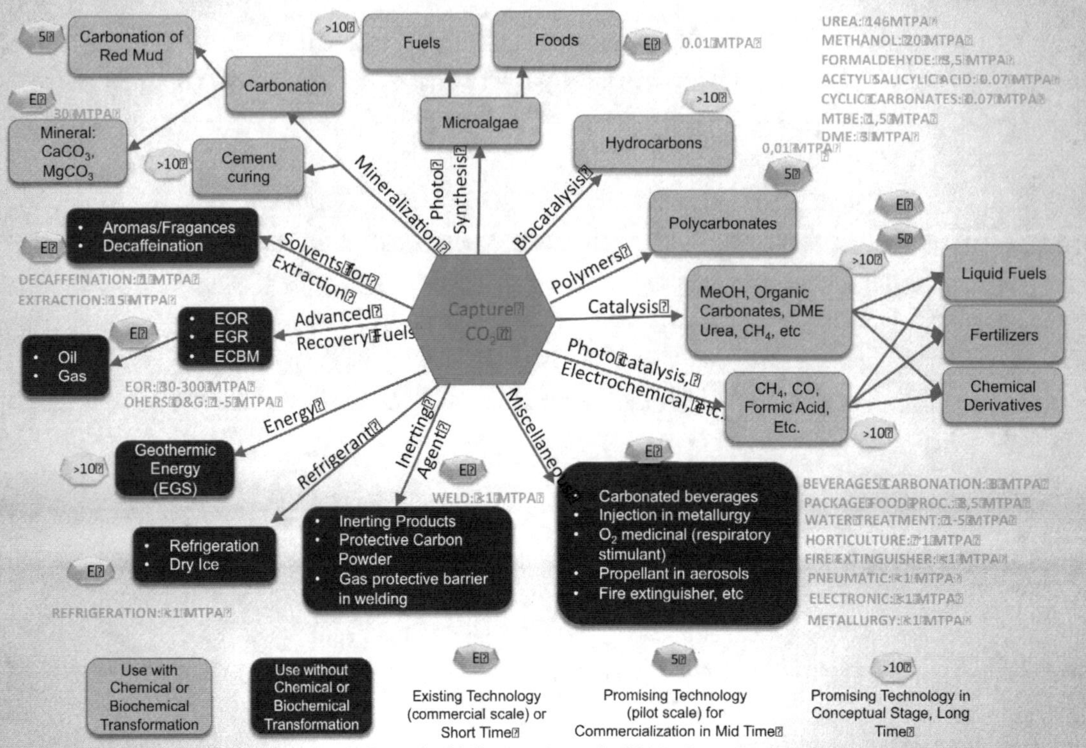

Built based on Gresser (2011), Bucin-Dumitriu et al. (2013), Araujo et al. (2014), Gouveia et al. (2008), Gürtler (2012).

* CCU - direct use

* GEOTHERMAL ENERGY GENERATION

Rather than using water to extract heat from deep underground to produce geothermal energy, it uses carbon dioxide.

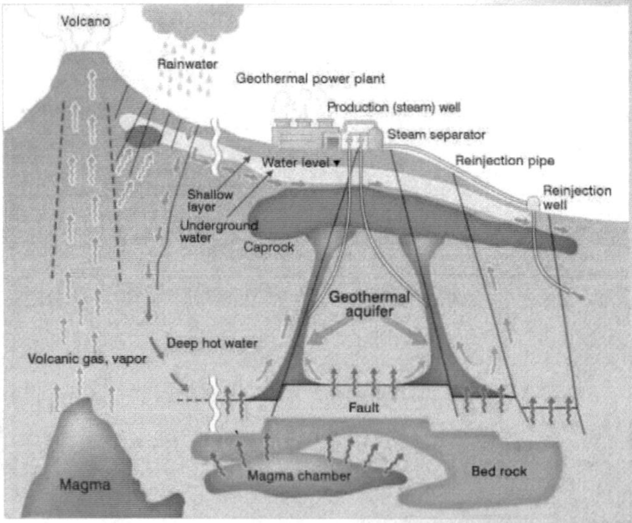

Source: Geothermal Energy, Serial No. 87 (July 1999)

* CCU- CO_2 conversion to chemicals

* The strategic objectives include:
  * rather than treating CO_2 as waste, it can be regarded as a chemical feedstock for the synthesis of other chemicals;
  * use CO_2 to produce added-value chemicals useful to the industry and able to reduce the penalty asssociated to the costs of capture process;
  * use CO_2 as benign fluid for processing or a way to recover energy and reduce emissions;
  * use CO_2 recycle as energy renewable source to preserve the carbon sources, aiming sustainable development.

CCU should not be regarded as an alternative technology to CCS but as a complementary technology

* CCU- CO_2 conversion to chemicals

* The main barriers to CO_2 utilization are:
  * the costs associated to CO_2 capture, purification and transport.
  * the unfavorable thermodynamics of many conversions
  * the requirement of energy for chemical conversion of CO_2
    * that means that there will be an energy cost associated with utilization
  * limited size market
  * small investments/incentives and lack of industrial commitment to increase CO_2 based chemicals; and
  * lack of social and economic drivers.

*CCU- CO_2 conversion to chemicals

*Chemicals from CO_2

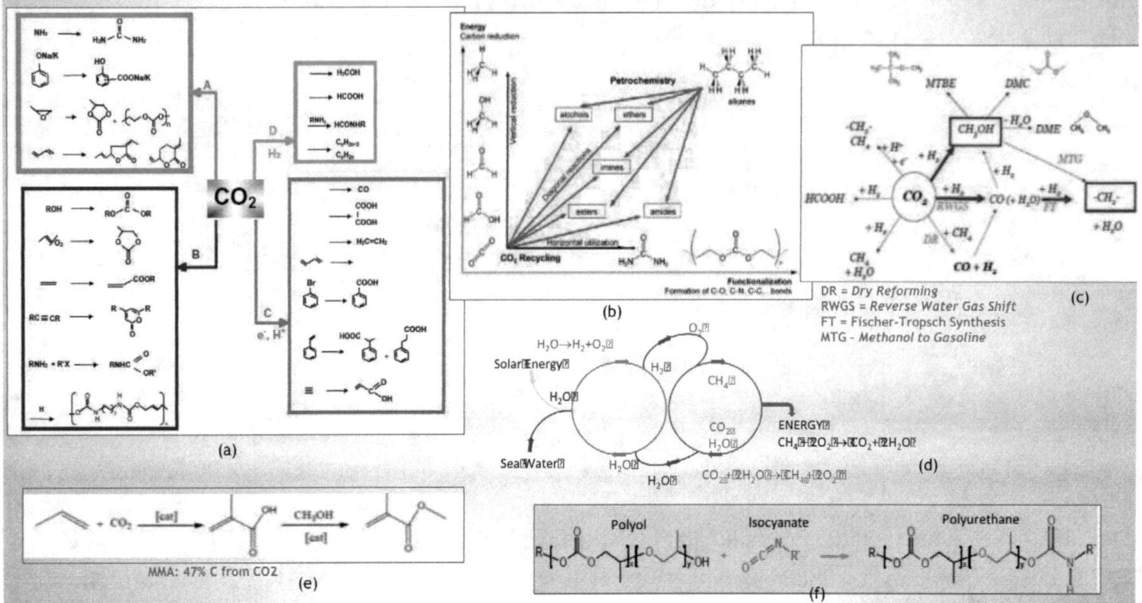

(a) Chemicals from CO_2, Aresta (2010); (b) "Diagonal Conversion", Gomes et al. (2012); (c) Synthesis of Liquid Fuels/Energy Storage, Hashimoto (2011); (d) Energy from H_2O, CO_2 and solar energy, adapted from Carrera (2011); (e) MMA (Wang, 2011); (f) Polyurethane, Gürtler (2012).

*CCU- CO_2 conversion to chemicals

*Chemicals from CO_2

Reagents and products generated by CO_2 reaction.

CO-REAGENT	CO_2 - products	CO-REAGENT	CO_2 - products
Alkane	*Syngas*, acids, esters, lactones	Substituted hydrocarbon [(c)]	Acids, esters, lactones, polycarbonates
Cycloalkane	Acids, esters, lactones	Alkyne	Lactones, organic carbonates, polycarbonates
Active H compounds	Acids, esters, lactones	Epoxy	Carbonates, (co)polymers (polycarbonates)
MonoAlkene	Acids, esters, lactones	NH_3 and Amine	Simetrical Ureias, amino acids, (co)polymers
Diene[(a)]	Acids, esters, lactones [(b)]	Diamine	Ureias, carbamates, (co)polyemers, (polyureias)
CycloAlkene	Acids, esters, lactones, (co)polymers	Imines	Carbamates, (co)polymers (uretane)

Source: Kargari e Ravanchi (2012)

*CCU- CO_2 conversion to chemicals

*CO_2 Converion to Synthetic Fuels

*Technological routes for CO_2 conversion to synthetic fuels (synfuel)

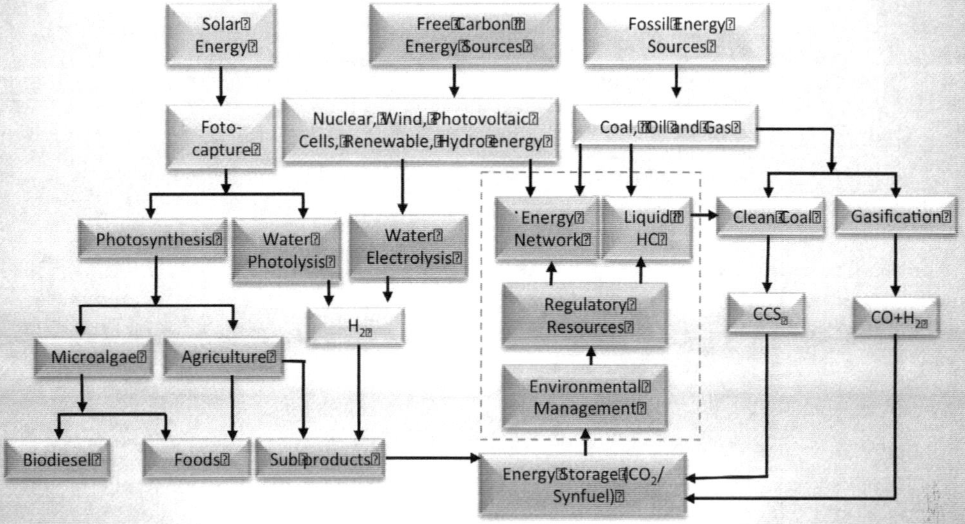

Source: Amoureux, Siffert e Massue (2011)

*CCU- CO_2 conversion to chemicals

*CO_2 Converion to Synthetic Fuels

*Solar route for Synthesis Gas from CO2

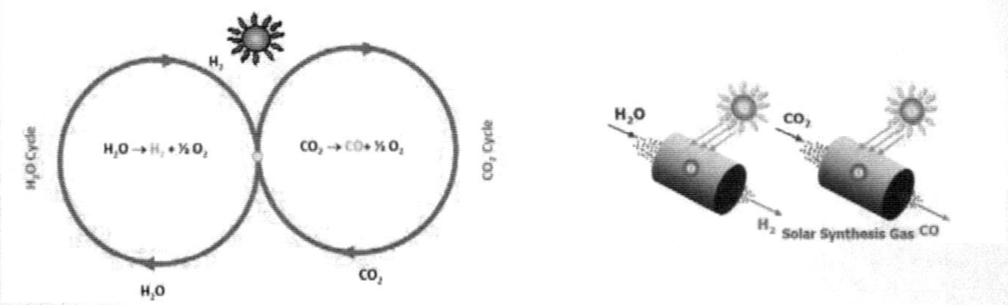

*Sustainable storage of H_2 e Carbon

$H_2 + CO \longrightarrow C_XH_Y$ Liquid Fuels – Fischer-Tropsch Process

$4H_2 + CO_2 \longrightarrow CH_4 + 2H_2O$ Gaseous Fuels - Methane/Sabatier Process

$H_2 + CO \longrightarrow C_XH_Y$ Plastics

Source: Konstandopoulos (2011)

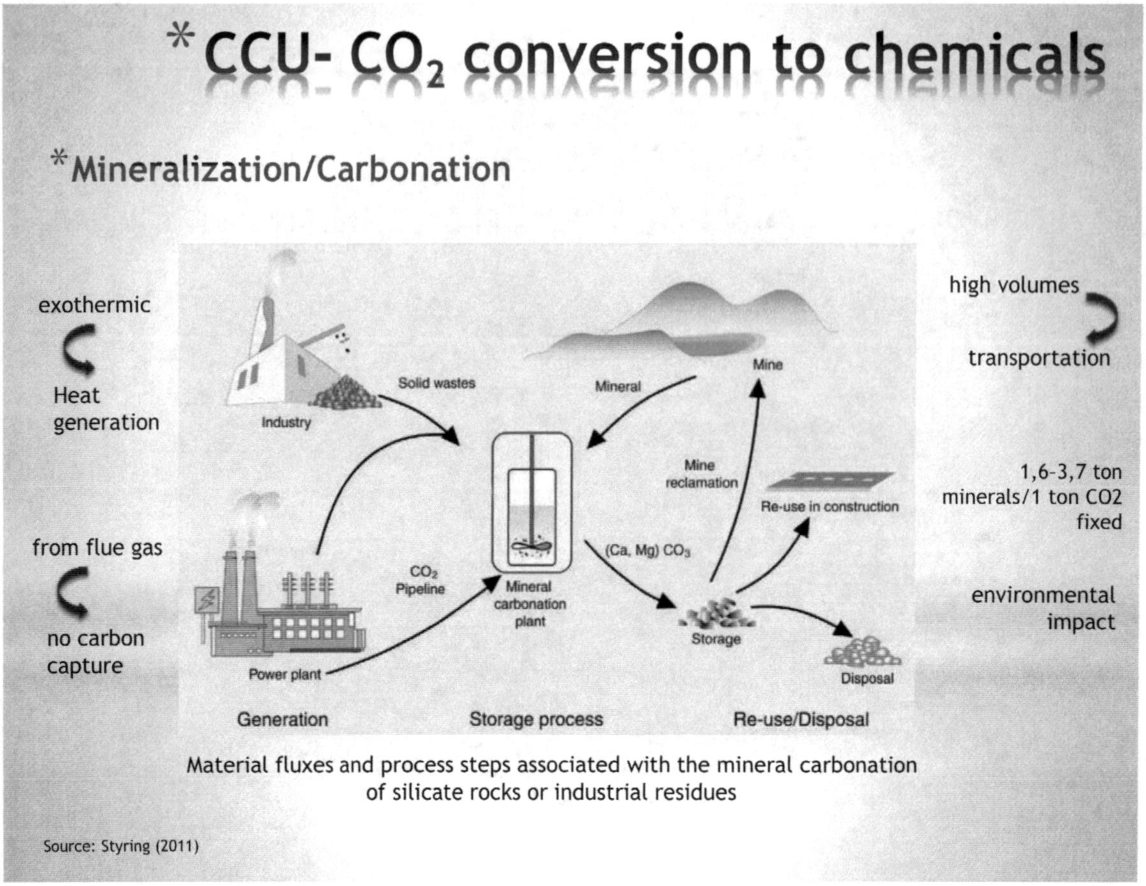
*CCU- CO_2 conversion to chemicals
*Mineralization/Carbonation
exothermic
Heat generation
from flue gas
no carbon capture
Industry
Solid wastes
Mineral
Mine
Mine reclamation
Re-use in construction
CO_2 Pipeline
Mineral carbonation plant
(Ca, Mg) CO_3
Storage
Power plant
Disposal
Generation
Storage process
Re-use/Disposal
high volumes
transportation
1,6-3,7 ton minerals/1 ton CO2 fixed
environmental impact
Material fluxes and process steps associated with the mineral carbonation of silicate rocks or industrial residues
Source: Styring (2011)

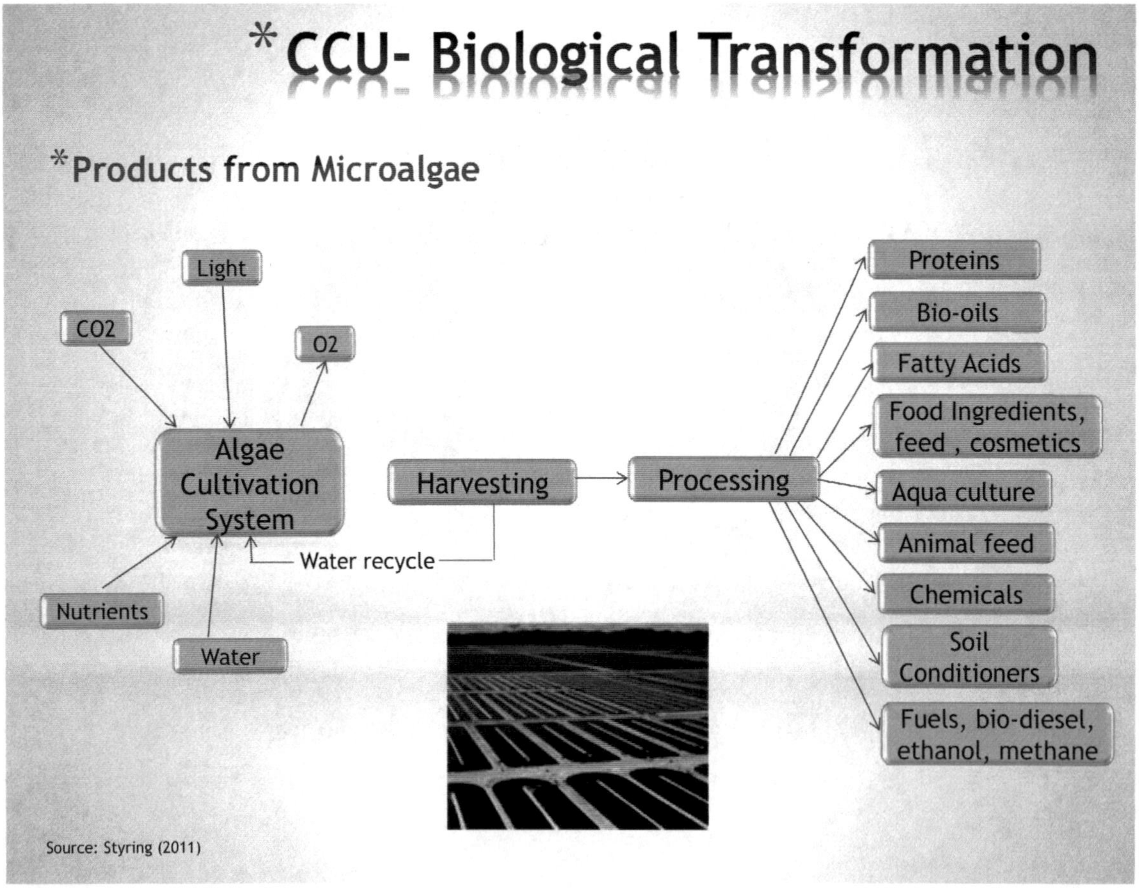
*CCU- Biological Transformation
*Products from Microalgae
Light
CO2
O2
Algae Cultivation System
Harvesting
Processing
Water recycle
Nutrients
Water
Proteins
Bio-oils
Fatty Acids
Food Ingredients, feed , cosmetics
Aqua culture
Animal feed
Chemicals
Soil Conditioners
Fuels, bio-diesel, ethanol, methane
Source: Styring (2011)

*CCU

*Technological maturity in CO_2 Utilization

CATEGORY	TECHNOLOGY	RESEARCH	DEMONSTRATION	ECONOMICALLY VIABLE	MATURE MARKET
CO2 TO FUELS	Hydrogen (Renewable Methanol)		■		
	Hydrogen (Formic Acid)	■			
	Microalgae (Biofuels)	■	■		
	Photocatalysis	■			
	Catalysts (nanomaterials)	■			
INCREASE PRODUCTIVITY	Power cycles with supercritical CO_2	■			
	Urea Boosting				■
ADVANCED RECOVERY	EOR				■
	EGR		■		
	ECBM		■		
MINERALIZATION	Cement Production		■		
	Concrete Healing			■	
	Carbonation Red Mud			■	
	Mineralization with Carbonate	■	■		
CHEMICALS	Sodium Carbonate				■
	Polymers	■			
	Other Chemicals (e.g., acetic acid)	■			
	Microalgae (for Chemicals)		■		

Fonte: Bocin-Dumitriu et al. (2013)

*Technological Roadmaps

* This work highlighted the state of art in technologies of *Carbon Capture and Utilization (CCU)* in order to
  * Identify the main drivers that will promote the development of technologies of capture, transport and utilization of CO_2
  * Identify the main technologies and classify them according to their technology readiness level (TRL)
  * Identify the Initiaves and technolgical perspectives
  * Build a technological roadmap able to associate short, mid and long term milestones to the technological solutions and organize them in a timeline.
  * Allow a broad vision about the most promising technologies in the short, mid and long term.

WHY, WHAT, WHEN & WHO

*Technological Roadmaps

*Roadmap Structure:

*Timeline

- *Current (or Today) represents technologies already in use
- *1 to 5 years includes the technologies in advanced stage of development or scale almost commercial
- *5 to 10 years are projects in demonstration or pilot plant scale and which efficiency has been demonstrated.
- *10 to 15 years are projects still in test or lab scale, which efficency hasn't been yet proved. As the case of researches recently published in literature.

*Technological Roadmaps

*Roadmap Structure:

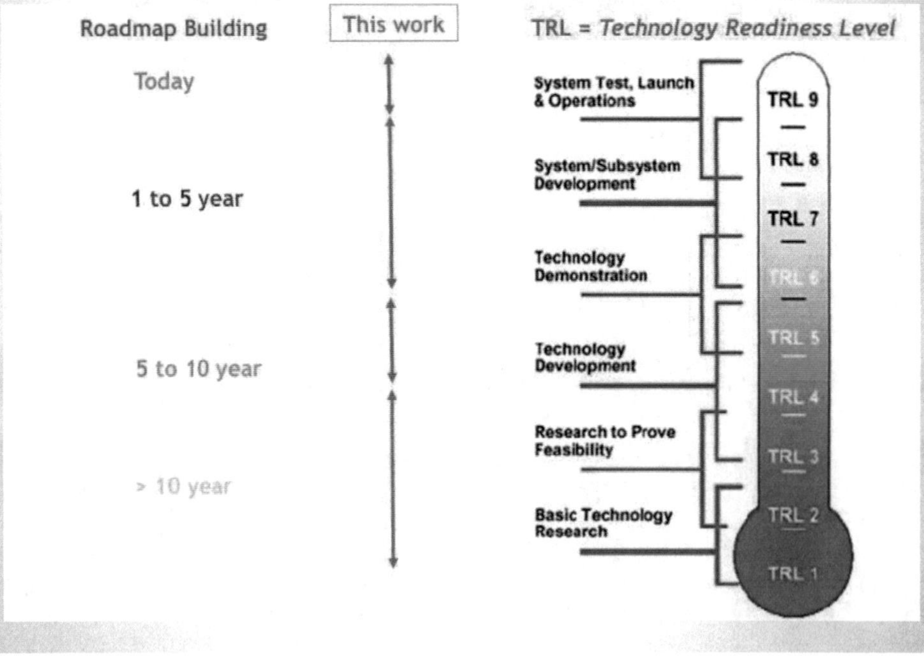

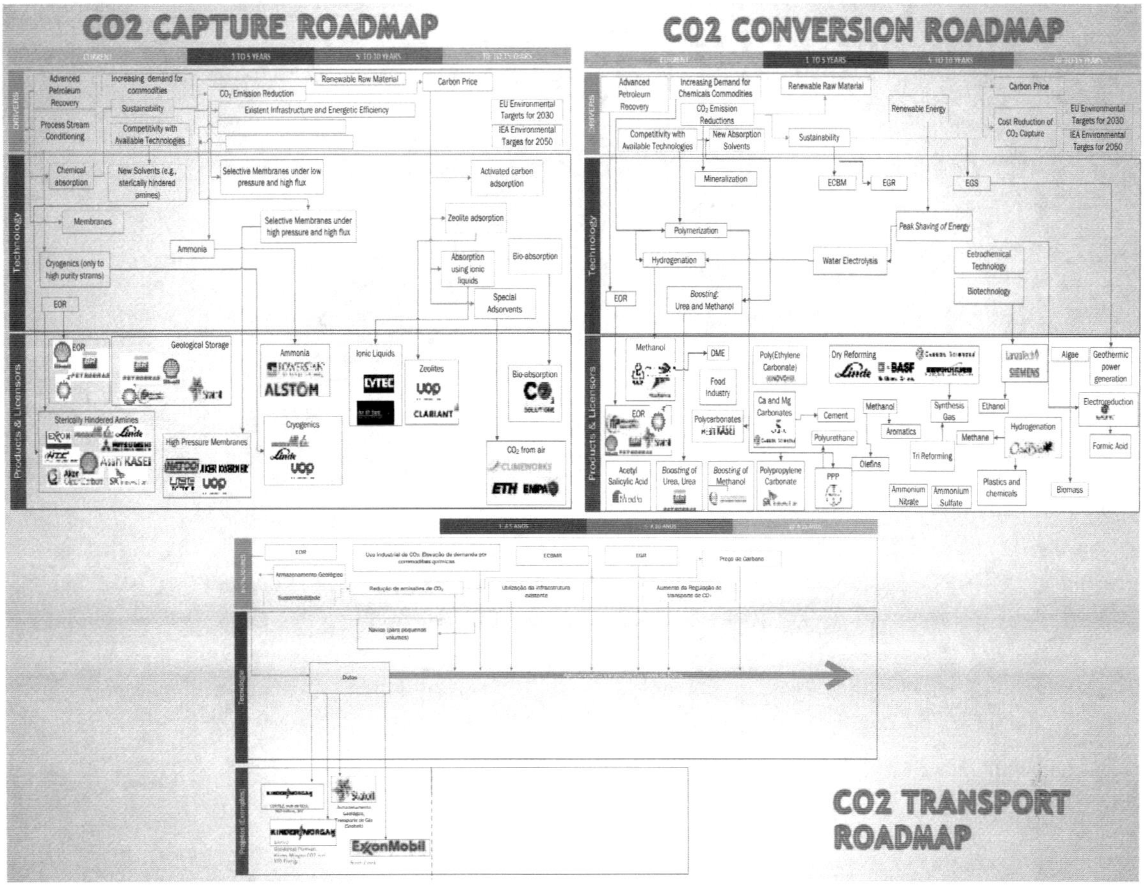

*Technological Roadmaps -Conclusions

*The cost of CO_2 capture, storage and utilization technologies should be analyzed to evaluate the project feasibility.

*The most promising technology for CO_2 capture is chemical absorption with amines.

*Regarding transportation technologies there is no doubt that the ducts are the present and the future. They are a safe, and they have already significant maturity in behavior and structure (due to similarities with pipelines for gas) and they are cheaper when compared to other forms of transportation.

*Researches on technologies for CO_2 utilization indicate that the EOR and geological storage destinations are the methods for which we have more knowledge and they are / will be able to absorb large amounts of CO_2, having a significant effect on reducing CO_2 emissions.

*Technological Roadmaps -Conclusions

*Concerning CO_2 conversion, the roadmap shows scenarios with opportunities for next decades.

*An important conclusion is that the driver for CO_2 utilization technologies is not the mitigation of emissions but the increasing in price of fossil raw materials, which would promote technologies based on CO_2 as a raw material - cheap and abundant due to emissions regulation.

*On the other hand, CCU alone cannot realistically remediate all emissions because of the volumes involved and the potential markets for the individual products.

Acknowledges

Braskem

CAPES

Contacts

rmbalves@usp.br

ofelia@eq.ufrj.br

*Thank you!

390952 Effects of Contaminants to Physical Solvents for Pre-Combustion CO2 Capture

Abstract:

The pre-combustion capture process involves converting fossil fuel into syngas, primarily a mixture of carbon monoxide and hydrogen, usually by gasification. During pre-combustion, coal is first transformed into hydrogen and CO_2. In a CO_2 separation unit, the carbon dioxide is separated from the hydrogen using a solvent, such as ionic liquids (ILs). This process has the potential of capturing and removing 90 percent of CO_2 emissions from power plants - major contributors to global warming.

ILs are salts with many advantages over conventional pre-combustion CO_2 capture materials, including low melting points, longer lifetimes and higher stability over a range of process conditions. While ILs for CO_2 capture are a popular research topic, less attention has been focused on the impact of contaminants, i.e., water and coal derived *contaminants* (primarily H_2S), on ILs for pre-combustion carbon capture process. We explore the use of physical solvents for pre-combustion carbon capture. Four solvents have been under investigation, including PDMS, hybrid-PDMS, and two types of ILs, i.e., P888 allyl Tf_2N and allyl pyridinum Tf_2N, with different initial viscosity. A 1-liter Continuous Stirred Tank Reactor (CSTR) equipped with a 2-liter gas buffer tank has been used to measure CO_2 isotherms and CO_2 breakthrough curves at different temperatures. For CO_2 solubility and kinetics measurement without contaminants, a dynamic pressure-step method was applied, while a continuous flow operation was used for CO_2 breakthrough measurement in the presence of H_2O and H_2S. Effects of H_2O and H_2S on these solvents physiochemical properties will be discussed. This study has established an understanding of the influence of precombustion contaminants, i.e., H_2O and H_2S, on solvents physiochemical properties, including CO_2 solubility, CO_2 absorption kinetics, viscosity, and so on.

Using PAT to Characterize Scale-up of Mixing Elbows Injectors for Continuous Crystallization of APIs

High shear mixing elbow injectors are used for crystallization processes to improve control of particle size (Polster 2013). Elbow injectors are similar to t-mixers and can be configured in a variety of ways to achieve mixing requirements. Poor mixing can lead to uncontrolled crystallizations with increased secondary nucleation. These mixing elbows can provide rapid controlled micromixing. Previous studies have been performed using the Fourth Bourne Reaction (Bourne 2003) to characterize mixing times (Singh, Spencer et al. 2005) in mixing elbow injectors. This work aims to further characterize mixing times in mixing elbows using the Fourth Bourne Reaction focusing on scale up and the configuration of the mixing elbow. This process was also used to further understand the role of different types of mixing within the elbow as the process is scaled up. A comprehensive lab scale study of a ¼" elbows was undertaken using the Fourth Bourne Reaction as well as Computational Fluid Dynamics. The possibility of using inline PAT, such as ATR-FTIR to monitor acetone concentration resulting from various conditions was investigated in order to increase the efficiency of the study as the scale of the process increases to 3/8" and ½" mixing elbow injectors. Inline PAT such as FBRM and ATR- FTIR is also essential for monitoring the process during case studies of model crystallization systems.

References

Bourne, J. R. (2003). "Mixing and the Selectivity of Chemical Reactions." ChemInform **34**(40): 471 - 508.

Polster, C. (2013). "Design and Scale-Up of the Continuous Reactive Crystallization of an API" AIChE Annual meeting, Pharmaceutical Discovery, Development and Manufacturing Forum (466c)

Singh, U. K., et al. (2005). "Incorporation and Characterization of a Mixing Elbow on the Pilot Plant Scale for a Mixing Sensitive Crystallization of an API." Industrial & Engineering Chemistry Research **44**(11): 4068-4074.

383588 How to Monitor Crystallization Processes in Manufacturing

Abstract:

The *in situ* measurement and imaging of crystals and crystal structures has become ubiquitous in many industrial laboratories during the development of crystallization processes. Standard tools such as ParticleTrack and PVM (formerly Lasentec) provide scientists with deep process understanding allowing them to develop higher quality processes for transfer to manufacturing.

In many cases it is desirable to apply the same tools at the manufacturing scale in order to compare performance from the lab to the plant and to continuously monitor and control the process for variability. Such an approach can also act as an insurance policy should a process fail, allowing a data-based root cause analysis to be performed and corrective measures to be implemented

This presentation will address some of the most common concerns and challenges associated with applying probe-based particle measurement techniques in large scale manufacturing.

- Building a business case for implementation of PAT in manufacturing
- Ensuring representative crystallization monitoring under large scale mixing conditions
- Working with plant managers to ensure effective installation, maintenance and data transfer
- Case studies in industrial crystallization monitoring

384304 Applications of Process Analytical Technologies in the Solid State Development of Pharmaceutical Compounds

Abstract:

In the recent years, process analytical technology (PAT) has been proven to be very useful on the pharmaceutical development and manufacturing for better process understanding and control. For example, the information of the in-situ particle size distribution by Lasentec ® FBRM is invaluable on the crystallization development and scale-up. This paper discussed some less common but very important PAT application cases on the solid-state development of pharmaceutical compounds, for example, 1) the use of on-line mass spectrometer to monitor the form conversion of a de-solvated solid form, 2) the use of the combination of IR and Raman to construct the phase diagram for the crystallization development of pharmaceutical co-crystals, 3) crystallization control using FTIR or Raman in the solvent swap crystallization.

379742 On-LINE Optimal Control of Crystal Properties in NON-Isothermal Antisolvent Crystallization

Abstract:

Controlling crystal properties has been a challenging issue in many crystallization operations due to inevitable disturbances, occuring during the process. This contribution proposes an online strategy to control crystal size distribution in joint cooling antisolvent crystallization systems.

As a first step in the study, digital image texturing analysis in the form of fractal dimension and energy signatures is applied and automated to track crystals' properties at different time of the experiment. The one dimensional Fokker-Planck equation (FPE) represents the dynamic characteristics of the crystal growth. A map of iso-mean and iso-variance curves are reported in an antisolvent flow rate-temperature plane based on evaluating the model at asymptotic condition. This allows identifying the region of input multiplicities in the system so punctual information regarding stability of the process can be obtained. Using controllability tools, regions of singularities needed to be avoided for proper control of the process is determined and it is further shown that the system is ill conditioned in the whole operating map. The problem is investigated experimentally where a multi-loop PID structure is tuned by pairing crystals' mean size with antisolvent feed rate and manipulating temperature to control the size variation. The proposed controller is demonstrated to be incapable of bringing the system into the desired set points and an offset in both crystals' mean size and variance is observed at the end of the batch. In the next step, a feed forward approach is developed in which temperature and flow rate trajectories are obtained from a static optimization procedure. The sensitivity of feed forward control strategy is assessed in the case of disturbances in antisolvent feed rate. Finally, in order to achieve a better performance a two stage controller is set up by combining the feed forward approach with a feedback controller which is designed based on IMC *(Internal Model Control)* algorithm. It is indicated that a perfect control over crystals' mean size can be obtained in this case by renouncing a strict control over size distribution coefficient of variation. Experimental validation of the strategies are carried out for the ternary system of water-ethanol-sodium chloride.

FORMULATION OF A NETWORK FOR THE SUSTAINABLE REDUCTION OF CO_2 EMISSIONS

R. Frauzem[a], P. Kongpanna[b], K. Roh[c], J. H. Lee[c], V. Pavarajarn[d], S. Assabumrungrat[b], R. Gani[a]

[a] CAPEC, Department of Chemical and Biochemical Engineering, Technical University of Denmark, Søltofts Plads, Building 229, Kongens Lyngby, Denmark
[b] Center of Excellence in Catalysis and Catalytic Reaction Engineering, Department of Chemical Engineering, Faculty of Engineering, Chulalongkorn University, Bangkok 10330, Thailand
[c] Department of Chemical and Biomolecular Engineering, Korea Advanced Institute of Science and Technology 291 Daehak-ro (373-1 Guseong-dong), Yuseong-gu, Daejeon 305-701, Republic of Korea
[d] Center of Excellence in Particle Technology, Department of Chemical Engineering, Faculty of Engineering, Chulalongkorn University, Bangkok 10330, Thailand

November 19, 2014

Overview

- Introduction & Objective
- Workflow
- Stage 1 - Network
- Stage 2 – Base Case Design
  1. Dimethyl carbonate (DMC) production
  2. Methanol synthesis via syn gas
  3. Methanol production via direct hydrogenation
- Stage 3 – More sustainable Design
  1. Dimethyl carbonate (DMC) production
  2. Methanol synthesis via syn gas
  3. Methanol production via direct hydrogenation
- Conclusions & Future Work

Energy Scenario

Increasing population & industrial development

↓

Increasing energy demand

↓

Increasing emissions

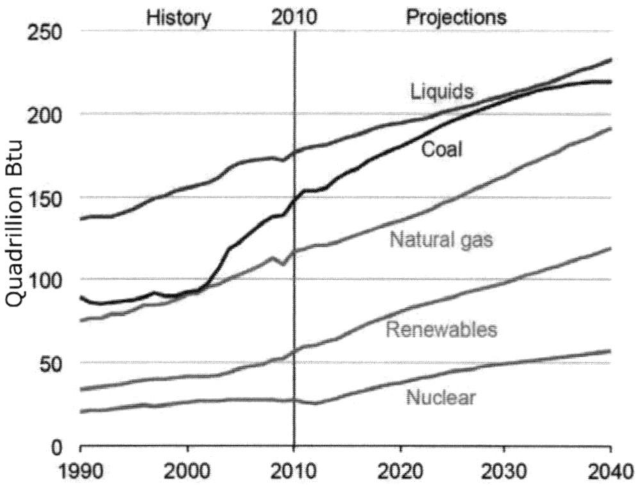

EIA, 2013. International energy outlook 2013, US Energy Information Administration (EIA .... EIA, Washington, DC.

Greenhouse Gases

- Greenhouse gases are considered primary contributors to global warming
- Global warming – the apparent increase in global temperature
- Most prominent is $\mathbf{CO_2}$

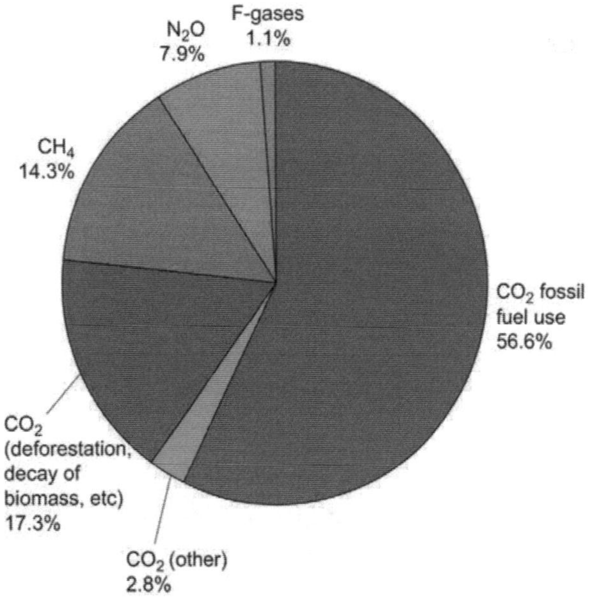

IPCC, 2007. Climate Change 2007: Mitigation of Climate Change. Cambridge University Press, New York.

CO_2 Mitigation

- 2 Methods of long-term reduction: Carbon Capture and Storage (CCS) and **Conversion**
- 2 targets for improvement of CCS
  1. Capture technology- reduction of additional energy, solvents or raw materials
  2. Storage methods- early stages of development
- Conversion involves use of chemical reactions to convert CO_2 to valuable product

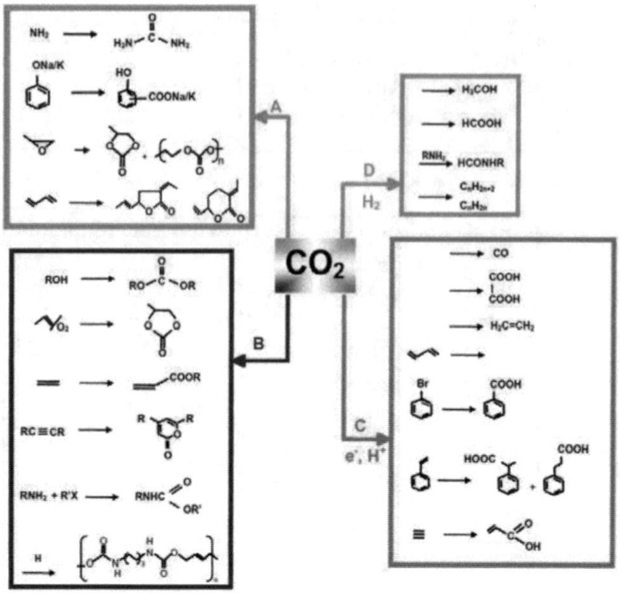

Aresta, M., 2010. Carbon Dioxide as Chemical Feedstock, Carbon Dioxide as Chemical Feedstock. doi:10.1002/9783527629916

Objective

- Environmental concerns show the need for a portfolio of CO_2 reduction methods
- Focus on conversion
- Formation of a conversion network subject to constraints
  1. Thermodynamically feasible reactions
  2. Non-trade-off solutions: reduced costs, increased profit, improved sustainability
  3. Overall CO_2 reduction

$$\text{Net } CO_2 = \Sigma CO_{2,generated} - \Sigma CO_{2,utilized}$$

  - Aim for negative value (more utilized in reaction that produced via emissions or energy requirements)

Workflow

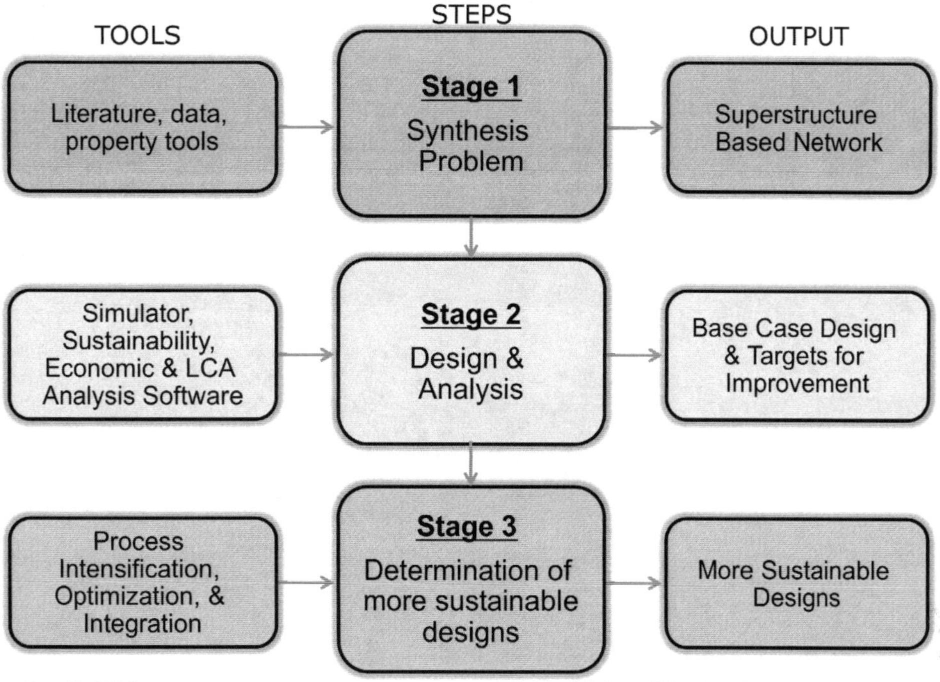

Superstructure based design

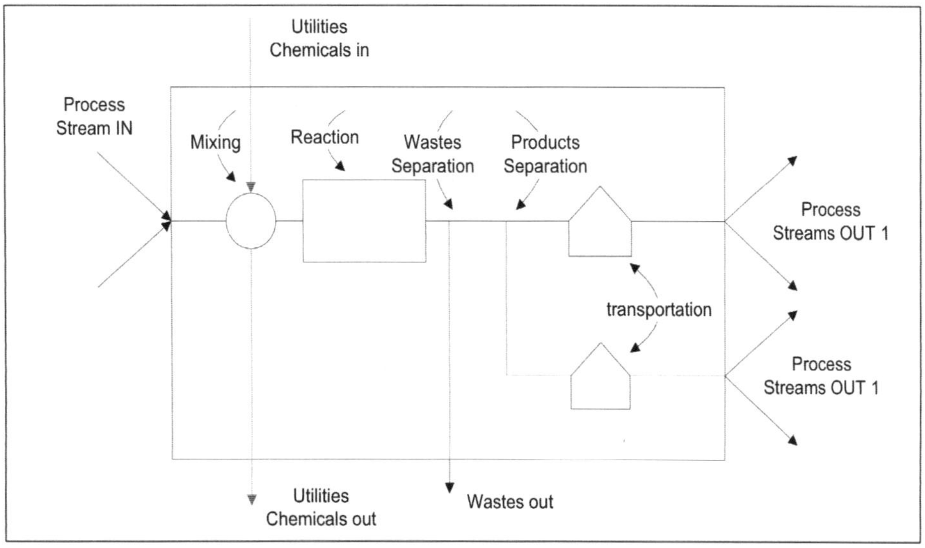

Quaglia, A., Sarup, B., Sin, G., Gani, R., 2012. Integrated business and engineering framework for synthesis and design of enterprise-wide processing networks. Comput. Chem. Eng. 38, 213–223. doi:10.1016/j.compchemeng.2011.12.011

Superstructure based design

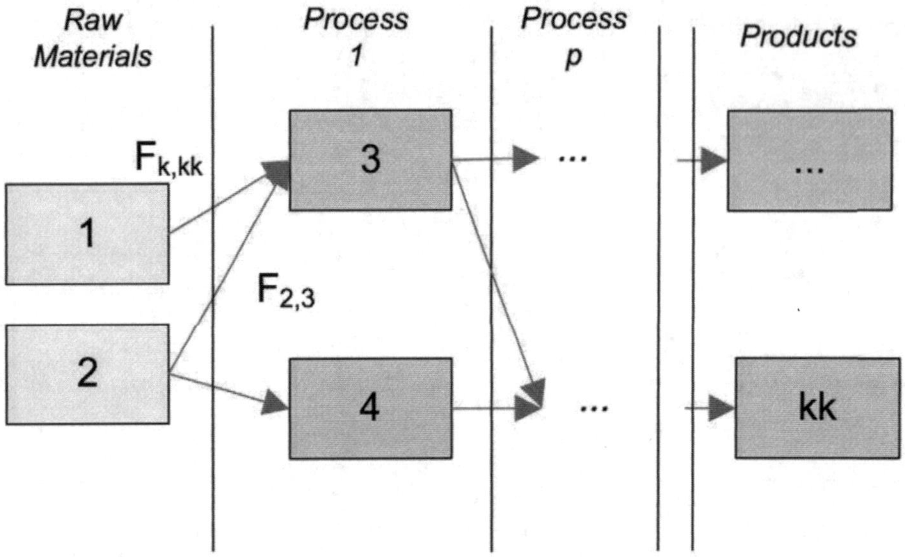

Quaglia, A., Sarup, B., Sin, G., Gani, R., 2012. Integrated business and engineering framework for synthesis and design of enterprise-wide processing networks. Comput. Chem. Eng. 38, 213–223. doi:10.1016/j.compchemeng.2011.12.011

Network of conversion processes

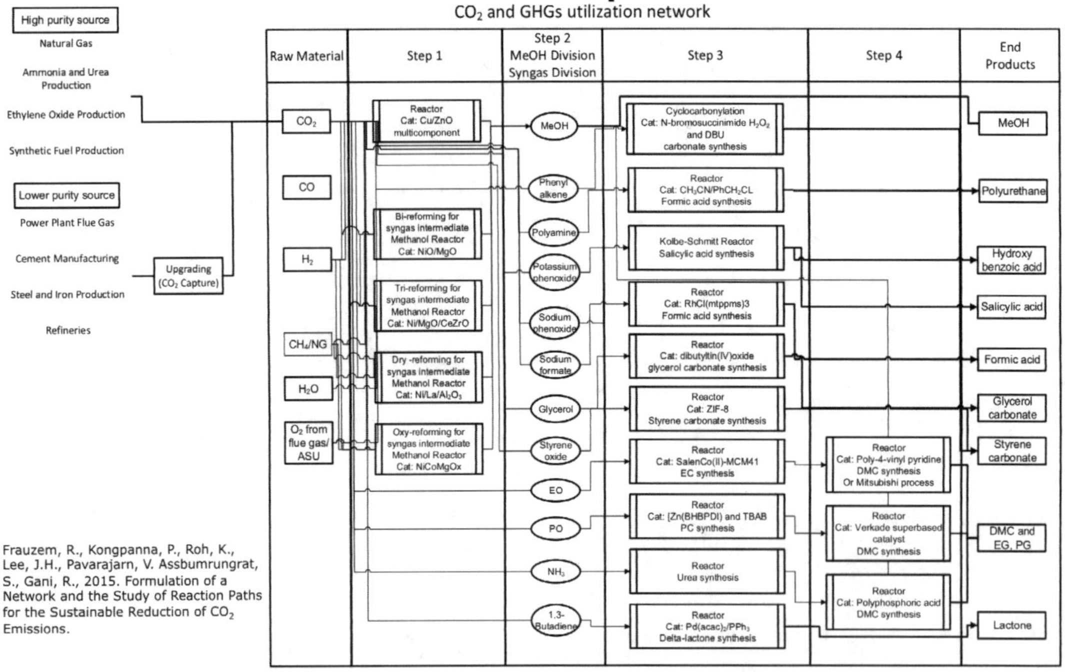

Frauzem, R., Kongpanna, P., Roh, K., Lee, J.H., Pavarajarn, V. Assbumrungrat, S., Gani, R., 2015. Formulation of a Network and the Study of Reaction Paths for the Sustainable Reduction of CO_2 Emissions.

Base Case Design

- This stage can only be done case-by-case
- Focus on 3 base cases
  1. Dimethyl carbonate (DMC) production
  2. Methanol synthesis via syn gas
  3. Methanol production via direct hydrogenation

1. DMC Production

- DMC is a secondary processing step in the network
  - Involves inputs from previous steps along with CO_2
- Base case from thermodynamic analysis – DMC via urea synthesis

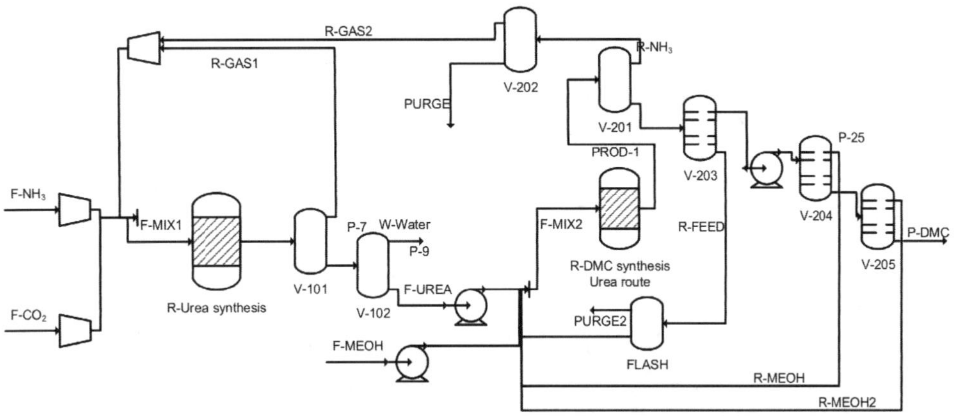

Kongpanna, P., Pavarajarn, V., Gani, R., Assabumrungrat, S., 2014. Techno-economic evaluation of different CO_2-based processes for dimethyl carbonate production. Chem. Eng. Res. Des. doi:10.1016/j.cherd.2014.07.013

1. DMC Production

- Analysis of simulation reveals a hot spot in the downstream separation
  - Process intensification performed
  - Targeted energy of downstream separation
  - Alternatives:
    1. Pervaporation
    2. Membrane reactor
    3. Reactive distillation

Babi, D.K., Woodley, J.M., Gani, R., 2014. Achieving More Sustainable Designs through a Process Synthesis-Intensification Framework. Comput. Aided Chem. Eng. 33, 31–36. doi:10.1016/B978-0-444-63456-6.50006-5

2. Methanol Synthesis via Syn Gas

- Methanol currently synthesized via synthesis gas
  - Two step process: reforming and synthesis
- 4 types of reforming possible: bi-reforming, tri-reforming, dry reforming, and oxy-reforming
- For this, base case is a bi-reforming (combined/mixed steam and dry reforming of methane) based process

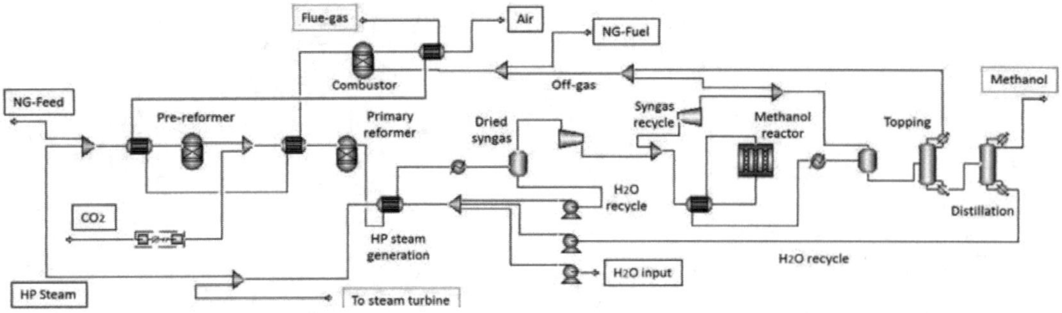

Holm-Larsen, H., 2001. CO_2 reforming for large scale methanol plants - an actual case. Stud. Surf. Sci. Catal. 441–446.

2. Methanol Synthesis via Syn Gas

- Reforming unit is energy intensive
  - Target optimization of reformer unit
  - Feed manipulation and operating conditions varied for reformer performed
    - Inject less steam, more CO_2
    - Increase reforming temperature
    - Find point for optimal pressure because of trade-off

Introduction | Workflow | Stage 1 - Network | **Stage 2 – Base Case Design** | Stage 3 – More Sustainable Designs | Conclusions

3. Methanol Production via Direct Hydrogenation

- Classical methanol synthesis via 2-stage process often results in high emissions (reforming requires high temperatures)
- Single-stage conversion promising
- Direct hydrogenation of CO_2 over a Cu/ZnO multicomponent catalyst

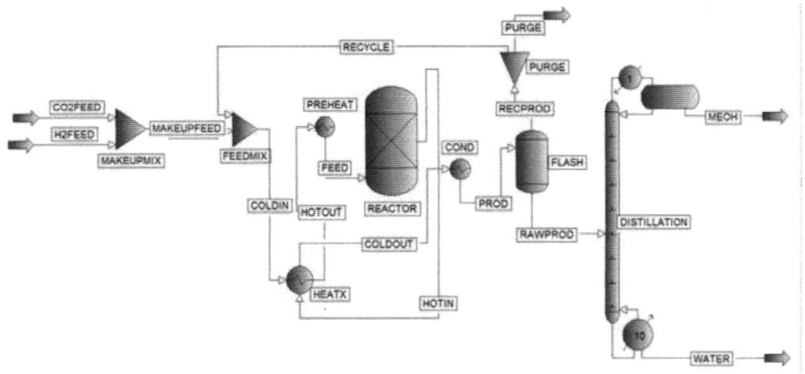

Ushikoshi, K., Mori, K., Kubota, T., Watanabe, T., Saito, M., 2000. Methanol synthesis from CO_2 and H_2 in a bench-scale test plant. Appl. Organomet. Chem. 14, 819–825. doi:10.1002/1099-0739(200012)14:12<819::AID-AOC90>3.0.CO;2-#

3. Methanol Production via Direct Hydrogenation

- For high selectivity, have low conversions resulting in a large recycle stream that is energy intensive
- Manipulation of feed, variation of operating conditions and integration of heat flow performed
- Heat intensification shows a **-1.125 $kgCO_{2,net}/kg_{MeOH}$** net CO_2

More Sustainable Design

- As with stage 2, stage 3 only possible for individual cases
- Take results from base case design and improve sustainability
- 3 cases:
  1. Dimethyl carbonate (DMC) production
  2. Methanol synthesis via syn gas
  3. Methanol production via direct hydrogenation

1. DMC Production

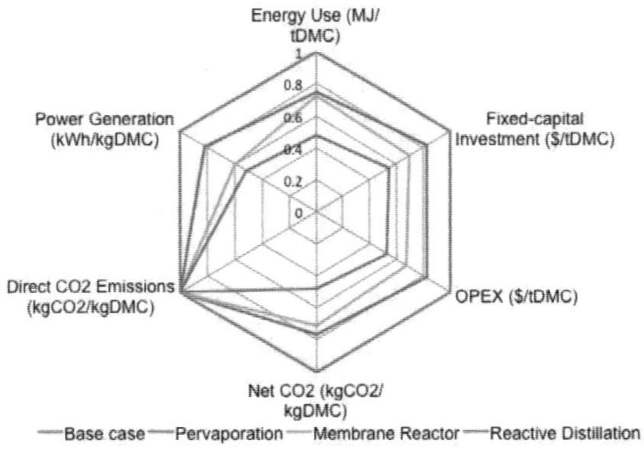

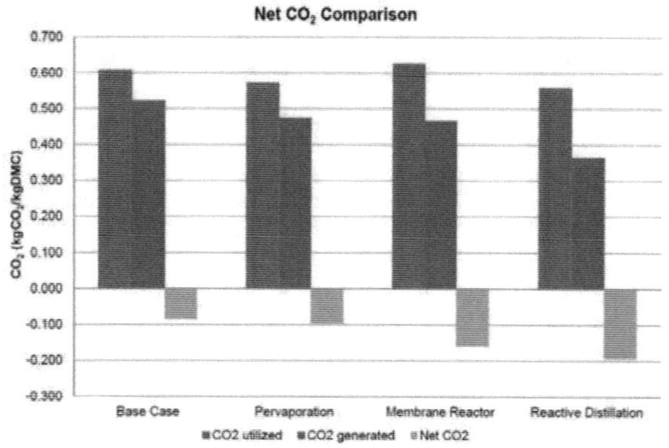

Kongpanna, P., Pavarajarn, V., Gani, R., Assabumrungrat, S., 2014. Techno-economic evaluation of different CO_2-based processes for dimethyl carbonate production. Chem. Eng. Res. Des. doi:10.1016/j.cherd.2014.07.013

2. Methanol Synthesis via Syn Gas

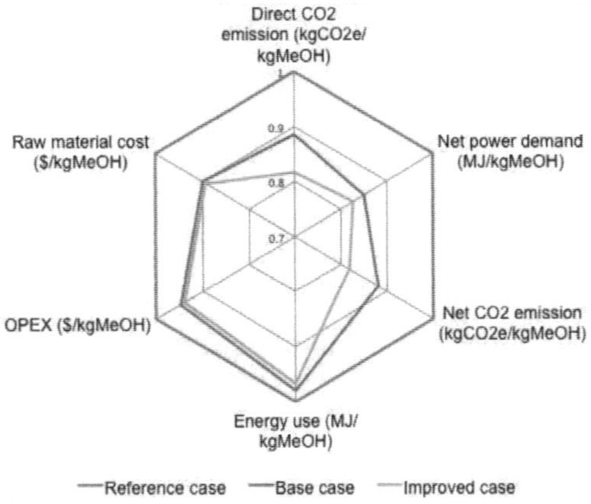

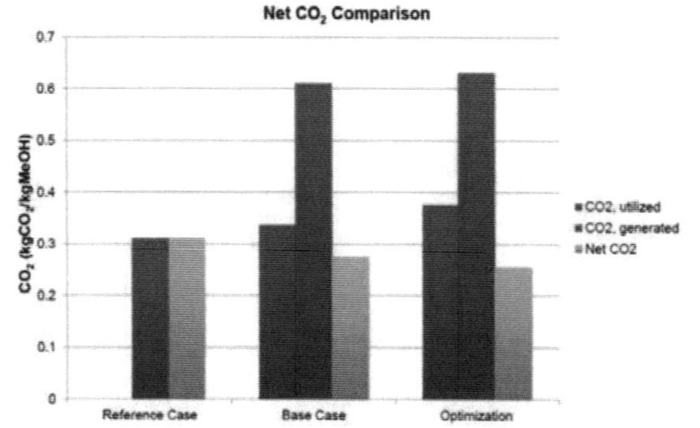

- In order to further improve sustainability, more work is being performed to create a zero or negative net CO_2 process

3. Methanol Production via Direct Hydrogenation

- Net CO_2:
  - Decreased with increasing impurities
  - Increased with increasing pressure and decreasing temperature

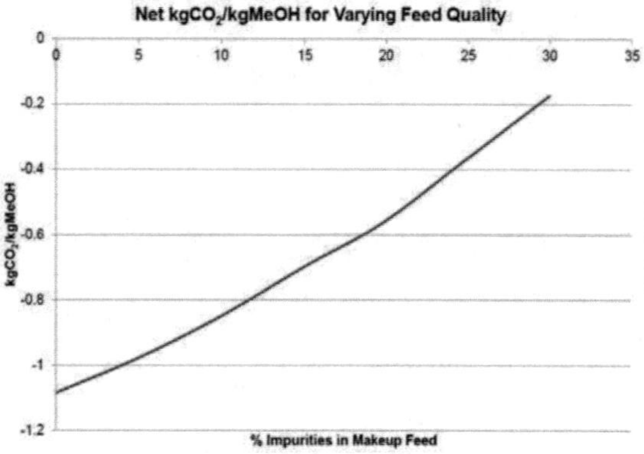

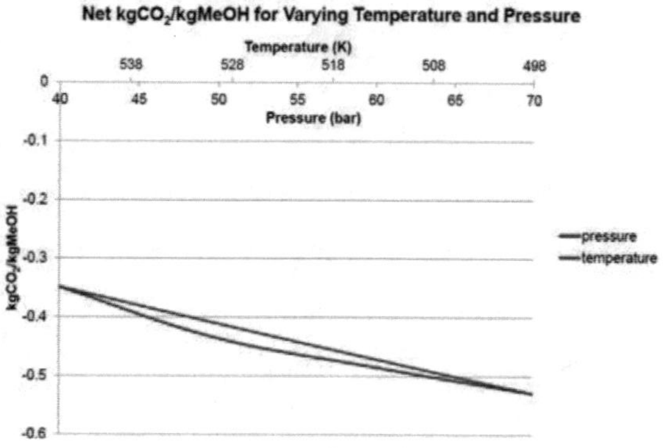

Conclusions & Future Work

DMC, intensified ($kgCO_2/kgDMC$)	MeOH from Syn Gas, optimized ($kgCO_2/kgMeOH$)	MeOH via Direct Hydrogenation ($kgCO_2/kgMeOH$)
-0.194	0.255	-1.125

- Following a superstructure based method, a network of CO_2 conversion processes is formed
- This network shows potential in reducing the amount of CO_2 emitted while producing a portfolio of products
- For the second and third steps, case studies have been performed on 3 processing steps in the superstructure based network
- Further work is still necessary:
  - Elaboration of the network
  - Continuation of base case simulations and sustainability improvements for the processing steps

Works Cited

- Aresta, M., 2010. Carbon Dioxide as Chemical Feedstock, Carbon Dioxide as Chemical Feedstock. doi:10.1002/9783527629916
- Aresta, M., Dibenedetto, A., Angelini, A., 2013. The changing paradigm in CO_2 utilization. J. CO2 Util. 3-4, 65–73. doi:10.1016/j.jcou.2013.08.001
- Babi, D.K., Woodley, J.M., Gani, R., 2014. Achieving More Sustainable Designs through a Process Synthesis-Intensification Framework. Comput. Aided Chem. Eng. 33, 31–36. doi:10.1016/B978-0-444-63456-6.50006-5
- Dincer, I., 2010. Global Warming, Green Energy and Technology. Springer US, Boston, MA. doi:10.1007/978-1-4419-1017-2
- EIA, 2013. International Energy Outlook 2013, US Energy Information Administration (EIA). Washington, DC.
- Frauzem, R., 2014. Techno-economic evaluation of a CO_2 -conversion process.
- Frauzem, R., Kongpanna, P., Roh, K., Lee, J.H., Pavarajarn, V. Assbumrungrat, S., Gani, R., 2015. Formulation of a Network and the Study of Reaction Paths for the Sustainable Reduction of CO_2 Emissions. In progress.
- Gani, R., 2004. Chemical product design: challenges and opportunities. Comput. Chem. Eng. 28, 2441–2457. doi:10.1016/j.compchemeng.2004.08.010
- Holm-Larsen, H., 2001. CO_2 reforming for large scale methanol plants - an actual case. Stud. Surf. Sci. Catal. 441–446.
- IPCC, 2005. Carbon Dioxide Capture and Storage, Intergovernmental Panel on Climate Change (IPCC). Cambridge University Press, New York.
- IPCC, 2007. Climate Change 2007: Mitigation of Climate Change, Intergovernmental Panel on Climate Change (IPCC). Cambridge University Press, New York.
- Knudsen, J.N., Jensen, J.N., Vilhelmsen, P., 2007. First year operation experience with a 1 t / h CO_2 absorption pilot plant at Esbjerg coal-fired power plant, in: European Congress of Chemical Engineering (ECCE-6). Copenhagen.
- Kongpanna, P., Pavarajarn, V., Gani, R., Assabumrungrat, S., 2014. Techno-economic evaluation of different CO_2-based processes for dimethyl carbonate production. Chem. Eng. Res. Des. doi:10.1016/j.cherd.2014.07.013
- Kubota, T., Hayakawa, I., Mabuse, H., Mori, K., Ushikoshi, K., Watanabe, T., Saito, M., 2001. Kinetic study of methanol synthesis from carbon dioxide and hydrogen 121–126.
- Last, G., Schmick, M., 2011. Identification and Selection of Major Carbon Dioxide Stream Compositions. Pacific Northwest National Laboratory, Oak Ridge.
- Olah, G.A., Goeppert, A., Prakash, G.K.S., 2009. Beyond Oil and Gas: The Methanol Economy. Wiley-VCH, Weinheim.
- Quaglia, A., Sarup, B., Sin, G., Gani, R., 2012. Integrated business and engineering framework for synthesis and design of enterprise-wide processing networks. Comput. Chem. Eng. 38, 213–223. doi:10.1016/j.compchemeng.2011.12.011
- Saito, M., Fujitani, T., Takeuchi, M., Watanabe, T., 1996. Development of copper/zinc oxide-based multicomponent catalysts for methanol synthesis from carbon dioxide and hydrogen. Appl. Catal. A Gen. 138, 311–318. doi:10.1016/0926-860X(95)00305-3
- US Department of State, 2010. U.S. Climate Action Report, US Department of State (US DOS). US DOS, Washington, DC.
- Ushikoshi, K., Mori, K., Kubota, T., Watanabe, T., Saito, M., 2000. Methanol synthesis from CO_2 and H_2 in a bench-scale test plant. Appl. Organomet. Chem. 14, 819–825. doi:10.1002/1099-0739(200012)14:12<819::AID-AOC90>3.0.CO;2-#
- Wilcox, J., 2012. Carbon Capture. Springer, New York. doi:10.1007/978-1-4614-2215-0
- Zakkour, P., Cook, G., 2010. CCS Roadmap for Industry : High-purity CO_2 sources. Carbon Counts Company (UK) Ltd.
- Zondervan, E., Nawaz, M., de Haan, A.B., Woodley, J.M., Gani, R., 2011. Optimal design of a multi-product biorefinery system. Comput. Chem. Eng. 35, 1752–1766. doi:10.1016/j.compchemeng.2011.01.042

383332 Spatial Decomposition Techniques for Strategic and Tactical Design of Biomass-to-Co-Combustion Supply Chains

Abstract:

Spatial decomposition techniques for strategic and tactical design of biomass-to-co-combustion supply chains

Mar Pérez-Fortes, José Miguel Laínez-Aguirre, Luis Puigjaner

Department of Chemical Engineering, Universitat Politècnica de Catalunya, Av. Diagonal, 647, E08028 Barcelona, Spain

The electricity generation sector needs to reduce its environmental impact and fossil fuel dependence, principally from coal. In terms of time to deployment and cost of the solution, co-combustion of biomass and coal is presented as a transition alternative towards a carbon-neutral energy sector. However, the use of biomass in large centralized systems requires the establishment of supply channels to provide the desired feedstock with the necessary attributes, at the right time and place. Hence, there is the need to evaluate the effective introduction of co-combustion projects in the current electricity production share. The purpose of this work is to investigate the application of an optimization-based supply chain (SC) design-planning approach to process seasonal and highly distributed biomass for raw material supply to an existing park of coal power plants. The problem is formulated as a multi-objective mixed integer linear program (MO-MILP); we integrate economic and environmental criteria: the net present value (NPV), measured in $€_{2010}$, and Impact 2002+ methodology with its corresponding metric for a life cycle assessment (LCA).

Special consideration is given to biomass intrinsic heterogeneity (due to its different origins), high moisture content, low bulk density, low heating value (LHV), degradation and seasonality. Such characteristics make it difficult its direct use as a coal substitute. However, biomass may be improved when needed, using a pre-treatment in order to enhance its stability in front of degradation and moisture content variations. Torrefaction, torrefaction combined with pelletization, pelletization, fast pyrolysis and fast pyrolysis combined with char grinding compose the superstructure of pre-treatment technologies considered in this work.

The starting point is a previous mathematical model for the biomass SC (Peréz-Fortes et al. 2014), which has been solved for a planning horizon of 10 years, with yearly strategic decisions and monthly tactical planning decisions. The model considers the following decisions: (i) selection of SC nodes and links among them (a node can be represented by processing or distribution activities), (ii) selection of the most cost-effective or environmentally friendly pre-treatment technologies, (ii) the corresponding capacity to install each planning period, (iii) the capacity utilization levels, (vi) fraction of coal replaced by biomass, and (vii) the amount of matter distributed along the different SC nodes.

The amount of potential suppliers and facility locations, the different technological alternatives, biomass properties characterization and the model granularity, can significantly increase the size of the optimization problem. Preliminary results demonstrate that this computational expense can be significantly reduced by using a decomposition approach. To tackle more efficiently this problem, we propose an adapted Lagrangian relaxation based approach, the Optimal Condition Decomposition (OCD), proposed in Conejo et al. (2002). One of the advantages of the OCD is that the procedure to update the multipliers is embedded in the solution of the sub-problems, thus avoiding the solution of a master problem and allowing the algorithm convergence in fewer iterations. These promising characteristics are evaluated and compared using a real large scale problem of retrofit design of biomass-to-co-combustion supply chains in Spain (Peréz-Fortes et al. 2014). Significant results obtained are discussed and future development perspectives underway are also presented.

Keywords: Bio-based supply chain, biomass pre-treatments, biomass seasonality, co-combustion, MILP, Lagrangian relaxation.

References

Conejo, A. J.; Nogales, F.J., 2002, A decomposition procedure based on approximate Newton directions Mathematical Programming. Mathematical Programming, Series A, 93, 495—515.

Gómez, A., Zubizarreta, J., Rodrigues, M., Dopazo, C., Fueyo, N., 2010, An estimation of the energy potential of agro-industrial residues in Spain. Resources, Conservation and Recycling, 54, 972-984.

Pérez-Fortes, M., Laínez-Aguirre, J.M., Bojarski, A.D., Puigjaner, L., 2014, Optimization of pre-treatment selection for the use of woody waste in co-combustion plants, Chemical Engineering Research and Design, article in press: http://dx.doi.org/10.1016/j.cherd.2014.01.004

Techno-Ecological Synergy: A Framework for Sustainable Engineering

Bhavik R. Bakshi
The Ohio State University

Guy Ziv
University of Leeds

Varsha Gopalakrishnan
The Ohio State University

May 12, 2014

Abstract

Even though there is little agreement on a definition of sustainability, defining unsustainability is much easier: it is a state where the impact of human activities exceeds nature's carrying capacity. Avoiding such a situation is even included in many definitions of sustainability. However, this carrying capacity is rarely considered in methods for sustainable engineering, including life cycle assessment, and methods for designing sustainable chemical processes. Ecological footprint does account for biocapacity, but in a highly aggregated manner that is not very useful for guiding decisions. Given the degradation of ecological processes highlighted in several studies [1, 2], accounting for the dependence and impact of human activities on ecosystems is an urgent need.

We propose a novel framework for Sustainable Engineering that explicitly accounts for the demand for ecosystem services generated by human activities, and the supply of ecosystem services available from nature at multiple spatial scales. The demand for ecosystem services is quantified by information about resource consumption and emissions. For example, the demand for the water provisioning ecosystem service is quantified by the volume of water withdrawn for an activity, while the demand for the carbon sequestration ecosystem service is quantified by the quantity of CO2 emitted. The supply of ecosystem services is determined from knowledge about available ecosystems at the selected spatial scale. Thus, forest cover and species determine the ability to regulate air quality by taking up pollutants such as nitrogen and sulfur oxides and carbon monoxide. Information about tree cover, soil quality and management, etc. can quantify the supply of the carbon sequestration ecosystem service. Models for quantifying various services are available from ecologists at multiple scales, including local, regional and global. They form the basis of the proposed framework.

In the TES framework, each technological system is represented as $T_{i,j}$ which consists of the i-th technological process at the j-th scale. These could be individual manufacturing processes or average processes obtained from a life cycle inventory database, or an economic sector in a regional or national economic input-output model. Each technological system is nested inside ecosystems in its vicinity, that is $T_{i,j} \in E_{i,j}$, where $E_{i,j}$ represents ecological systems in the vicinity of $T_{i,j}$. Thus, if the smallest technological scale includes manufacturing processes, then the smallest ecological scale could be the plant site or the corporate campus; if the technological system is an single-family home, the smallest ecological scale would be the yard around the house. This smallest scale is then expanded to consider regional, national, and global scales. The largest ecological scale that should be considered depends on the ecosystem service being analyzed. A service such as carbon sequestration is global in nature because of the global flow of CO_2, which means that for sustainability analysis, its assessment needs to be at the global and smaller scales. In contrast, a service such as Pollination is local and determined by the range covered by pollinators. Thus, this service is not relevant at larger geographical scales.

Results from a TES study consists of a pair of numbers $\{D_{i,j}, S_{i,j}\}$ representing demand and supply for each ecosystem service, $i = 1, \ldots, I$, and at each ecological scale, $j = 1, \ldots, J$.

These demand and supply numbers may be used to define sustainability metrics to compare alternatives and as objectives for designing sustainable systems. The degree of overshoot for the i-th service may be defined as,

$$V_{i,j} = \frac{S_{i,j} - D_{i,j}}{S_{i,j}} \tag{1}$$

A necessary, but not sufficient condition for absolute sustainability is,

$$V_{i,J} \geq 0, \ \forall i \tag{2}$$

Thus, for each ecosystem service, the demand cannot exceed the supply at the largest scale, $j = J$. This condition is not sufficient for sustainability since other factors such as dynamics and complexity also matter but are not captured in the current framework. If demand does exceed the supply, that is, if $V_{i,J} < 0$ then such a situation usually results in symptoms such as depletion of ground water, accumulation of CO_2 in the atmosphere, or of nutrients in water bodies. The criterion given by Equation 2 may also be applied at smaller scales. If

$$V_{i,j} \geq 0, \ j < J \tag{3}$$

then, it means that dependence on the i-th ecosystem service is *locally* sustainable at scale j. It could happen that Equation 3 is satisfied, while Equation 2 is violated. For example, a process may emit less CO_2 than can be sequestered by its surroundings to be locally sustainable, but the electricity it buys from outside the process boundary may emit more CO_2 than can be sequestered by the ecosystems at the larger scale.

This framework may require allocation of a regional ecosystem services among multiple activities that demand it. Two approaches are explored: allocation based on quantities such as monetary value of the activity or its emissions, and no allocation. The first approach is analogous to what is used in LCA and is subjective. The second approach is easier to use. It implies that if at the selected scale use of an ecosystem service is unsustainable, then all activities in that region that rely on the selected service will be considered to be unsustainable. For example, because the global demand for carbon sequestration far exceeds the global supply of this service, all activities that demand this service are considered to be unsustainable at the global scale.

Some benefits of the TES framework include the following. 1) The goal of reducing ecological overshoot for individual ecosystem services at multiple scales will encourage enhancements in technological efficiency, along with ecological protection and restoration. Existing methods mainly encourage the former. 2) If human activities demand services that ecosystems simply cannot provide, then the only way to satisfy the TES goal of $V_{i,j} \geq 0$ is by technological efficiency that eliminates this demand. This will discourage use of nonrenewable resources and encourage cycling of molecules that do not occur or are rare in nature. 3) The TES approach favors systems that have a smaller overshoot with respect to ecosystems at the selected scale. This approach encourages integrated design of technological and ecological systems such as that of a building and its surrounding land [3], and of a manufacturing process and ecosystems on available land [4]. 4) If $V_{i,j} < 0$, then the impact of emissions and resource use will need to be determined by various impact assessment methods. In contrast, if $V_{i,j} \geq 0$ then the system is locally sustainable and there is no need for applying methods such as LCA. Thus, a company could focus on achieving local sustainability, and only including locally sustainable manufacturers in its life cycle. In such a situation, methods such as LCA become unnecessary. These and other characteristics of the TES framework will be demonstrated with the help of a case study.

References

[1] MA. *2005 Millennium Ecosystem Assessment (MEA) , Ecosystems and Human Well-being: Synthesis,*. Island Press, 2005. `www.maweb.org`, Accessed December 23, 2013.

[2] J. Rockstrom, W. Steffen, K. Noonel, A. Persson, F. S. Chapin, E. F. Lambin, T. M. Lenton, M. Scheffer, C. Folke, H. J. Schellnhuber, B. Nykvist, C. A. de Wit, T. Hughes, S. van der Leeuw, H. Rodhe, S. Sörlin, P. K. Snyder, R. Costanza, U. Svedin, M. Falkenmark, L. Karlberg, R. W. Corell, V. J. Fabry, J. Hansen, B. Walker, D. Liverman, K. Richardson, P. Crutzen, and J. A. Foley. A safe operating space for humanity. *Nature*, 461(7263):472–475, 2009.

[3] R. A. Urban and B. R. Bakshi. Techno-ecological synergy as a path toward sustainability of a North American residential system. *Environmental Science & Technology*, 47(4):1985–1993, 2013.

[4] Varsha Gopalakrishnan, Bhavik R. Bakshi, and Guy Ziv. Innovative and sustainable process design by incorporating ecosystems as unit operations. In *AIChE Annual Meeting*, Atlanta, GA, November 2014.

383884 Multistage Fuzzy Decision-Making for Sustainability Performance Improvement

Abstract:

Sustainability is a vital issue in pursuing the short-to-long-term development of industrial organizations and systems. One of the main challenges in developing development strategies and plans is how to deal with uncertainties that could appear in various stages of a decision making process. The uncertainties can be in general classified into two types: aleatory and epistemic. Aleatory uncertainty refers to the inherent variation associated with the physical system or the environment under consideration, and it is objective and irreducible. By contrast, epistemic uncertainty is caused by the lack of knowledge or information, and it is subjective and reducible. Fuzzy logic based arithmetic appears an effective method to handle uncertainties associated with various types of sustainability problems. Fuzzy logic theory is a mathematical system in which rigorous logical mathematics are used to deal with uncertain information and data that are difficult to compute using conventional mathematics.

In this paper, we introduce a mathematical framework for multistage hierarchical fuzzy logic-based decision-making when solutions need to be derived for sustainability performance improvement. It aids in the determination and selection of decisions in different decision stages that must be implemented in order to achieve an improved state of sustainable development in the future. The methodology can be applied in two scenarios: forward and backward decision making. In forward decision making, for a given state of sustainability and known resources available, it is to determine how to achieve the most sustainable state at multiple time intervals (i.e. multistage) into the future. In backward decision making, for a sustainability goal set to achieve in a target time frame, it is to determine the best development pathway from a given current stage of sustainability at the lowest cost. This methodology is computationally very efficient, suitable for analyzing the sustainability status of any system, or predicting the system short-to-long-term behavior in the scope of sustainability under uncertainty. The methodology has been applied to an automotive focused industrial zone case study to quantify and analyze the state of industrial sustainability, and determine recommendations for future development that would increase the values of the indicators identified as promoting or decrease the values of those identified as impeding sustainability can be made.

Bio-based Bottles – Recyclable but more Ecological?

Merten Morales[a,*], Sven Linster[a], Sara Badr[a], Stavros Papadokonstantakis[a], Konrad Hungerbühler[a]

[a]Institute for Chemical and Bioengineering, ETH Zurich, CH-8093 Zurich, Switzerland

merten.morales@chem.ethz.ch

1. Background + Aims

The introduction of a new generation of biorefineries into the market presents a multi complex problem, where questions like environmental sustainability, economic feasibility and conceptual exploration need to be answered beforehand. The biorefineries of the second generation need to tackle the objective of how chemicals can be produced from biomass, and not compete with the food production. Lignocellulosic biomass presents a raw material with a high potential to replace corn or sugarcane. Innovative production routes from wood to chemicals have to be included within the biorefinery concept. The quest of finding the optimal biorefinery concept is becoming even more complex by considering besides the conventional also innovative processes for the production of chemicals. The approach for finding the optimal concept is that generic optimization algorithm, developed by EPFL in Lausanne, is able to integrate different processes in order to find a biorefinery concept under the objective of profit maximization and environmental impact minimization.
A major part of the research project is the modelling of the different production paths to produce either platform chemicals (e.g. Glucose, Ethylene, Succinic acid) or high value added chemicals (Polyurethanes, Polyesters, Vinyl polymers). For the production routes conventional concepts as well as innovative concepts are considered for finding the optimal production routes. The concepts are evaluated through environmental indicators (Global Warming Potential, Ecoindicator 99, Cumulative Energy Demand) and economic indicators (Profit, Operational costs, Investment cost).

2. Methods + Results

The usual approach for a biorefinery concept is the replacement of petrochemicals by chemicals, which can be produced from renewable resources. Another approach is the production of entirely new monomers based on bio based materials. The monomer 2,5-furandicarboxylic acid (2,5- FDCA) can serve as an replacement for terephthalic acid. The promising polymer poly(ethylene furandicarboxylate) (PEF) can be produced from 2,5-FDCA. PEF is seen as a possible replacement for PET. Regarding to thermal properties, PEF is more interesting than PEF as it has higher thermal stability combined with lower processing temperature. Furthermore PEF is recyclable with the common machines, which are used for PET. For realizing the production of PEF a key chemical is 5-hydroxymethylfurfural (HMF). For the production of PEF different processes were modelled and optimized:

1. Glucose production from wood by acidification
2. Fructose production by Glucose isomerization [1]
3. HMF production by dehydration of Fructose [2]
4. 2,5-FDCA production through oxidation of HMF
5. PEF production by polymerization of 2,5-FDCA with ethylene glycol

The different production steps were evaluated to answer the question if plastic bottles made of PEF are more ecological friendly or economical feasible than from PET.

3. Summary/Conclusion

In this research project the production from alternative plastic bottles made of PEF were evaluate from an environmental, economic as well as from a safety point of view. As an outcome of this study the main environmental burden in the production route was pointed out.

References
[1] Method of Producing High Fructose Corn Syrup from Glucose using Noble Gases, 5,593,868, 1997
[2] Method and Apparatus from preparing Hydroxymethylfurfural, US 2012/0016141 A1, 2012

358083 New Systematic Approach Using Combined Constraints Logic Propagation and Mathematical Programming Techniques for Energy Efficient Synthesis of Eco-Industrial Parks

Abstract:

Industrial symbiosis adaptation on the road to having real Eco-industrial parks (EIP) can serve a significant role in realizing economic, environmental and social benefits to both industrial and non-industrial communities.

Even though, Eco-industrial Parks are often publicized as a mean of reducing environmental damage through reduced waste; the major driving factor for the creation of most eco-industrial parks has been to attain financial gain. Industrial clusters played a significant role in the economic growth of many countries. Industrial ecology concept may promote a new path of local development through the transition from industrial clusters to EIPs (eco-transformation). It is the best exploitation of the geographic proximity of time-dependent and non-dependent plants and the non-industrial community of malls, hospitals, hotels, housing compounds, schools and so on, that make the eco-transformation of industrial complexes financially warranted. Implementing Eco-Industrial Parks' principles in an existing industrial cluster represents a significant opportunity for its revitalization. IEP could exploit synergies from industrial clusters and non-industrial activities to easily create new production models in which the economic and environmental dimensions are symbiotic.

Previous EIPs key success has been a sequence of independent economically driven actions. Such evolutionary pattern followed by countries like Denmark for instance, may not be easily transferred to other eco-transformation potentials. The authors in this paper present a holistic approach in addressing the problem, using new concept of sharing, method, and retrofit tools.

The literatures show that it is very difficult, with the current state-of-art methods and tools, to manufacture EIPs to work from scratch. First, there should be the basic ingredients in place, namely the desire of firms and communities to actively participate and the correct membership as well as structure of firms. These basic ingredients can then be enhanced and improved upon, with correct support structure. A major factor that can enhance the success of EIP synthesis or an industrial complexes eco-transformation is the presence of a large firm which acts as a magnet for other ones. From mass and energy integration points of views, the key to developing a successful EIP or eco-transformation applications is to determine the best connections among different industrial plants and its surrounding communities. Specifically it is the material and energy flows relationship among the different members in the alliance of plants or firms which permit establishing optimal linkage to form a fruitful inter-dynamic structure. If such structure does not exist a successful EIP would not be realized and the eco-transformation of an industrial complex would not be worth exploring. This paper addresses the energy component of EIP via novel system approach.

The current approach for the eco-transformation or the planning of new EIPs for energy efficiency maximization is the export of waste heat from one plant to a nearby another. In other words, each firm in the eco-industrial park alliance allows the usage of its waste energy to be another adjacent firm. The energy waste of one company is used partially or totally in another adjacent one. It is merely a sort of cooperation through the waste integration to save transportation cost and energy degradation during transit.

The current technical sharing methods are essentially three. The first one is the ad hoc method of sharing in which an obvious waste heat stream from a power plant, for instance, is used in a nearby process. This method is not systematic and not efficient where huge amount of energy efficiency opportunities will be left untouched.

The second method is the total site based pinch technology where the waste heat from all processes can be theoretically considered as a source of heat in other processes. The waste heat sources are converted first to steam, and then passed to processes that are in heat deficit through the steam system infrastructure. To identify the external heating and cooling requirements of a group of individual plants from the central utility system, the temperature/enthalpy data from individual plants are first required to be extracted from the plant after the thermal integration of its hot streams to be cooled and cold streams to be heated using individual plant' grand composite curve. This curve defines each plant' thermal heat deficiency and thermal heat surplus after intra-plant heat integration. The collection of grand composite curves of the whole site are then used to graphically add all thermal deficiencies to draw the total site heating demand curve, and add all thermal surpluses to draw the total site cooling demand curves. Such two curves are superimposed on one graph with the existing and/or suggested steam generation levels and steam supplying levels to find the minimum total site combined heating and power (CHP) external energy utilities requirement and naturally best indirect inter-plants thermal integration. In this method, intra-integration is done first, which means only waste heat of one plant is shared with other parks' members (which is not proactive form of cooperation). In this method, the mismatch in number of steam levels required for EIP users in generation and utilization result in energy loss. This method while in literature addresses in its application both time non-dependent and dependent sites processes, the use of time as an optimization variable for hybrid inter-time-inter-systems energy integration is ignored. It also ignores the integration of the combined cooling and power generation/consumption (CCP) system with the total site. The third method uses mathematical programming but with many assumptions to be able to model the whole city' industrial and non-industrial processes without having mathematically an intractable problem. It can theoretically speaking find best mass and energy integration linkage among the EIP members and design the optimal energy utility systems. However, up to our knowledge there is no public domain literature for any application of such method in the retrofit and/or planning of energy efficient EIPs.

From the EIP synthesis and/or eco-transformation tools, software, we do not have much. The most famous software that can be partially used is the aspentech company total site commercial software. Other US EPA' decision support program products/software for Eco-industrial Park planning in general, namely "FaST"; "DIET" and "REaLiTy", are data base software with linear programming capability. These software are essentially focused on material exchange and the very obvious waste heat exchange; where a waste heat stream in one plant/power station is used in other eco-industrial park' plants.

The well planning of new EIP or industrial complexes eco-transformation can brings significant value to energy efficiency, it is about more than twenty percent consumption saving. However, the problem is mathematically huge multi-variable multi-dimensional optimization problem in which the eco-industrial park network depends on as small as stream condition and as big as the whole park functionality. Integration among many industrial plants and non-industrial ones, in adjacent geographical locations, can bring in more degrees of freedom to optimize the "waste energy recovery problem" and consequently presents new horizon to the radical energy-based GHG emissions reduction to levels never thought of before.

It is true that we still face many arguments against the eco-transformation applications such as the processes are having different start up and shut down times. Besides the processes can work at partial loads and the processes can have seasonal changes in its conditions. From capital point of view we all know that in eco-transformation cases utility systems, heaters and heat exchangers network, coolers capital will not be reduced. On the processes control side the disturbance in one process can propagate to another one if they are integrated which make the process difficult to control and the distance-time known as velocity lags affect the controllability of processes. Last but not least, the geographical distances among firms will cost us energy in pumping or compression and capital in piping, pumps and compressors beside safety issues due to the travel of a fluid from one area to another as well as the fear of leakage. However, many of the practical reasons hindering the application to date, even though they are valid, most of it can be addressed in a novel cost-effective way.

This paper presents new concept, method and tool for total site energy integration which includes many plants and utilities system to attain new levels of energy saving and GHG emissions reduction. The paper uses hybrid, direct and indirect, method that systematically looks to all options together and find best combinations out of the available solutions package while considering both inter-time-zones-inter-systems energy integration. This paper is a continuation-in-part to previous pinch-based methods for enhancing energy efficiency via systematic inter-processes integration as a preliminary screening tool. However, our new method includes new steps for systematically identifying hybrid direct and indirect inter-systems-inter-time-zones matching solutions through selecting best energy efficient routes; generating technically viable energy efficient eco-industrial parks alternatives, identifying best generation and allocation of energy utilities and synthesizing the utility systems which satisfies the eco-park demands during each time zone as well as rendering its best operating scenario at each specific time-zone. The paper identifies using a simple constraints logic propagation model the best and the second best and so on matching among plants. The new systematic method for membership uses constraints logic propagation (CLP) and the mathematical programming model is used to find best cost-effective detailed solutions. The method and the models are illustrated by two industrial scale case studies. The first one is a typical state-of-art oil refinery and the second one is an integrated multi-generation gasification-based facility.

***References:**

1. Dhole V.; Linhoff B., "Total site targets for fuel, cogeneration, emissions, and cooling" Computers and Chemical Engineering, 17,(suppl.):S101-9, 1993
2. Terry Tudor, Emma adam, and Margaret Bates, "Drivers and Limitations for successful development and functioning of EIPs (eco-industrial parks): A Literature review", Ecological Economics, 61, (199-207), 2007
3. P.Varbanov and J.Klemes,"Total site integrating renewables with extended heat transfer and recovery", Heat Transfer Engineering, 31(9), (733-741), 2010
4. Tom Maes, et al.,"Energy management on industrial parks in Flanders", Renewable and sustainable Energy Reviews, 15,(1988-2005), 2011
5. Seok Jung, et al., "A novel approach for evaluating the performance of eco-industrial park pilot projects", Journal of cleaner Production, 39, (50-59), 2013

364033 Self-Emulsification of Alkaline-Dissolved Clove Bud Oil By Whey Protein, Gum Arabic, Lecithin, and Their Combinations

Abstract:

Essential oils (EOs) are well-known for their antioxidant and antimicrobial activities and have been widely used as functional ingredients in food, pharmaceutical and cosmetic products. However, poor water solubility and high volatility during processing are two major obstacles for utilizing EOs as preservatives in food matrices or as sanitizing agents. In the present study, a novel self-emulsifying process was studied to prepare clove bud oil (CBO) emulsion without any specialized equipment or organic solvents. In this method, the major CBO component (eugenol) was deprotonated at alkaline conditions, which was accelerated by heating. Heating 10% v/v CBO in 3 M NaOH at 110 °C for 10 min was observed to be optimal conditions to obtain a transparent CBO solution. The alkaline-dissolved CBO was mixed at an overall CBO concentration of 1% v/v in aqueous neutral solutions containing 1% w/v naturally-occurring emulsifiers - whey protein concentrate (WPC), gum arabic, lecithin, and their combinations. Emulsions were obtained after neutralization to pH 7, during which eugenol became re-protonated and encapsulated in situ by emulsifiers. CBO extracted from emulsions had identical UV-visible absorption spectrum as native CBO, and the peak in reversed-phase HPLC had a same elution time and peak area as the native eugenol. These results verified the maintained structure of eugenol after self-emulsification. The freshly prepared emulsions had hydrodynamic diameters in the range of 160–540 nm, but their dimension and stability were dependent on the emulsifiers. The sample self-emulsified with WPC only had the smallest particles, followed by the combination of all three emulsifiers and the WPC-lecithin mixture, while that with gum arabic alone had the largest particles. The emulsifier blend generally resulted in better stability of the emulsion particle size than individual compounds. Emulsions were evaluated for their antimicrobial activities against *Escherichia coli* O157:H7 ATCC 43895, *Listeria monocytogenes* Scott A and *Salmonella* Enteritidis. Emulsions generally exhibited better activity than same amounts of ethanol-dissolved CBO, showing lower minimum inhibitory and bactericidal concentrations and stronger inhibition of bacterial growth in tryptic soy broth. Our findings indicated that this novel self-emulsifying process may find promising applications in preparing EO emulsions using low cost food ingredients for use as antimicrobial food preservatives.

OIL EXTRACTION FROM SPENT COFFEE GROUNDS USING ADVANCED TECHNIQUES

Javier A. Dávila, Moshe Rosenberg, Gonzalo Taborda, Carlos A. Cardona
Universidad Nacional de Colombia, Manizales. University of California, Davis, Universidad de Caldas

2014 AIChe Annual Meeting

1. INTRODUCTION

SPENT COFFEE GROUNDS (SCG)

Coffee is one of the most important agroindustrial products in Colombia generating appreciable amounts of residues

654,000 T/Year

- Accounts the 10% of the total weight of the grain
- Contains bioactive compounds (Chlorogenic acids)
- Oil content (4% - 25%)
- 44% of fiber (Cellulose, hemicellulose)

2

1. INTRODUCTION

SPENT COFFEE GROUNDS (SCG)

<u>Products</u>

- Cellulose
- Hemicellulose

- Alcohols
- Biofuels
- Bioproducts
- Furfural
- PHB
- Xylitol
- Acids

OIL

- Food
- Pharmaceutical
- Biofuels

(Rodriguez, Cenicafe, 2011)

1. INTRODUCTION

CHEMICAL COMPOSITION OF SCG

Component	(%)
Nitrogen	1.64
Protein	10.30
Ash	0.59
Fiber (Hollocelulose)	39.72
Lipids	25.83
Extractives	21.92

Value Added Product

1. INTRODUCTION

EXTRACTION TECHNIQUES

Most common techniques used:

- Mechanical extraction (High pressure)
- Solvent extraction (Hexane)
- Supercritical extraction (CO2 at supercritical conditions)

6

1. INTRODUCTION

HOW TO EXTRACT OIL FROM SCG

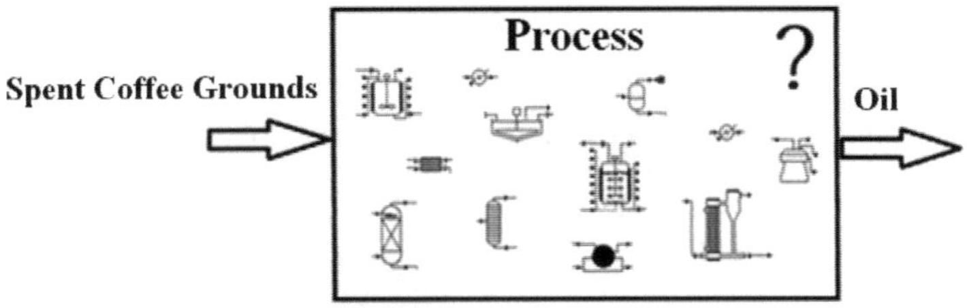

Yields?
Cost of production?

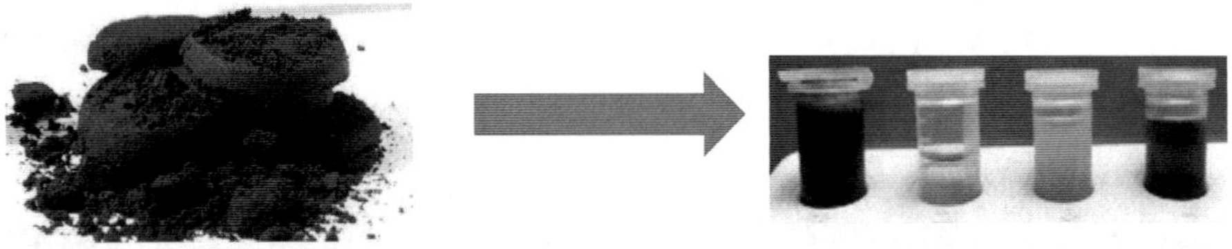

7

1. INTRODUCTION

INTEREST OF THIS WORK

Oil from Spent Coffee Grounds

Motivation:

Extract oil from SCG using three different techniques and obtain a preliminary techno-economic analysis of the extractions

8

2. METODOLOGY

METHODOLOGY

- Mechanical extraction
- Solvent extraction
- Supercritical extraction

Experimental

Process Simulation

- Construction of PFD
- Mass and energy balances

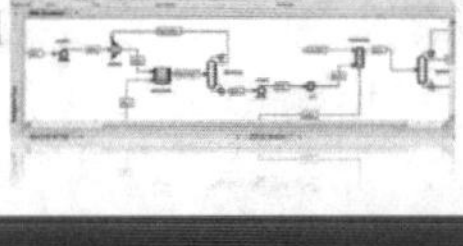

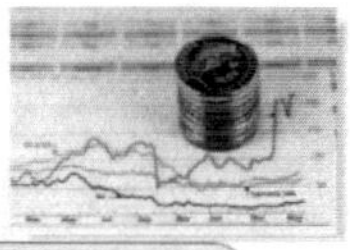

- Economic parameters (Colombian context)
- Economic evaluation

Techno-economic Analysis

9

2. METODOLOGY

MECHANICAL EXTRACTION

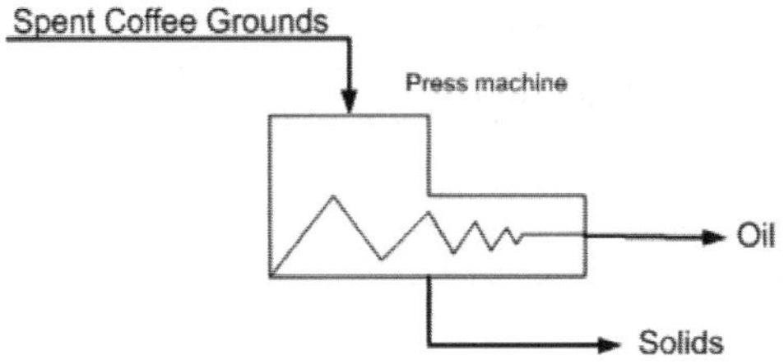

MECHANICAL EXTRACTION

- Pressure: 60 bar
- Temperature: 50, 70, 90 °C

2. METODOLOGY

SOLVENT EXTRACTION

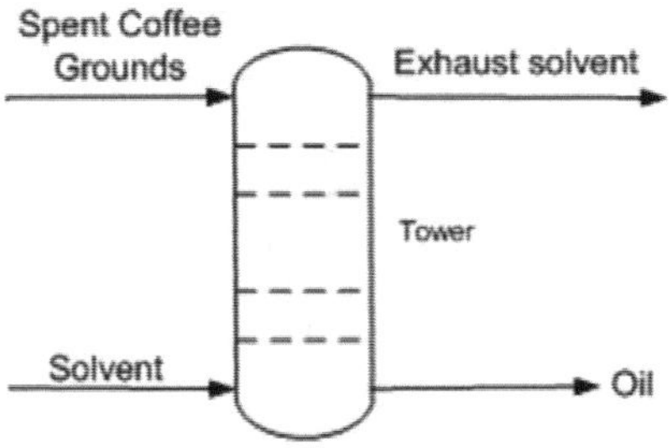

SOLVENT EXTRACTION

- Solvent: Hexane
- Temperature: 90, 120, 160 °C
- Time: 24 h

2. METODOLOGY

SUPERCRITICAL EXTRACTION

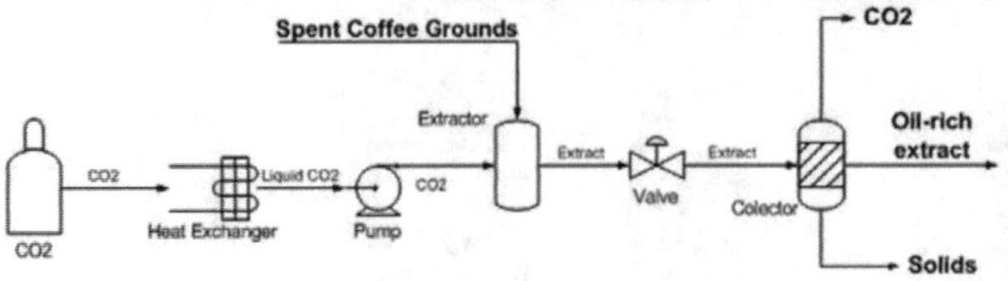

SUPERCRITICAL EXTRACTION

Supercritical Extraction

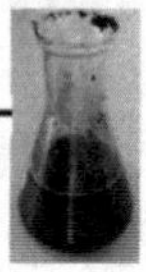

- Solvent: CO_2
- Temperature: 45 °C, pressure 300 bar
- Time: 5 h

12

2. METODOLOGY

PROCESS SIMULATION

Characterization of SCG → Simulation of the extraction processes → Mass and Energy Balances (ASPEN PLUS v8.0 software) → Economic Evaluation (Aspen Process Economic Analyzer) → Flows, yield, energy and cost results → Analysis of the process

13

3. RESULTS

EXPERIMENTAL RESULTS

Yields

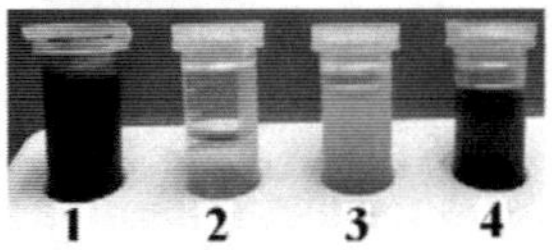

Oil 1: 60 bar, 50, 70, 90 °C
Oil 2: Hexane, 90 °C
Oil 3: Hexane, 120 or 160 °C
Oil 4: Hexane, 120 or 160 °C evaporating inmediately. Supercritical extraction

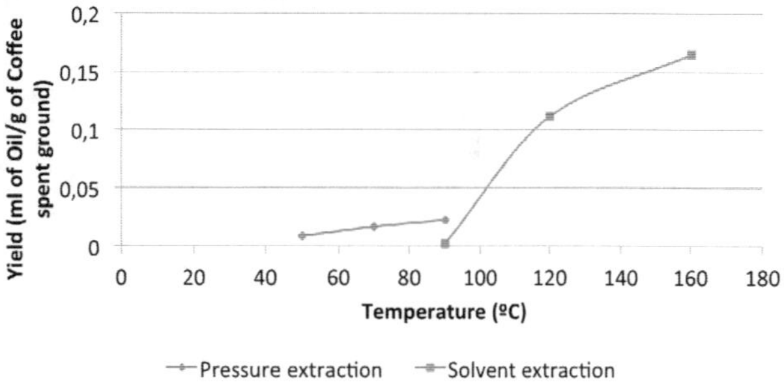

Melanoidins
Liposoluble dyes

Technology	Oil (g/g)	Yield (%)
Solvent extraction	0.165	63.88
Mechanical extraction	0.02	8.96
Supercritical extraction	0.2066	80

14

3. RESULTS

ECONOMIC ANALYSIS

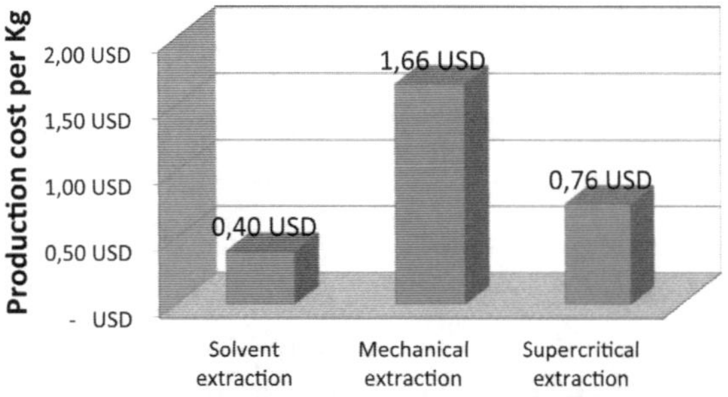

Market price of Palm Oil 0.64 USD/kg

(Index Mundi, 2014)

15

3. RESULTS

ECONOMIC ANALYSIS

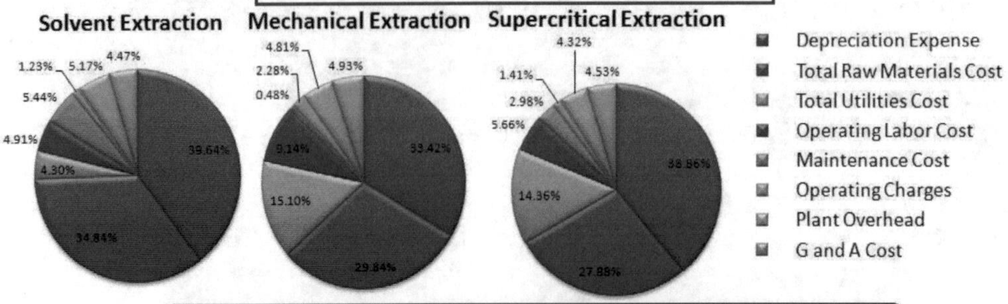

PRODUCTION COST PER YEAR

Solvent Extraction (1.048 million USD/Year)

Mechanical Extraction (0.617 million USD/Year)

Supercritical Extraction (1.211 million USD/Year)

16

3. RESULTS

OIL EXTRACTED

Oil extracted from SCG

Free Fatty Acids (FFA) were measured according to $AOCS\ Ca\ 5a - 40$ official method (Rukunudin et al., 1998)

Technology	Acid Index (% FFA)
Solvent extraction	3.26
Mechanical extraction	3.08
Supercritical extraction	2.95
Palm Oil	2.4 (Rincón et al., 2009)
Coffee Oil from SCG	2.99 (Ramírez, 2008)

Acid Index values ranges from 0.1 % to 4.7 % for biodiesel production.

17

4. CONCLUSIONS

CONCLUSIONS

1. Oil extracted from SCG is a potential raw material to be for other purposes.
2. Solvent extraction is the most interest technique to obtain oil from SCG.
3. Acid Index indicates that oil from SCG can to be used for biodiesel production.

18

REFERENCES

REFERENCES

1. Rodriguez Valencia Nelson. 2011. Recent experiences on the use of byproducts from coffee. CENICAFE 2011.
2. Price Indexes. Palm oil price market. 2014. Available in www:indexmundi:com
3. Rincon M. Sandra Milena, Martinez C. Daniel Mauricio. Analysis of the properties of palm oil in the industries development. 2009. Palmas Vol 30 No. 2.
4. Ramirez Ovalle Lourdes Mara. Assessment of the yield extraction and characterization of oil from coffee. 2008. Bachelor thesis, Guatemala Universidad de San Carlos de Guatemala.

19

Microencapsulation of Blackberry antioxidants. Modeling and simulation

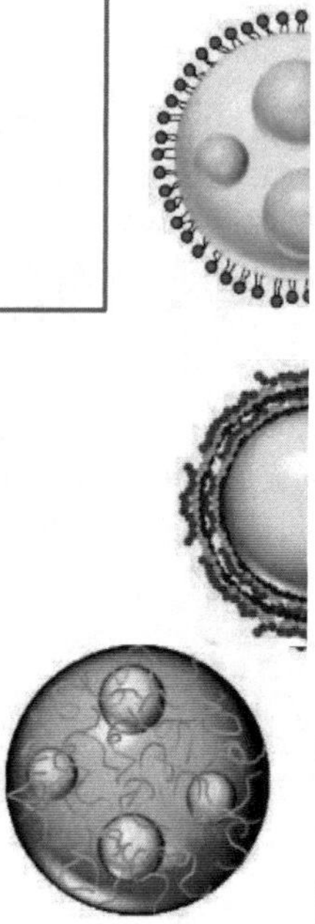

Miguel A. Rojas[1], Javier A. Davila[1] Moshe Rosenberg[2], Carlos A. Cardona[1*]
[1] Biotechnology and Agribusiness Institute, Department of Chemical Engineering
Universidad Nacional de Colombia Sede Manizales, Manizales -Colombia
[2] University of California Davis, Food Science & Technology Department
marojasre@unal.edu.co; jadavilar@unal.edu.co;
mrosenberg@ucdavis.edu; *ccardonaal@unal.edu.co;

14AIChE
Annual Meeting, Atlanta, GA

1. INTRODUCTION

The Interest in Designing functional foods

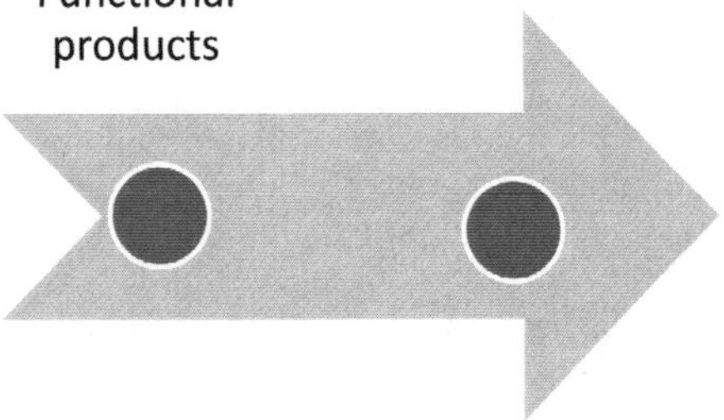

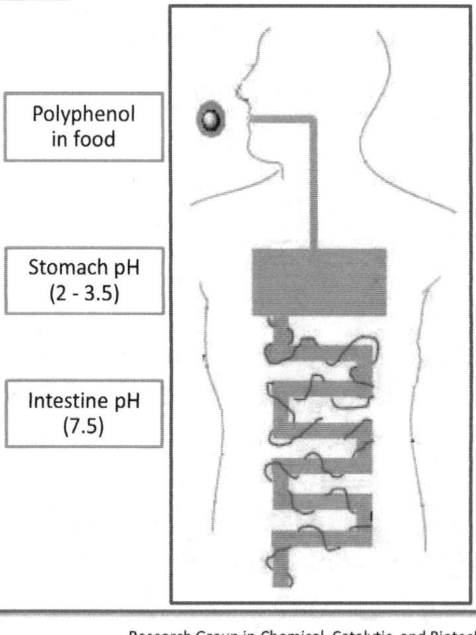

1. INTRODUCTION

Foods rich in polyphenols

- Red Wine
- Green tea
- Coffee
- Pomegranate
- **Blackberry**
- Nuts
- Prune
- Cholocate

Development of natural products with high contents of benefical compounds

Significant market for functional foods

14AIChE Annual Meeting, Atlanta, GA

3

Research Group in Chemical, Catalytic and Biotechnological Processes
Universidad Nacional de Colombia - Sede Manizales

1. INTRODUCTION

Foods rich in polyphenols

- Red Wine
- Green tea
- Coffee
- Pomegranate
- **Blackberry**
- Nuts
- Prune
- Cholocate

Blackberry (*Rubus glaucus benth*) is an exotic tropical fruit that have high content of natural antioxidants, phenolic compounds and anthocyanins.

In 2013, Colombia produced 105,285 tonnes of blackberry from 11,986 ha.

14AIChE Annual Meeting, Atlanta, GA

4

Research Group in Chemical, Catalytic and Biotechnological Processes
Universidad Nacional de Colombia - Sede Manizales

1. INTRODUCTION

After the extraction of active ingredients they need to be immediately protected against biological activity

Food manufacturers

- Polyphenols
- Antimicrobial
- Flavors
- Preservatives

Problems

Chemically and physically unstable

They are incompatible with the product matrix

Lack the appropriate functional attributes

1. INTRODUCTION

After the extraction of active ingredients they need to be immediately protected against biological activity

Food manufacturers

- Polyphenols
- Antimicrobial
- Flavors
- Preservatives

Challenges

Overcome by incorporating the active ingredients into some kind of delivery system before it is introduced into the final product

1. INTRODUCTION

Microencapsulation is a potential solution to the loss of biological activity and maintain the stability of polyphenols during food processing and store

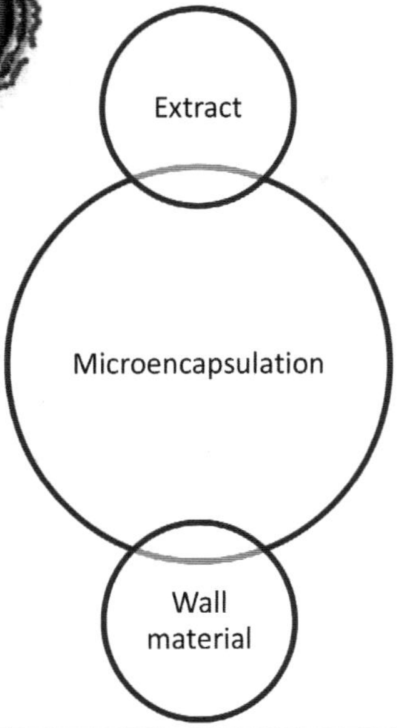

Shell Materials used for microencapsulation purposes

- Polysaccharides: Maltodextrin, Dextrin, Cyclodextrin, Cellulose, gum Arabic, Alginate, Pectin
- Protein: Soy, Wheat, Corn, Gelatin, Casein, Whey protein Isolate, Caseinate
- Polymers: Chitosan, Ethyl cellulose, Polyethylene glycol
- Lipids: Hard fat, Glycerides, Phospholipids
- Waxes: Beesswax, carnuba wax

1. INTRODUCTION

Microencapsulation is a potential solution to the loss of biological activity and maintain the stability of polyphenols during food processing and store

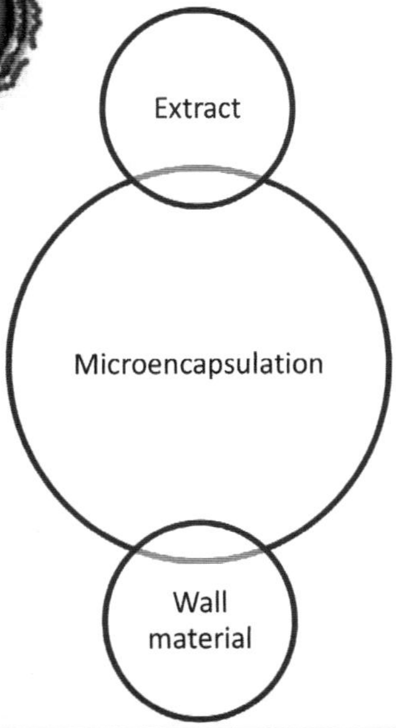

Microencapsulation methods used within the food industry.

Physical	Chemical
☐ Spray drying	☐ Phase separation
☐ Spray cooling	☐ Solvent evaporation
☐ Extrusion	☐ Interfacial polarization
☐ Coextrusion	☐ Liposomes

2. OBJECTIVE

- The objective of this study was to evaluate the influence of microencapsulation by spray drying of blackberry extract on the antioxidant activity and phenolic content and also to obtain the economic analysis of the process

3. METHODOLOGY

Extraction Process → Sample Preparation → Spray Drying → Chemical Analysis → Modeling and Simulation → Economic Evaluation

Supercritical fluid extraction

This Blackberry extract need be protected

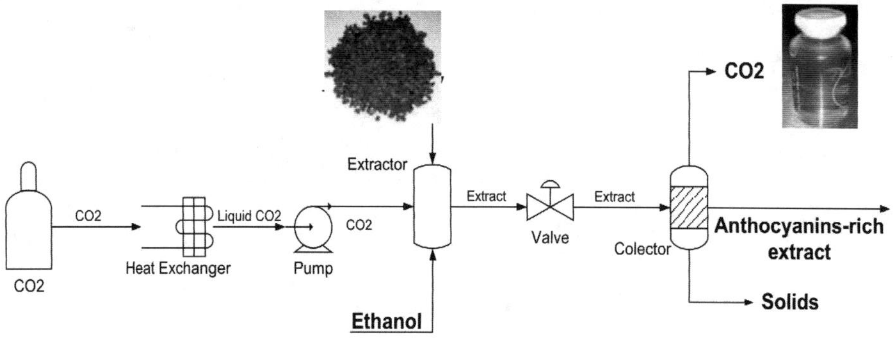

- Presion: 150 – 300 bar
- Temperature: 45 °C

3. METHODOLOGY

- Extraction Process
- Sample Preparation
- Spray Drying
- Chemical Analysis
- Modeling and Simulation
- Economic Evaluation

Delivery System — Designed to encapsulate, protect and release the active ingredients

Blackberry extract + Maltodextrin 20 DE (Solution 6% w/w) → Core in wall solution

3. METHODOLOGY

- Extraction Process
- Sample Preparation
- Spray Drying
- Chemical Analysis
- Modeling and Simulation
- Economic Evaluation

Microencapsulation Method.

Conditions:

Flow rate 0.49 kg/h.

Inlet air temperature:160 and Outlet air temperature: 80ºC

Nozzle atomizer: 0.7mm diameter

Drying flow rate: 0.36 m3/h

Aspirator flow rate: 35 m3/h.

Total Solid content: 30%

Initial phenolic compounds [mg GAE/ 100g]: 267.21

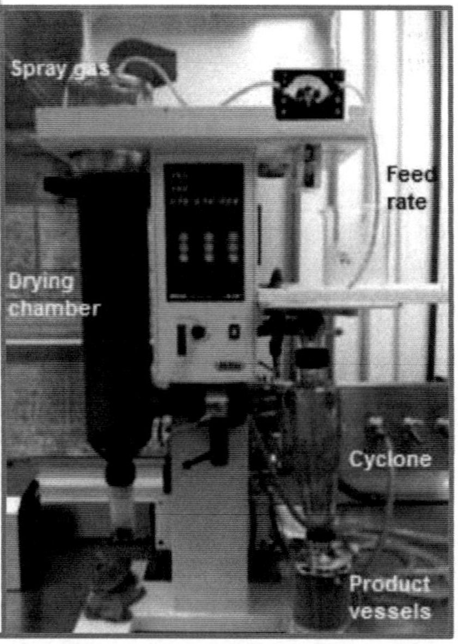

3. METHODOLOGY

Extraction Process → Sample Preparation → Spray Drying → Chemical Analysis → Modeling and Simulation → Economic Evaluation

Characterization of the blackberry extract microencapsulated:

Moisture content: (AOAC)

Water activity:(Hygrometer)

Total phenolic compounds (TPC): Folin Ciocalteu method

Total phenolic retention (TPR):

$$TPR = \frac{TPC\ in\ BEM\ (mg/100g)}{TPC\ in\ feed\ (mg/100g)}\ x\ 100$$

Antioxidant capacity: (DPPH method)

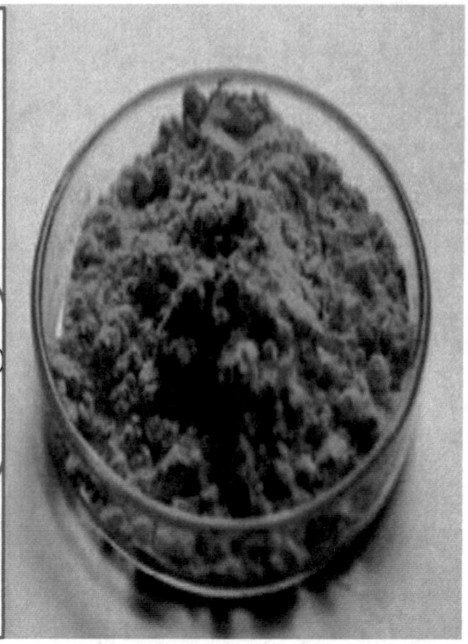

3. METHODOLOGY

Extraction Process → Sample Preparation → Spray Drying → Chemical Analysis → Modeling and Simulation → Economic Analysis

Blackberry Extract → Core in Wall solution

Biopolymer → Hydration → Core in Wall solution

Core in Wall solution → Final emulsification → Spray drying → Encapsulated powder

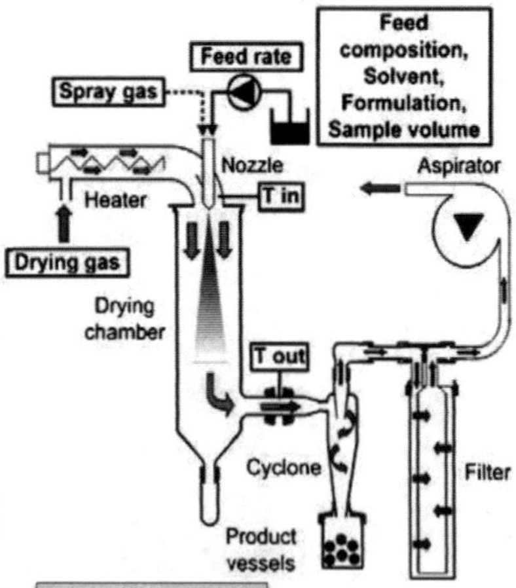

3. METHODOLOGY

Extraction Process → Sample Preparation → Spray Drying → Chemical Analysis → Modeling and Simulation → **Economic Evaluation**

The economic analysis was performed by calculating the production cost per unit of product (e.g. kg). The total production cost considers the total of raw material, utilities, operating labor, maintenance costs, operating charges, plant overhead and general administrative costs. The costs of personnel labor, water and electricity were obtained from typical prices in Colombia.

4. RESULTS AND DISCUSSION

4.1 Chemical Analysis of Blackberry Extract (BE) and Blackberry Extract Microencapsulate [BEM]

Chemical analysis	[BE]	[BEM]
Moisture content [%]	92.47	3.42
Water activity [aw]	0.97	0.26
Phenolic compounds [mg GAE/ 100g]	267.21	253.30
Total phenolic Retention [%]	N.A	94.79
Antioxidant activity DPPH [EC_{50} mg sample /mg DPPH]	21.33	30.23

N.A= Not applicable

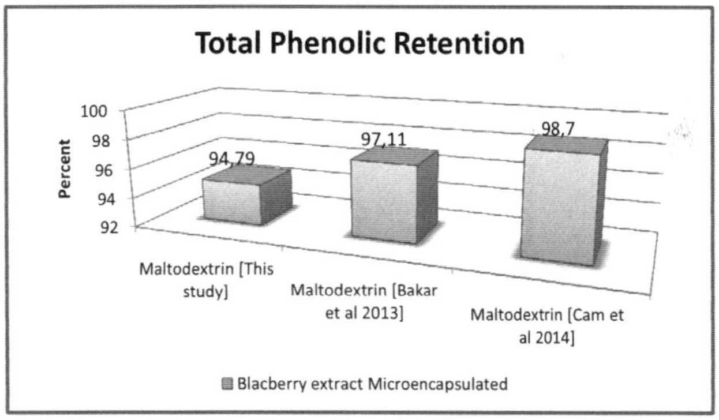

A delivery system should be capable of incorporating an appropriate amount of the active ingredient into the final product

It is important to obtain a high retention of the phenolic compounds within the deliver system until they are released at the required site of action.

Chemical Analysis of Blackberry Extract (BE) and Blackberry Extract Microencapsulate [BEM]

4. RESULTS AND DISCUSSION

Chemical analysis	[BE]	[BEM]
Moisture content [%]	92.47	3.42
Water activity [aw]	0.97	0.26
Phenolic compounds [mg GAE/ 100g]	267.21	253.30
Total phenolic Retention [%]	N.A	94.79
Antioxidant activity DPPH [EC_{50} mg sample /mg DPPH]	21.33	30.23

N.A= Not applicable

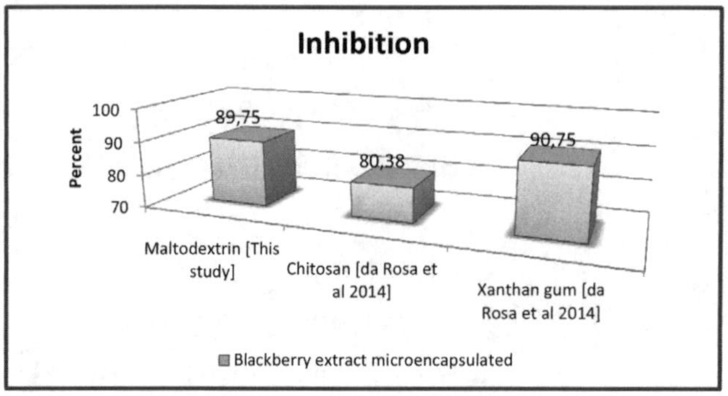

The results from the antioxidant activity studies indicated that the encapsulation procedure had no untoward effect on phenolic compounds.

The phenolic compounds present in the blackberry extract microencapsulated showed a greater antioxidant activity with a high inhibition of 89.75%

Modeling and Simulation

4. RESULTS AND DISCUSSION

Distribution Cost in the production of BEM

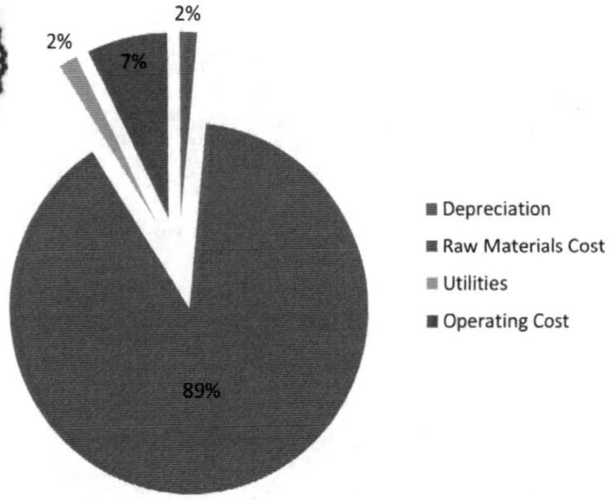

In this context, it can be observed that the price of the raw material has a great influence on the distribution cost in the production of BEM.
The utilities had a low percentage of the total cost

Economic parameters used for BEM estimation cost

Raw material		
Blackberry Extract[a] (Solvent system CO_2 + EtOH)	USD /kg	7.63
Maltodextrin[c]	USD/ kg	0.38

[a]Previous work
[b]http://www.icis.com/

4. RESULTS AND DISCUSSION

Economic evaluation

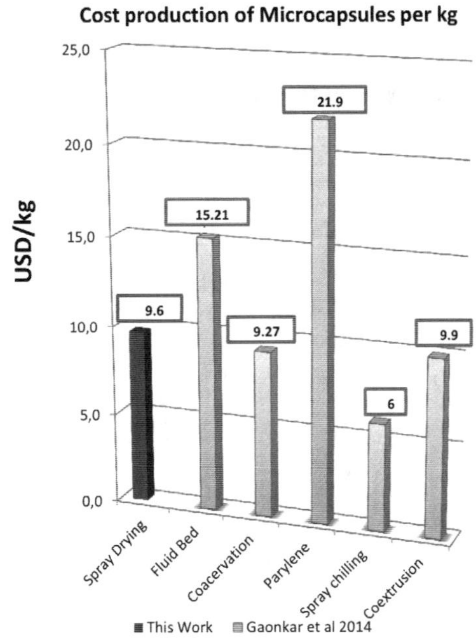

The process economic evaluation largely depends on some factors that must be taken into account before claiming that the BEM process is not economically viable.

Factors:

- Cost and content of the active ingredient to entrap
- Operating conditions
- Water content in the initial feed mixture
- Small scale and high scale

5. CONCLUSIONS

- The BEM obtained by spray drying process had an high retention of 95% of the total phenolic compounds and high inhibition of 89% of free radicals
- The use of polyphenols can not be considered without the protection provided by microencapsulation
- The benefits gained with microencapsulation of blackberry extract should outweigh the additional cost associated with the process of encapsulation

6. ACKNOWLEDGEMENTS

- The authors express their acknowledgments to the Universidad Nacional de Colombia at Manizales
- Research group of Chemical, Catalytic and Biotechnological Processes.

7. REFERENCES

- T. Egan, J.-C. Jacquier, Y. Rosenberg, and M. Rosenberg, "Cold-set whey protein microgels containing immobilised lipid phases to modulate matrix digestion and release of a water-soluble bioactive.," *J. Microencapsul.*, vol. 31, no. 2, pp. 184–92, Jan. 2014.
- T. Egan, J.-C. Jacquier, Y. Rosenberg, and M. Rosenberg, "Cold-set whey protein microgels for the stable immobilization of lipids," *Food Hydrocoll.*, vol. 31, no. 2, pp. 317–324, Jun. 2013.
- C. C. Ferrari, S. P. M. Germer, and J. M. de Aguirre, "Effects of Spray-Drying Conditions on the Physicochemical Properties of Blackberry Powder," *Dry. Technol.*, vol. 30, no. 2, pp. 154–163, Feb. 2012.
- Da Rosa, C. G., Borges, C. D., Zambiazi, R. C., Rutz, J. K., da Luz, S. R., Krumreich, F. D., ... Nunes, M. R. (2014). Encapsulation of the phenolic compounds of the blackberry (Rubus fruticosus). *LWT - Food Science and Technology*, *58*(2), 527–533. doi:10.1016/j.lwt.2014.03.042
- Veggi, P. C., Cavalcanti, R. N., & Meireles, M. A. a. (2014). Production of phenolic-rich extracts from Brazilian plants using supercritical and subcritical fluid extraction: Experimental data and economic evaluation. *Journal of Food Engineering*, *131*, 96–109. doi:10.1016/j.jfoodeng.2014.01.027
- Santos, D. T., Veggi, P. C., & Meireles, M. A. a. (2010). Extraction of antioxidant compounds from Jabuticaba (Myrciaria cauliflora) skins: Yield, composition and economical evaluation. *Journal of Food Engineering*, *101*(1), 23–31. doi:10.1016/j.jfoodeng.2010.06.005
- Anilkumar Gaonkar, Niraj Vasisht, Atul Khare and RS. Microencapsulation in the food industry. A practical Implementation Guide. (Elsevier AP in an imprint of, ed.). San Diego, CA; 2014.

THANK YOU FOR YOUR KIND ATTENTION

Corresponding Author: Carlos Ariel Cardona Alzate.
Km 9 Vía al Aeropuerto La Nubia, Manizales, Caldas
tel. +57 68879400 ext 55354
e-mail: ccardonaal@unal.edu.co
Universidad Nacional de Colombia at Manizales

383760 Liquid-Phase Electrical Discharge Plasmas Rapidly Inactivate Pathogenic and Spoilage Microorganisms in Water

Abstract:

This work seeks to reduce food wastage caused by microbial spoilage through the development of an energy efficient and effective food sterilization technology applicable to the liquid foods manufacturing sector.

We have developed an energy efficient plasma technology that can rapidly achieve 5-log or greater reduction of food spoilage organisms in water. The technology, which creates plasma directly in a liquid, sterilizes microorganisms by combinations of chemical (hydrogen peroxide and oxidative oxygenated species), physical (shockwaves and UV light), and electrical effects.

Through the manipulation of plasma chemistry, we were able to reduce the treatment time for selected microorganisms from 30 min to ~ 50 sec while achieving greater than 5-log reduction, and have noted differences in optimal plasma chemistries for different organisms of interest. The extent of sterilization appears to be determined by the plasma reactor design, input power, plasma electrode material, and electrode polarity. The work also discusses the mechanism of sterilization and investigates the possibility of using the technology to treat fruit juices.

Modeling the Drying Kinetics of Carrot Particles in a Closed Chamber Following Forced Convection and Electric Field Treatment

A.B. Jemai[a], L. Khezami[b], M. HadjKali[a], and E. Vorobiev[c]

a. King Saud University, College of Engineering, Chemical Engineering Dept., P.O. Box 800 Riyadh 11421- KSA.
b. Al Imam Mohammed Ibn Saud University, Faculty of Sciences-Department of Chemistry. P.O.Box: 90950 Riyadh: 11623 Kingdom of Saudi Arabia.
c. University of Technology of Compiegne, Chemical & Industrial Process Engineering Department

The impact of forced convection and Pulsed Electric Field pretreatment on the drying kinetics of grating type carrot particles, placed in a closed chamber, have been studied based on continuous relative humidity measurements ($\Delta\Phi$) of circulating air. The particles were placed in a laboratory-scale closed chamber (30g particles capacity) which can be used for the subsequent pressing operation and drying air circulation. The kinetics of moisture content loss was indirectly deduced from a normalized form of the relative humidity difference between the entering and leaving air streams. Using two thermo-hygrometers, the continuous measurements of relative humidity difference between entering and exiting air were used in a normalized form to study the kinetics of drying for various air speed. The effect of an electric field pre-treatment of the particles was also illustrated using this experimental setup. About 20% reduction in drying time could be achieved following PEF pretreatment.

Keywords: Agro-food particles processing, Relative humidity, Modeling Drying kinetics, Forced convection, Electric fields.

383942 Antimicrobial Peptide Segments from Soy Protein for Use in Food Safety

Abstract:

Antimicrobial peptides (AMPs) kill microbial cells through insertion and damage/permeabilization of the cytoplasmic cell membranes and has applications in food safety. Soy protein may be such an attractive, cost-saving candidate for commercial consideration because the protein subunits have amino acid sequences that contain several α-helix or 3-10 helix domains which possess characteristics of AMPs. A methodology for identification of AMPs from soy protein is proposed. They were identified from soy β-conglycinin (7S) and glycinin (11S) based on (i) number of amino acids, (ii) positive charge, (iii) hydrophobicity and (iv) hydrophobic moment. Explicit solvent molecular Dynamics (MD) simulation was employed to assess the secondary conformation of these peptides in POPC/POPG bilayers to mimic their permeation action on the cell membrane of microorganism. The effects of number of peptides, their orientation and hydrophobic moment on the deformation of membrane and formation of water channel were investigated. A mathematical model for the prediction of minimum inhibitory concentration for deactivation (antimicrobial activity) was proposed accounting for free energy of formation of a pore by an aggregate of peptides. The antimicrobial activity was tested by using synthetic selected peptides against *Listeria monocytogenes* and *E. Coli* and compared with predictions.

James E. Van Wyk P.E.

Process Development for Sustainable Food and Biochemical Production

AIChE Annual Meeting Atlanta GA

19 June 2015

Making Food Safe in Light of New Demands

1. Introduction

Food products are non-Newtonian. This means that there are few if any imperial formulas that can be used to predict heat transfer properties. The most reliable data is from laboratory data and field results. This is compounded by the fact that if a new ingredient is added to an existing product, the properties can change dramatically.

The concern in the design of a food heat exchanger system is to insure that all potentially harmful bacteria are destroyed. When a new ingredient is added it can change the heat properties to an extent that a previous safe product becomes unsafe.

There are numerous examples that can be shown. In this discussion we will focus on two products which have become popular in the current market

1. Orange juice which has now other juices added such as tropical fruits
2. Tomato soup which has products added such a bits of potatoes, carrots and other vegetables

While admittedly there could be many more these two were chosen because of their time in the market as a stand-alone product and how changes drastically changed the way they are processed.

2. Product Particulars

   a. Orange Juice- Orange juice has been on the market for a number of years. The trend since 1990 has been for NFC or not from concentrate. This actually represents the same product as you would get if you squeezed it yourself. Since the orange juice is blended form various types, the consumer gets a consistent product that always tastes the same

      In the early 2000's the demand for single strength orange juice had peaked. In fact the per person consumption of orange juice has dropped. Added is the fact that major producers were attempting to move orange juice from an only breakfast drink to a drink consumed any time of the day.

      To achieve these objectives marketing people added tropical juices to orange juice to add variety of flavor. Many different juices were tried. The one that appeared to be the best taste wise was cranberry.

      From a thermal point cranberry changed the characteristics tremendously. The juice no longer behaved as orange juice with an additive. It behaved as an entirely different product.

      The way we learned was due to spoilage. Orange which is pasteurized has a shelf life of 60-80 days. This product had a shelf life of 10 days. It should be noted that no human became ill from the product. It simply spoiled very quickly

   b. Tomato Soup

      Tomato soup has been available in cans for a long time. The can are retorted (Cooked in can). This actually causes overcooking and can give a burnt flavor to the product. Since 1980 there has been a market demand for a fresher flavor. In order to do this the soup has to be pasteurized and then aseptically filled in a container.

Plain tomato soup sales were either flat or slightly declining. In order to increase volume it was decided to add ingredients. Small diced potatoes and carrots were added.

The issue became how can be we be sure that the potatoes and carrots are adequately heat treated. These products constitute an added mass that not only affects the heat transfer properties but also insurance that they have been heat penetrated to insure safety.

3. Laboratory Results
    a. Orange Juice. We first did a side by side calculation to see what properties might have changed.
        i. Viscosity- No change
        ii. Thermal Conductivity No change
        iii. Heat Capacity-Significant change A decrease of almost 20 %
    b. Tomato Soup We first did a side by side calculation to see what properties might have changed.
        i. Viscosity- A decrease in dynamic viscosity of 29%
        ii. Thermal Conductivity No change
        iii. Heat Capacity-Significant change A decrease of almost 20 %
    c. Laboratory Conclusions-
        i. The design of the heat exchanger had to be modified to meet the change in heat capacity. We either had to add surface or lower the processing rate

ii. With the tomato soup we found that the particles were not heated sufficiently. This meant that the soup had to be exposed to the heating for a longer period. The resonance time went from 10 min to almost 18 minutes. This meant that either we add surface or lower the processing rate,

4. Field Results

   a. Orange Juice-The juice processing rate is closely coupled to the rate of filling. Specifically the processing rate is 100 gpm to coincide with a ½ gallon filling rate of 80 gpm. This meant the only option for us was to increase the heat transfer surface. Since we were using a plate heat exchanger this was accomplished with almost no difficulty. We increase the area to what the laboratory test would indicate. This however proved to be not enough. We still had to increase by an additional 5% to achieve the pasteurization temperature.

   b. Tomato Soup- This proved to be a more difficult fix. We could not increase the surface area due to space limitations and cost. This was being processed in a tubular heat exchanger. We therefore had to slow the flow rate. The actual processing rate was about 30% less. From an engineering point of view this was acceptable. The economics at running at a lower rate proved not the case. Once the product was established in the market an new line as built for the soup

5. Conclusions

   a. Assume you know nothing- You cannot simply add an ingredient and assume that it will behave as before. Since the majority if not all food products are non-Newtonian, laboratory testing must be done each time.

b. Look at economics-Can you add surface area with a minimal amount of cost? Adding this cost will the product still be profitable? In some cases this may kill a project.

c. Never put a product out for public consumption until you are 110 % sure

386751 Low Temperature Batch Conversion of Cellulosic Biomass

Abstract:

The project aims at developing small, affordable batch reactors that communities in developing countries can use to produce clean cooking gas from biomass. Our approach extracts clean combustible gases such as methane and hydrogen using low cost locally available catalytic ingredients such as sodium chloride and sodium carbonate. Compared to uncatalyzed biomass conversion, the onset of pyrolysis/gasification shifts to significantly lower temperatures in the presence of catalysts. One possible explanation is that the ions in solution migrate in between cellulosic strands/fibers and disrupt the interstrand hydrogen bonding. Upon drying, the salt nucleates and crystallizes in between the fibers, leading to mechanical strain and loss of structural integrity of the fibers. The recovered biochar can be used as a supplemental source of fuel, while the gas evolved can be collected inside the pressure vessel and connected to a gas burner. This allows the use of a wide range of biomass sources without forcing a community to radically change their way of living, thereby increasing the probability of acceptance of this new method.

380169 Environmental and Techno-Economic Analysis for Ethylene Production from Shale Gas

Abstract:

Environmental and techno-economic analysis for ethylene production from shale gas

Andrea Paulina Ortiz Espinoza [a], Arturo Jiménez Gutiérrez [a], Mahmoud M. El-Halgawi [b]

[a] *Departamento de Ingeniería Química, Instituto Tecnológico de Celaya, Celaya, Gto., Mex., 38010*

[b] *Chemical Engineering Department, Texas A&M University, College Station, TX 77843, USA*

Ethylene is a light olefin that is used as a large building block for many petrochemicals. It works as the raw material in the manufacture of several products such as plastics, resins, fibers, etc. Ethylene is typically produced from thermal cracking of naphtha or ethane and propane from natural gas. In the thermal cracking, valuable by products including propylene, butadiene, and benzene are also produced in significant amounts. Besides the conventional process used to produce ethylene, alternative methods have been developed, some of them using methane as a feedstock.[1] These particular processes have been gaining importance due to the lower prices of natural gas and most recently shale gas discoveries.

One of these methods is the Oxidative Coupling of methane, which involves direct conversion of methane to ethylene by using a catalytic reactor. This process has been investigated for about 30 years and has not been industrially implemented due to its medium conversion and low selectivity. Despite this, the process is a very promising alternative for the production of ethylene from methane that is worth of consideration.[2] Another process that has been developed for the production of ethylene is the Methanol to Olefins process. In this process, methane is converted to syngas through a reforming reactor, and then the syngas is converted to methanol in a catalytic reactor. Once the crude methanol is obtained it is converted to ethylene.[3]

This work presents a techno-economic analysis and comparison of the above mentioned alternatives to produce ethylene. An environmental assessment for each process is also carried out in order to show the advantages and limitations of these alternatives. First, Aspen Plus® simulations are developed to obtain mass and energy balances. Then, along with data reported in literature, return on investment is calculated to show the profitability of each process. Finally CO_2 emissions are calculated to assess the environmental impact for the processes. The results identify the compromise between economics and environmental implications for these production alternatives.

1. Sundaram, K.M.; Shreehan, M.M.; Olszewski, E.F. 2010. Ethylene. Kirk-Othmer Encyclopedia of Chemical Technology, 1-39.

2. Godini, H.; Xiao, S.; Jašo, S.; Stünkel, S.; Salerno, D.; Son, N.; Song, S.; Wozny, G., 2013. Techno-economic analysis of integrating the methane oxidative coupling and methane reforming processes. Fuel Processing Technology, 106, 684–694.

3. Ren, T.; Patel, M.K.; Blok, K. 2006. Steam cracking and methanol to olefins: energy use, CO_2 emissions and productions costs. Energy, 33, 817-833.

381139 Study of Flare Minimization Opportunities during an Ethylene Plant Shutdown

Abstract:

It has always been beneficial for an ethylene plant to reduce flaring during its shutdown because the reduction of burned-off gases reduces pollutant generation, saves raw materials, and also lowers the utility/energy costs. An ethylene plant shutdown generally includes three stages of operations: feed and inventory reduction, facility shutdown, and unit decommissioning (involving liquid discharge, vapor discharge, and nitrogen (N_2) purge). In this paper, a dynamic simulation model has been built to study flare minimization opportunities of a front-end de-ethanizer ethylene plant shutdown operations. In its normal shutdown procedure, three of the five furnaces will be firstly cut off, and the charge gas compressor (CGC) will be reduced to a small operating load while internal recycle lines in CGC section will be kept open for safety operation. Secondly, the liquid levels of various vessels are reduced to their minimum operating values. Thirdly, the left furnaces will be brought off-line when CGC shutdown is started. Finally, the residual liquid and vapor inventories will be sent to the flare system and nitrogen will then be injected into each vessel for purge operations.

To minimize the shutdown flaring and smooth the plant shutdown operation, various operational improvements have been initiated and examined in this paper via the developed dynamic simulation model. For example, such an idea has been tested that shutting down all furnaces first and then recycling vapors from downstream columns to the CGC inlet to necessitate CGC running meanwhile enable flare source recovery. After that, there are much less hydrocarbons left in downstream vessels after CGC is shut down and before the nitrogen purge operation. Compared with base case, significant flare emissions have been saved. With the help of plant-wide dynamic simulations, other flare minimization initiatives have also been tested and quantified.

379650 A Novel Energy-Efficient Process for Isolation and Purification of a Sustainable Industrially Relevant Polymer

Abstract:

The current process used by SABIC's Innovative Plastics business for an industrially relevant polymer purifies & isolates the resin from the reaction mixture dissolved in reaction solvent by separation of the aqueous phase and subsequent precipitation using an anti-solvent. This process efficiently removes most of the impurities, however is associated with a significant energy cost in the downstream operation due to the large quantity of the anti-solvent used. Also, the final product, obtained as a fine powder, can pose dust explosion risks and potential transfer and feeding issues during compounding.

This presentation summarizes the technical feasibility of a novel energy-efficient direct isolation process design based on liquid-liquid extraction as an alternative to anti-solvent precipitation for this standard polymer purification. After the non-volatile impurities are efficiently removed by the new process, the polymer can be isolated from the solution using standard devolatization equipment. The effect of different extraction process parameters on the Efficiency, Yield, Solvent Load and Recovery, Product Specifications, Impurities, Odor Body and Color Removal is presented. The new process has a significant impact on energy cost in the downstream operations related to solvent recovery and recycle. Finally, an ASPEN model is developed to demonstrate a significant energy reduction in the polymer isolation operation compared to anti-solvent precipitation.

Process of Making Post-Consumer Upcycled Polybutylene Terephthalate (PBT) Based Engineering Thermosplastic Resin Enabling Improved Hydrostable Product

Husnu Alp Alidedeoglu*, Tianhua Ding

In recent years there has been a growing social concern regarding the effect of industrial development and global population growth on the long-term availability of the world's natural resources, primarily fossil fuels (mineral oil, gas, and coal), and the earth's ecosystem. The United Nations Environmental Program's Global Environment Outlook estimated that global society demands more than 1.5 planets' worth of productive biological capacity.[i] Therefore, there has been a growing public interest in the sustainable resins and products which are recycled, consume less energy and yield less carbon dioxide emissions in their manufacturing than traditional resins.

SABIC has introduced the VALOX iQ™ PBT resin to foster sustainability and help mitigate these environmental concerns without compromising engineering performance properties. VALOX iQ* PBT resin is designed to emit 34% less greenhouse gases and save 47% energy compared to traditional DMT-based PBT resin.[ii] The VALOX iQ* PBT resin produced with modified catalyst delivers competitive hydrostability performance while enabling lower environmental impact by up cycling post-consumer recycle polyethylene terephthalate (PCR PET) waste.

This work represents a chemical upcycling process which converts PCR PET into a sustainable PBT copolymer using a modified titanium based catalyst. This process comprises four main steps: 1) PET bottle bales are first cleaned and processed into clean PET flakes; 2) The clean PET flakes are then depolymerized into oligomeric units; 3) These oligomeric units are then re-polymerized to create PBT copolymer in a suitable batch reactor using a modified titanium based catalyst; 4) Finally this PBT copolymer is then further compounded into finished products.

By modifying the catalysts in-situ during the polymerizations process we are able to simplify the formula and create a sustainable PBT compound with improved performance in outdoor applications. This presentation will cover the optimization of the in-situ catalyst modification process and its effect on the final compounded PBT product produced from PCR PET.

---

[i] http://www.unep.org/geo/pdfs/geo5/GEO5_report_full_en.pdf, pg 144
[ii] http://www.pod-sabic-ip.com/KBAM/Reflection/Assets/Thumbnail/20333_7.pdf

381779 Design of Methane Hydrate Decomposition Process Combined with Conventional LNG Receiving Terminal

Abstract:

Recently, transportation of natural gas (NG) in the form of natural gas hydrate (NGH) is in interest because of its economic feasibility and safety. It also gives advantage of diversification of NG transport and import. Nevertheless, there is no study of converting NGH to pipeline natural gas (PNG), which is the most common distribution method in domestic transport.

In this research, the process of converting MH to PNG combined with existing LNG receiving terminal is designed. It is assumed that NG is transported in MH dry pellet form, composed of CH4 and H2O in the proportion of 1:6. Storage tank of MH is assumed to be 1 bar and -20 °C.

First, MH decomposition process (regasification process) is designed. Sensitivity analysis of the process using design variables, i.e. decomposition temperature, is performed. Gas stream from decomposed MH goes through the glycol dehydration unit to meet water content specification of PNG (less than 120 ppm). Then, the dehydrated gas is compressed in multi-stages for use as PNG while cryogenic energy of LNG is integrated in the process to cool down the compressed gas stream. Assuming fixed PNG send out rate, optimal quantities of LNG and MH are determined for minimum operating cost.

Technical Risk Assessments:
A Tool for Process Development and Technical Risk Reduction – 470a

Daniel D. Euhus, Ph.D., P.E., Chevron Technology Ventures Co, Houston, TX
Steven F. Sciamanna, Ph.D., PolyDiamond Technologies LLC, Orinda, CA

Summary
A Technical Risk Assessment (TRA) is an engineering tool used for raising technical questions that should be addressed in order to reduce technical risk and thereby increase confidence in the technology (Fundamental Science, Engineering-Science and Engineering) upon which a commercial venture will be based. The outcome of the TRA is a reference registry of technical risks, after a systematic and guided exercise to identify those risks, which need to be reduced in order to mitigate their impact on the financial viability of a proposed technology. The TRA identifies areas for technical development in order to reduce technical risk. Alternatively, the decision makers agree to accepting known technical risk if the benefit of further risk reduction does not outweigh the cost of the development. The TRA is an "evergreen" register created during the early stages and then maintained throughout the technology development program. A TRA can also be used when introducing a technology that is new to your company or industry. Ideally, the TRA is conducted by members of the technology development team and outside subject matter experts (SME) who can provide an impartial technical perspective. The benefits of the TRA are: reducing project risk due to a technology failure or shortfall; provides guidance to the R&D effort so that time and money is spent wisely; and allow for early review by SMEs to avoid pitfalls and hopefully gain some insights. A simplified risk matrix can be used to quantify the risk associated with each technical concern raised during the assessment process. These concerns and questions are quantified and ranked enabling better use of project resources at various development stages to reduce the technical risks associated with the process technology.

Introduction – The "Process" of Process Technology Development & Commercialization
For the capital-intensive process technology-based industries, when it comes to turning "science and engineering" into a commercial success, wishful thinking and cutting corners to meet a schedule are not a viable strategy. More often than not these are just "separation" processes where investors are separated from their money. In the 2013 AIChE meeting, Dan Euhus et al. (2013) presented a paper entitled "*Pathway to Process Technology Commercialization*". This paper detailed the prescriptive steps and work process or pathway moving from the "Aha! or "Eureka moment" in the lab through to the commercial or full-scale deployment. The pathway to commercialization coincides with reducing technical risk (as well as other forms of risks). Managing and mitigating technical risk without stifling creativity or enthusiasm can be tricky but when done properly can enhance them by posing and answering key technical questions.

Types of Process Technology Development (Technical) Risk
In this paper we consider three types of technical risk that must be reduced or eliminated during the development and implementation of a new technology. As shown in Figure 1, these are; <u>fundamental-science risk</u>, <u>engineering-science risk</u> and <u>engineering risk</u>.

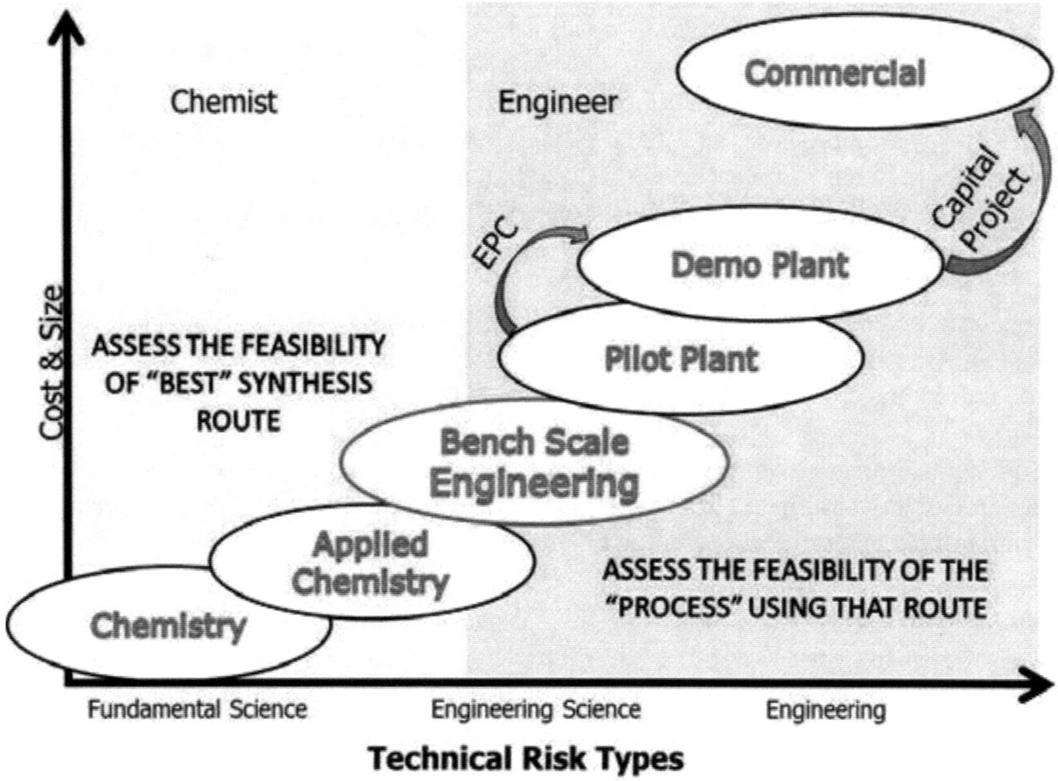

Figure 1: Technical Risk Types along the Process Development Path

Types of Process Technology Development (Technical) Risk
In this paper we consider three types of technical risk that must be reduced or eliminated during the development and implementation of a new technology. As shown in Figure 1, these are; fundamental-science risk, engineering-science risk and engineering risk.

Companies, investors and capital project managers have no appetite for fundamental-science risk because its resolution is not time-certain nor is it possible to estimate its cost. Process design engineers don't want it because their effort then starts by making critical assumptions after which their work is best described as "Imagineering". There is no substitute for devoting the time, effort and expertise to insure that there are no fatal flaws of a fundamental-science nature (e.g. chemistry or physics). Developing a process technology based on flawed science happens all too often and nothing good can be said about it.

The risks associated with the second type of technical risk, engineering-science (e.g. separations, reaction engineering, heat transfer, etc.), are addressed at the bench and pilot scale steps. Here the unit operations are shown to exist or are developed. Furthermore, they are shown to perform as per the process intent and that no surprises appear when, for example, recycle loops are considered. The reduction of the engineering-science risk also reduces engineering and operational risk by demonstrating, for instance, that a piece of equipment or unit operation can be built and continuously operated.

The third form of technical risk, engineering risk, addresses the design, construction and operability of new equipment, unit operation or novel integration.

A fourth risk-type of a technical nature worth mentioning here is "techno-economic" risk. An example of this risk type is one where the three previously described technical risk types have been successfully mitigated but "so what". That is, the business drivers (e.g. CapEx, OpEx, jobs created) fall short of the objective. It is imperative that a Techno-Economic Analysis (TEA) is performed at the earliest stage of a conceptual process and is refined as the technology development effort progresses. An evergreen TEA defines and redefines the boundaries and limits of what needs to be accomplished to insure that "there is a there there". Without it, technology development professionals and managers/decision-makers potentially expose themselves to one of the most embarrassing situations possible in their career.

The costs of reducing technical risk for the full-scale commercial facility increase as the project moves along the process development path. So it is imperative that technical risk be reduced at the smallest meaningful scale to use resources efficiently.

Technical Risk Assessments

Everyone in the process industries should be familiar with the HAZOP and a Process Hazard Analysis (PHA). In short, these are systematic ways to review the safety and environmental risks present in a process, and to evaluate the mitigation methods utilized to "de-risk" the hazards. These processes can be quantitative and usually involve a determination of both the consequence of an outcome from a potential hazard and the likelihood of its occurrence. Individual companies will likely have different levels of risk tolerance when it comes to process safety risk. These will be evident in the decisions made about what risks can be mitigated and to what level. Some risks may even be identified as intolerable no matter what level of mitigation is achieved. Because at its heart, risk is a psychological perspective and only qualitative, i.e. what feels risky to me may not be feel risky to you. One person may never fly on a plane because of their perception while another flies constantly. In both cases, the facts around the safety of flying are equally available to both parties, but their interpretation of the data and their perception of the risk are different. But, as technical people, we prefer quantitative results and analysis. So, a process risk assessment serves as a useful quantitative tool once the evaluation team has been calibrated as to a definition of acceptable and unacceptable risk – based on their company's philosophy.

A Technical Risk Assessment (TRA) is similar to a PHA in that its goal is a systematic approach to analyzing an aspect of a process (or any technology). It can also be used to evaluate a theoretical or conceptual design and determine its high-level technical risks. But, in some ways, a TRA differs from a PHA. For example, a TRA should be initially performed BEFORE the process design is finalized and then updated as necessary. Rather than looking at health, safety or environmental risks, a TRA is looking at the technological aspects of new processes, chemistries, unit operations, etc. The assessment also seeks to address the proper scale where risks to technical success should be reduced based on cost of the scale and validity of the results. At its core, a Technical Risk Assessment raises technical questions that require answers in order to have confidence in the Science, Engineering Science and Engineering as well as the Operability of the technology in question.

The primary purpose of a TRA is to reduce the risk of technical failure that would lead to a commercial failure for a technology. While so-called "black swans" can and do change commercial outcomes for projects utilizing both new and/or accepted technologies, a project should NOT fail due to a technical risk. Technical risk CAN and SHOULD be reduced during technology development!

A second, and highly valuable, purpose of a TRA is to provide insight and guidance to the research and development program. Anyone involved in managing research and development knows that it is critically important, especially in this day and age to spend time and money wisely. The results of the TRA will help the research and development effort to focus on answering the questions that matter, the questions that lead to key decisions. Even more importantly, the TRA will help to focus the answers to these questions at the smallest possible relevant scale where costs will be the lowest by intentionally introducing this discussion at the beginning of the work. The TRA also allows for early technology review by "cold-eye" technology experts, a step that is critical for success in technology development, by forcing outside experts and perspectives into the conversation.

A Technical Risk Assessment can be a very valuable tool in a number of different situations. First, it should be used for any new process technology, whether this is a new process or chemical pathway, a new unit operation (or piece of equipment in new service) or any combination of these. Second, it should be strongly considered when changes are being made to an existing process that changes the technology utilized such as a debottlenecking project or changes being made to a unit operation (perhaps tray changes in a column). It is NOT appropriate for a replacement in kind or replacement with an equivalent (NOTE that this MAY still require an MOC). Last, it can be used to help guide early concept research to help evaluate the likelihood of success. This helps avoid uncomfortable questions like, "did you ever consider that one this size has never been built?"

For a technical risk assessment to succeed requires that the right people be in the room together when conducting the assessment. The team should definitely include the various project members involved including such roles as chemists and industrial chemists, process engineers and specialists focusing on such areas as reactions, separations and heat transfer. But, the team should also include technical experts NOT involved in the project from the areas mentioned above as well as program managers from other areas. The leader of the TRA is likely to be the program manager (i.e. the person overseeing the larger technology development, rather than the individual stages – see Euhus et. al, (2013) for more details on technology development and team organizational structure).

As with a PHA, there are different ways to approach a TRA. A very formal approach would be to utilize a keyword study where the discussion is guided by keywords as in a PHA. This is a little more difficult, and often is not as fit for purpose as a simpler "what-if" analysis. The so-called "what-if" approach ask various what-if type questions about various parts of the technology to try and understand what is proven (accepted and utilized) technology, and what still needs development. One approach to conducting a TRA would proceed using the following outline:

1. Detailed presentation introducing and reviewing the proposed process. It should include the chemistry, catalysis (if appropriate) and unit operations (BOTH reactions and separations).
2. Next the group should brainstorm their technical issues, concerns and questions. Brainstorming may (and probably should) include both open and guided formats. Open brainstorming is where perhaps PFDs are attached on the walls and participants individually, or as small groups, document issues/concerns by posting them with sticky notes on the individual PFDs in proximity to the appropriate area. Guided brainstorming is where the group is led through each individual PFD like as is done with a P&ID in a HAZOP or PHA. Here the group, as a whole, can synergize and build on each other's thoughts.
3. "Offline" the TRA Leader compiles the discussion into a tabular form and begins binning the results (binning is the process of sorting which risks will be addressed in which stages of development, i.e. what is the appropriate scale for addressing each risk) and determining the status of the technology: is it proven or unproven, where has it been proven, what is the status of development, etc. For examples see Huyse (2013) and the US DOE Technology Readiness Guide (2009).
4. After completing the initial review of the TRA results, they should be presented back to the team for general agreement and discussion around the binning exercise. A significant role of the technical team members and advisers is to make sure the technical risk is being reduced at the appropriate stage. Another way to look at this is that not all questions need to be answered to move to the next stage, but the critical (go/no-go) type questions certainly need to be addressed.

The "product" of the TRA is an evergreen Technical Risk Register. This will most often be in the form of a spreadsheet.

Documenting the TRA in a Technical Risk Register

		Risk of Failure							Cost of Failure	Ranking	Risk Reduction Plan		
		Y/N/U	Y/N/U	Y/N/U	Y/N/U	Y/N	Y/N	Y/N			L,B,P,D	Y/N/U	Y/N/U
Process Unit/Area/Equipment	Technical Issue/Question/Concern	Used by Company	Used in Comparable Company	Used a Desired Commercial Scale in Industry	Used at Smaller Commercial Scale in Industry	Proven at Demonstration Scale	Proven at Pilot Scale	Proven at Bench Scale	NPC ($)	From Company Specific Graph	Minimum Scale for Technical Proof	Appropriate for Computational Proof	Engineering Design Risk
Moving Bed R.E.	How does catalyst flow through the system?	N	N	N	N	N	N	N	D	9	P/D	Y	Y
Moving Bed R.E.	What is the catalyst life (activity)?	Y	Y	Y	Y	Y	Y	Y	A	1	B/P	N	N
Moving Bed R.E.	Is the catalyst durable for moving bed?	Y	Y	Y	Y	Y	Y	Y	B	2	P	N	N
Moving Bed R.E.	What is liquid/liquid equilibrium behavior in packed bed section	U	U	N	N	N	N	Y	C	7	B/P	U	U

Figure 2: Example Technical Risk Registry w/ Ranking and Risk Reduction Plan

The form of the technical risk register is really up to the one utilizing it. Figure 2 provides a sample Technical Risk Register that is consistent with the rest of this paper where the Risk of Failure and the Cost of Failure lead to a ranking of the risk. Finally, the appropriate scale and method for risk reduction has been recommended. The technical risk assessment in Figure 2 is for a hypothetical process involving a heterogeneous catalyst in a moving bed reactive extraction (and is by no means complete). What can quickly be identified from the register is that catalyst movement and liquid/liquid equilibrium behavior are significant technical risks while catalyst life (or activity) is not as concerning based on their overall rankings. It is also clear that a minimum risk reduction plan will include piloting, and given the high risk for catalyst flow through the system – one might want to seriously consider demonstration as well.

Quantifying the TRA

	NPC>E							1
	D<NPC<E							
Net Present Cost (NPC)	C<NPC<D							
	B<NPC<C							
	A<NPC<B							
	NPC<A	10						
		Very Low Risk	**Low Risk**	**Low-Medium Risk**	**Medium Risk**	**Medium-High Risk**	**High Risk**	**Very High Risk**
		Used by Company	Used in Comparable Company (Industry)	Used at Desired Commercial Scale Industrially	Used at Smaller Commercial Scale Industrially	Proven at Demo Scale	Proven at Pilot Scale	Proven at Bench Scale

Figure 3: Evaluation Matrix for Technical Risk

At its core, all risk management is qualitative. This has to be so because individual perception of risk (and company perception of risk) is going to be different. As humans, our risk perception is colored by experience. This is one reason that diversity of thought and experience is so important when considering risk. Operating conditions that one company considers "highly risky" because of "high" temperatures and pressures may be regarded as normal by another company because that is where they operate every day. The underlying consequences of failure are the same for both companies, but the risk is perceived differently based on experience and ability to manage the risk. When managing technical risk, different individuals and different companies are likely to have different appetites. For this reason, a company should have a perspective on their own personal risk appetite that is clearly understood by the employees. An effort to remove individual bias to risk is reduced by then quantifying the risk assessment process and providing company guidelines – similar to the system used by many companies for process risk assessments through process hazard analysis.

Figure 3 has attempted to suggest one possible approach to quantifying the risks present without giving instruction as to what appropriate risk tolerance should be. On the bottom, the scale shows varying level of risk related to the development level of the technology (what scale has it been "proven" at) with least risky on the left and most risky on the right. Similarly, on the left, the cost of failure (Net Present Cost – assuming no benefit, i.e. the project fails) is displayed with least costly on the bottom and most costly on the top. This leaves a clear conclusion that the least risky corner is the bottom left and the most risky corner is the top right. But, how an individual company might decide to fill in the blanks is entirely up to them based on their own aversion to technical risk as it relates to a commercial project. This evaluation can then be used to help determine the best investment points for funding of development to reduce overall technical risk. Remember that the goal of development is to move the technology forward to key decision points about whether or not to keep developing the technology – not just to spend more development money!

Using the TRA to Manage & Reduce Technical Risks

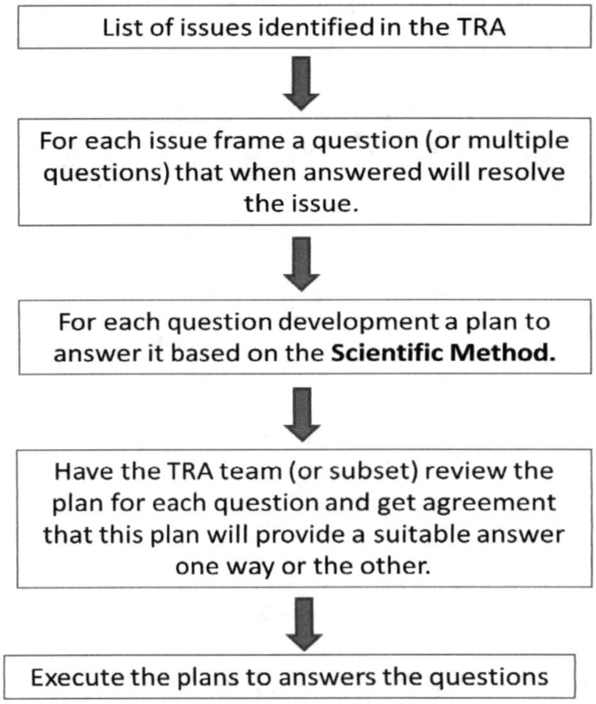

Figure 4: Resolving issues identified in the TRA

The output from the TRA is a list of issues that can be resolved into questions. These are grouped together based on similarity or because of common root issues and then ordered based on the TRA score. Questions whose answers obviate the need for answering others should be done first. If useful, questions can be simplified and/or converted into multiple questions. A plan is devised to answer a question unequivocally. A good practice is to design the plan based on the Scientific Method (2007 Lafferty). The goal here is to answer the question posed and not come up with new alternatives or necessarily solve the problem. A technical project decision to Proceed, Recycle or Stop must be data-driven based on the answered questions. Here is an opportunity to manage and mitigate technical risk without stifling creativity or enthusiasm.

If an answer uncovers a fatal flaw with the technology then the answer is clearly "Stop" this particular project. However, the new information might suggest an alternate approach or other ideas worth pursing possibly that are completely unrelated to the original technology. This could be the seed for a new development effort as a strategic research project. Alternatively, the answer might show that the uncertainties of the results are too high as indicated by error bars on the data. In this case, it's not that the proposed idea is fatally flawed but rather sometimes "it" works and other times not but we're not sure why. Another case is just that more information or data is needed, for instance, to further define the operating envelope of the process. These are an opportunity to put the forward project schedule on hold while the project development effort is in "Recycle". The duration of the Recycle time period depends on the situation. In the first case described above, the uncertainty of resolving why the desired results happen only part of the time could take a long time. In this case it may be better to stop the project and continue just the R&D program effort. The second case described above of getting more information is easier to time rate its duration. So in this case it may be prudent to not demobilize the project team. Alternatively if other opportunities are deemed nearer to fruition, then the project can be moved back into a research phase until a more significant "Aha" moment has happened.

These are business decisions made by the stakeholders informed by the technology development project team. Whether the decision is to Proceed, Recycle or Stop communications must be made with stakeholders and the management chain. This is a secondary but still very important benefit of the TRA. It provides transparency and documents the data-driven decisions which are made as a result.

Remember the following:

- Don't skip answering critical questions and rely on wishful thinking.
- Formulate plans to answer the questions using the principles of the Scientific Method.
- Focus work and resources on answering the key root questions first.
- Use the TRA as a communication tool with the team and stakeholders.

Conclusion

This paper has introduced the concept of a Technical Risk Analysis and discussed its use as a systematic method to manage and reduce process technology development risk. Three types of technology risks were described: *Fundamental-Science, Engineering-Science* and *Engineering*. The TRA is a systematic but flexible guided-exercise, the output of which is a register of issues to address and questions to be answered along with their relative importance based on the risk tolerance metrics of the stakeholders. The answers to these issues and questions will reduce the technological risk component and may suggest other development opportunities. A TRA is useful at any point along the development path but has a high impact for new technology development at the early concept stage. The review team should include both technical team members for the project as well as outside technical experts providing a "cold-eye" perspective. The TRA should be used in conjunction with a Techno-Economic Assessment (TEA) to increase the likelihood of commercial success. As is already the case for the HAZOP and PHA, the TRA methodology is a tool that should be included in the process technology developer's engineering toolbox.

References

Euhus, D.D., Sciamanna, S.F., Bryan, P.F., "*Pathway to Process Technology Commercialization*", Paper 115d, Proceedings of the 2013 AIChE National Meeting, San Francisco, CA

Huyse, L., "*Using Technology Classification and Qualification Status as a Tool for Strategic Technology Screening and Selection*", OTC-25419-MS, 2014 Offshore Technology Conference, Houston, TX

US Department of Energy "Technology Readiness Assessment Guide" https://www.directives.doe.gov/directives-documents/400-series/0413.3-EGuide-04, 2009

2007 Lafferty http://stem.aihec.org/Projects/airei/Energy%20Institute%20Materials/scientific-method-handout.pdf

AIChE November 2014 Annual Meeting

Presentation Abstract to Session 12C01: "Reducing Risk during Technology Transfer"

William G. Etzkorn
Mid-Atlantic Technology Research & Innovation Center
E-mail: William.etzkorn@matricinnovates.com
Work Phone: 304-720-8612 extension: 936
Mobile Phone: 304-552-1342

Knowledge Custodian – Incorporation of Technology Expert Role Enhances Effective Communication throughout Project Implementation

The importance of the effective and timely communication of know-how and technical information is highlighted in practically all project management guidelines and project critiques. An approach to project implementation that focuses on facilitating communication promises to streamline knowledge flow and directly influence efficiency throughout all stages of the project.

During the chemistry, conceptual, and process development in the early FEL-stages, fundamental knowledge of the process, pilot plant design, and operation is established. This process-specific knowledge, especially when combined with a breadth of chemical industry experience, creates a significant potential source of value that can be tapped during technology transfers throughout the project. In a typical project a FEL-1 or 2 provider turns over a package of technical data to a (FEL-3) process engineering firm, who, in-turn, passes a design package to an EPC firm, whom then finally turns over the reins to the client/owner. This approach leaves many opportunities for miscommunication, loss of implementation efficiency, etc. Creating an active project role for a "Knowledge Custodian" – a technology/chemistry/process expert that will follow the process through all stages of project development from conceptualization through to start-up and operation – presents opportunity to improve communication and increase project implementation success. Such a "Knowledge Custodian" will act as a process technology caretaker to significantly improve the probability of a successful project and secure work process efficiency saving.

Opportunity to improve knowledge transfer and facilitate communication achieved with a Knowledge Custodian include:

1. The perception of "ownership" by project technical personnel is important to drive the project to success. Project ownership is fostered by involving key technical people in the project from formative stages through to commercial implementation.

2. Know-how developed in FEL-0 and FEL-1 is effectively incorporated in the development of FEL-2, and is efficiently transferred to FEL-3 (Process Design) and then to EPC to optimize performance of the project and streamline the timeline for process engineering and operations implementation.

Mid-Atlantic Technology Research and Innovation (MATRIC) has established a reputation for developing early-stage chemistry and process technology fundamentals. Extending the availability of this knowledge during all phases of technology transfer ("Knowledge Custodian" role) over the course of the project significantly improves communication and streamlines the transfer of technology. More successful project implementation and shorter implementation times result by applying this approach.

While large chemical companies with significant in-house R&D resources have had opportunity to apply similar approaches to projects, this has been done only sporadically. We propose this approach as a routine, planned project implementation work process, and demonstrate how early-stage technology development resources such as at MATRIC can bring these project advantages to start-up and/or smaller-sized chemical company's projects.

370726 Treatment and Reuse of a Brine Waste Stream

Abstract:

The reaction to make polycarbonate generates a byproduct brine stream. The byproduct stream has retained value that can only be claimed if its quality is improved via the removal of various compounds including TOC (Total Organic Carbon). The current study investigates the treatment of such brine streams containing high ppm levels of organics. Examples of these compounds include sodium gluconate, BPA, TEA, sebacic acid, methylene chloride, and others.

The purification of the brine samples is done via the introduction of the solution on NORIT GAC 830R activated carbon bed.

Both batch and continuous modes of operation are presented in this work. The resulting brine stream is suitable for use in an electrolysis reaction to produce chlorine and caustic soda

392104 Analysis and Interpretation of a Process Risk Assessment

Abstract:

A case study will be presented on a late stage quality risk assessment carried out on a manufacturing process using Failure Mode and Effects Analysis (FMEA). The case study will focus on demonstrating methods and techniques that can be used to analyze the results of a risk assessment to extract the most learnings and value from the exercise. In particular, categorization of risk reduction actions and data visualization are leveraged to better understand the current risk associated with a process and determine the level of risk that may need to be accepted moving forwards. The outcome is a comprehensive framework for making process changes geared towards reducing risk on scale up.

Design and Operational Analysis of Integrated Gasification and Steam Methane Reforming

The demand for hydrogen in industries continues to rise with increased plant capacities and is currently met by steam reforming or gasification of carbonaceous feedstock. The most preferred feedstock for the steam reforming process has been natural gas, while for gasification it has been coal. Steam reforming of natural gas is a highly endothermic and catalytic process that requires external heat to drive the reforming reactions, orthodoxly supplied by combusting fossil fuels in a furnace chamber. On the other hand, gasification in entrained flow gasifiers is a high temperature process where the product synthesis gas exits the gasifier at a temperature of around 1300 °C to 1400 °C. The coal-derived synthesis gas, with a low molar H_2/CO ratio of 0.75-1.1, needs to be cooled down to 150 °C for downstream acid gas treatment. Conventional methods employ radiant coolers with tubes for steam generation and quench cooling to provide the required cooling duty. This work presents the model development and design options of a hybrid system that integrates gasification with catalytic natural gas steam reforming by utilizing the heat duty available from the hot coal-derived syngas to drive the endothermic reaction inside the tubes of the radiant cooler. The advantages of the proposed system are process intensification by integrating two independent processes, production of a more valuable product using available exergy, improving the molar H_2/CO ratio of coal-derived synthesis gas and ability of a gasifier to respond to transient demands of a polygeneration plant.

Systems level analysis by Adams and Barton [1] has shown that the new cooling strategy is both profitable and technically feasible. However, the design and operational stability of the proposed system has to be studied in detail under transient operating conditions. A multi-scale, dynamic, heterogeneous, two-dimensional model of the radiant syngas cooler and steam methane reforming has been developed that can track temporal and spatial variations in concentration and temperature at the catalyst particle level. This work will discuss model development and validation approach adopted for a new system without supporting experimental data, associated computational complexities and the procedure to design an integrated system that meets the operating constraints. Simulation results show that the process is feasible providing the required cooling duty to the coal-derived synthesis gas and at the same time, achieving a higher conversion of methane at par with conventional reforming reactors. Also, an analysis of different design configurations considered will be presented with optimal design parameters for each of the case.

REFERENCE:

[1] T. A. Adams II and P. I. Barton, "Combining coal gasification and natural gas reforming for efficient polygeneration," *Fuel Process. Technol.*, vol. 92, no. 3, pp. 639–655, Mar. 2011.

TECHNION
Israel Institute
of Technology

Demonstration and Design Principles of an Efficient Scaled Down Autothermal or Heated Membrane Reformer for Pure Hydrogen Production

Michael Patrascu, Moshe Sheintuch
Department of Chemical Engineering
Technion-IIT, Haifa, Israel

IETHyCo

New Energy World JU
fuel cells & hydrogen for sustainability

Outline

TECHNION
Israel Institute
of Technology

- Introduction

1. Heated (solar) membrane reformer
    - Reactor performance: Experimental and theoretical
    - Membrane inhibition
2. Autothermal membrane reformer
    - Thermodynamically limited performance
    - Distributed recycle
    - Experimental system and results

- Conclusions and future work

IETHyCo

Introduction – the scale down

TECHNION
Israel Institute of Technology

$$MSR: \quad CH_4 + H_2O \rightleftarrows CO + 3H_2 \quad \Delta H = 206 \ kJ/mole$$

$$WGS: \quad CO + H_2O \rightleftarrows CO_2 + H_2 \quad \Delta H = -41 \ kJ/mole$$

http://www.lindele.de/process_plants/hydrogen_syngas_plants/gas_products/hydrogen.php

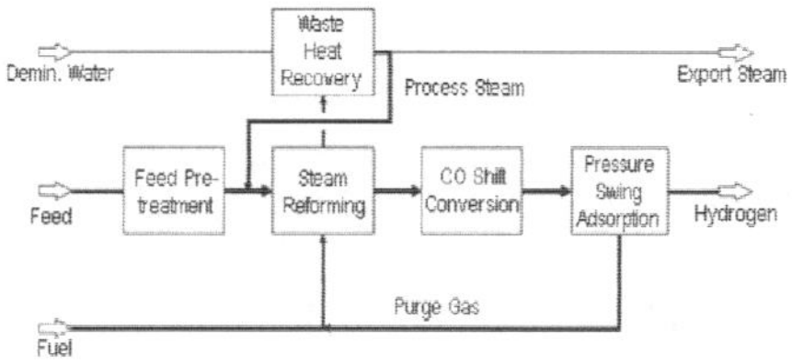

Main Methane Steam Reforming limitations:

1. Equilibrium
2. Endothermic
3. Purification of H2 is required

Scale Down:
10^3-10^5 → 10^{-1}-10^1 Nm^3/hr

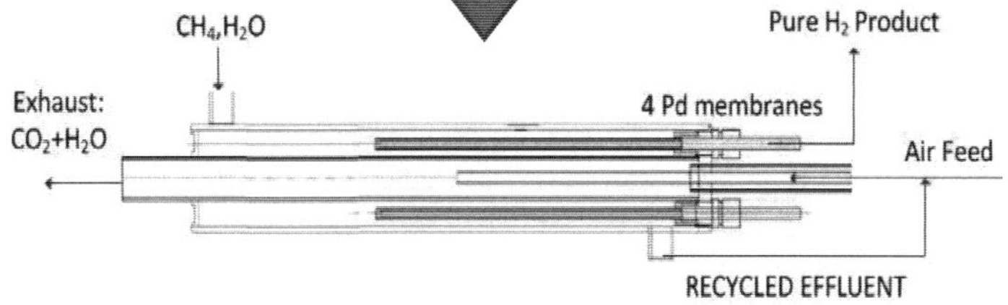

IETHyCo

Introduction - the Pd membrane effect

TECHNION
Israel Institute of Technology

$$MSR: \quad CH_4 + H_2O \rightleftarrows CO + 3H_2 \quad \Delta H = 206 \ kJ/mole$$

$$WGS: \quad CO + H_2O \rightleftarrows CO_2 + H_2 \quad \Delta H = -41 \ kJ/mole$$

$$combined: \quad CH_4 + 2H_2O \rightleftarrows CO_2 + 4H_2 \quad \Delta H = 165 \ kJ/mole$$

Stoichiometric feed, P=10bar

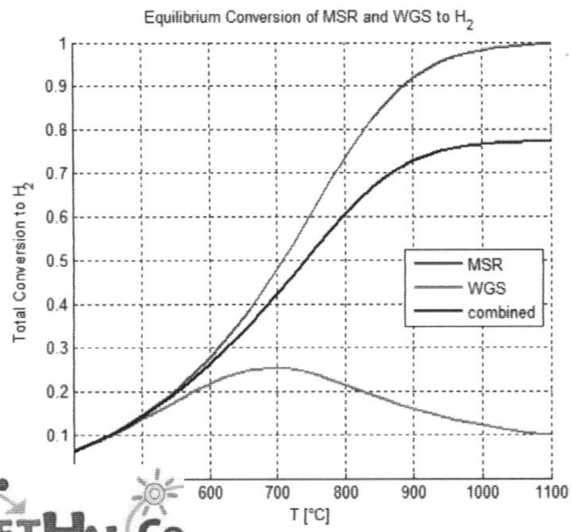

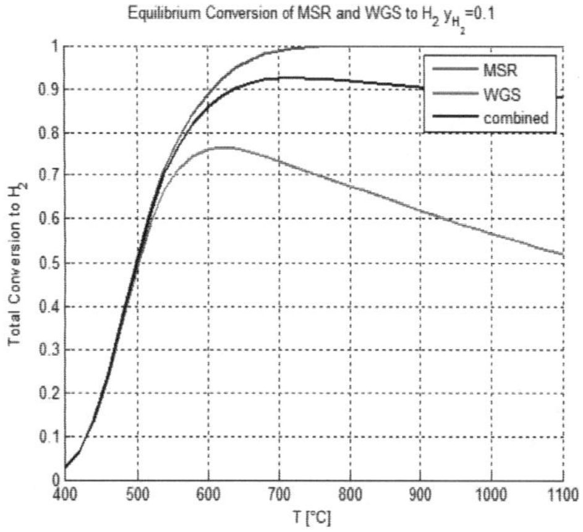

IETHyCo

Heated (solar) membrane reformer

TECHNION
Israel Institute
of Technology

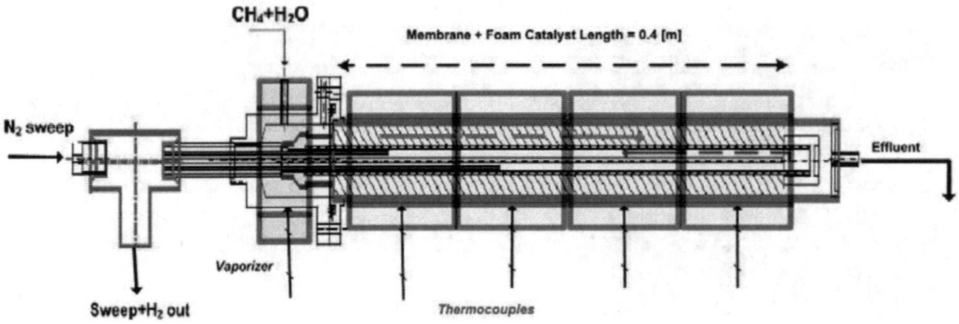

$$\frac{dy_i}{d\tau}=\frac{1}{Pe_c}\frac{d^2y_i}{d\zeta^2}-\left(\frac{d\tilde{F}}{d\zeta}y_i+\frac{dy_i}{d\zeta}\tilde{F}\right)+\frac{1}{M_f}\sum_{j=1}^{3}\alpha_{i,j}r_j-\kappa_i y_{CH_4,0}Pe_m^{-1}\left(\sqrt{y_i}-\sqrt{P_r y_i^{perm}}\right)$$

$$\frac{t_T}{t_R}\frac{\partial\theta}{\partial\tau}=\frac{1}{Pe}\frac{\partial^2\theta}{\partial\zeta^2}-\frac{\partial\theta}{\partial\zeta}+\frac{1}{E_f}\sum_j(-\Delta H_j)r_j-U(\theta-\theta_w)$$

IETHyCo

Effect of flow rate

TECHNION
Israel Institute
of Technology

$$Pe_m=\frac{\text{CH}_4\text{ molar feed flow}}{\text{max H}_2\text{ transmembrane flow}}=\frac{F_{CH_4,0}}{S_m J_m\sqrt{P}}$$

$$J_m=A_m\exp\left(-E_m/RT_{ref}\right)\left[mol\left(H_2\right)/m^2s\sqrt{bar}\right]$$

$$\text{Sieverts law flux}: Q_m=J_m\left(P_{H_2}{}^{retentate}-P_{H_2}{}^{permeate}\right)$$

GHSV= 50 – 1100 [1/hr]

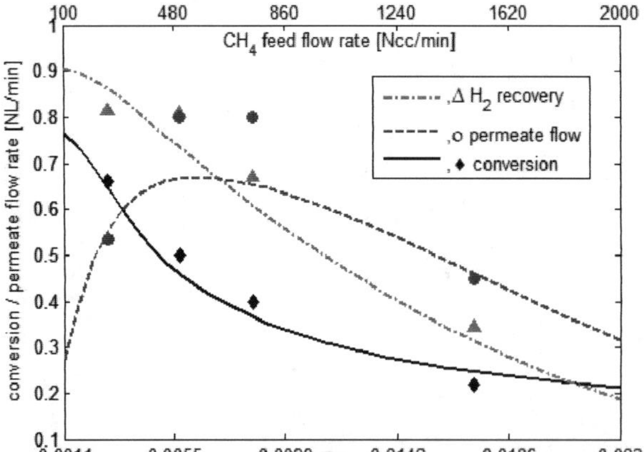

Patrascu at al., Chem. Eng. J., accepted (October 2014)

Conversion, permeate flow rate, and H_2 recovery vs feed flow rate (T_w=525°C, P=10bar ,S/C=3, no sweep). Lower abscissa indicates Pe_m number and the upper one indicates the corresponding CH_4 feed flow rate (Ncc/min).

IETHyCo

Effect of pressure and sweep

TECHNION
Israel Institute
of Technology

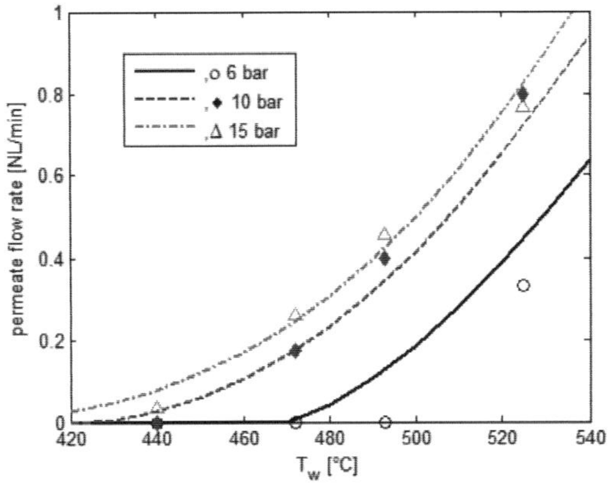

Permeate Flow rate vs. wall temperature . Effect of reaction (retentate) pressure. CH_4 feed flow rate = 0.5 NL/min, no sweep.

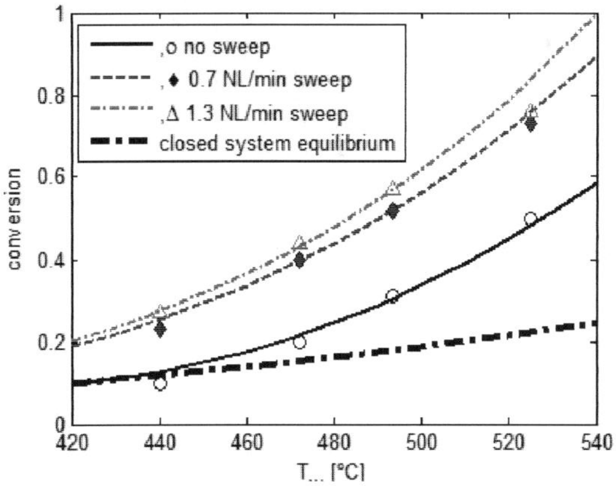

Conversion vs. wall temperature . Effect of sweep flow rate. CH_4 feed flow rate = 0.5 NL/min, P=10 bar

IETHy Co

Patrascu at al., Chem. Eng. J., accepted (October 2014)

Effect of composition – evidence of permeance inhibition

TECHNION
Israel Institute
of Technology

All at same H_2 partial pressure

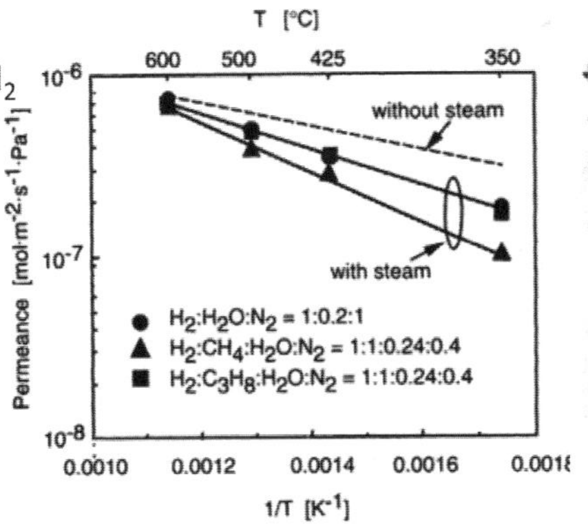

Jung et al, Journal of Membrane Science 170 (2000) 53–60

T = 400 °C, *P* = 20 bar,

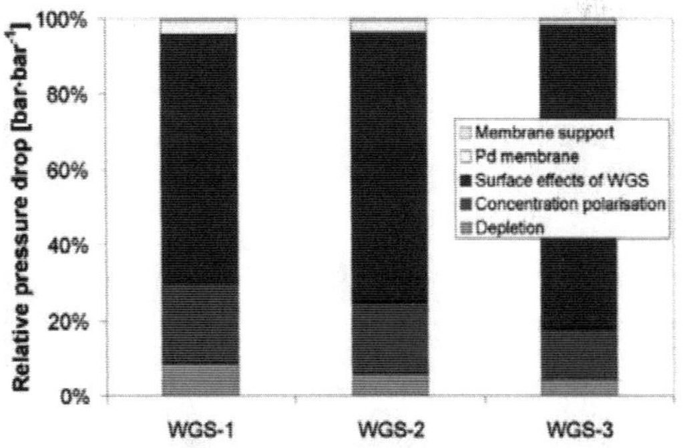

Gas	WGS-1 (%)	WGS-2 (%)	WGS-3 (%)
H_2	60.2	30	10
CO_2	19.2	40.5	60
H_2O	15.4	25	25.5
CO	4	2	2
CH_4	1.2	2.5	2.5

T.A. Peters et al. , Journal of Membrane Science 316 (2008) 119–127

IETHy Co

Measurements in unpacked annular system

Effect of pressure - revised

TECHNION
Israel Institute
of Technology

$$J_m = \Theta A_m \exp\left(-\frac{E_m}{RT}\right)\left[\frac{mol}{m^2\, s\sqrt{bar}}\right]$$

$$\Theta = \frac{1+\sqrt{K_H P_{H_2}}}{1+\sqrt{K_H P_{H_2}} + K_{H_2O}P_{H_2O} + \left(K_{CO}P_{CO}\right)^{1/3} + \left(K_{CO_2}P_{CO_2}\right)^{1/2} + K_{CH_4}P_{CH_4}} \quad (*)$$

Permeance in pure H_2 was measured to be $\sim 90\left[m^3/m^2 hr\sqrt{bar}\right]$ at 500 ∘ C(**). Results of reactor tests reflect ~80% in permeance reduction.

The cause might be competition by species formed by **reaction** on the Pd surface.

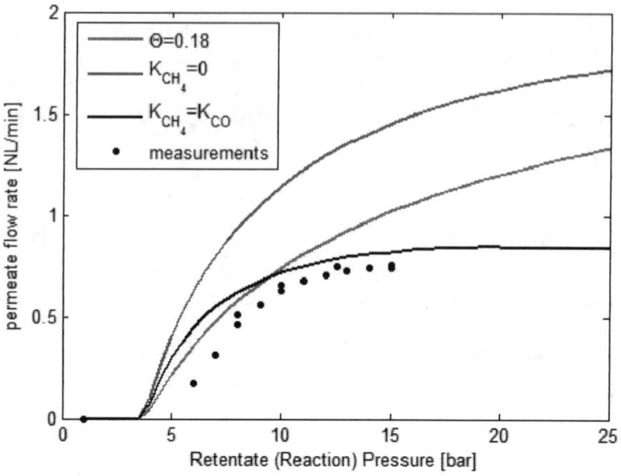

IETHyCo

(*) values adopted from: Israni S.H and Harold M.P, Ind. Eng. Chem. Res. **2010,** *49,* 10242–10250
(**) as reported by ECN, the membrane producer.

Autothermal membrane reformer

TECHNION
Israel Institute
of Technology

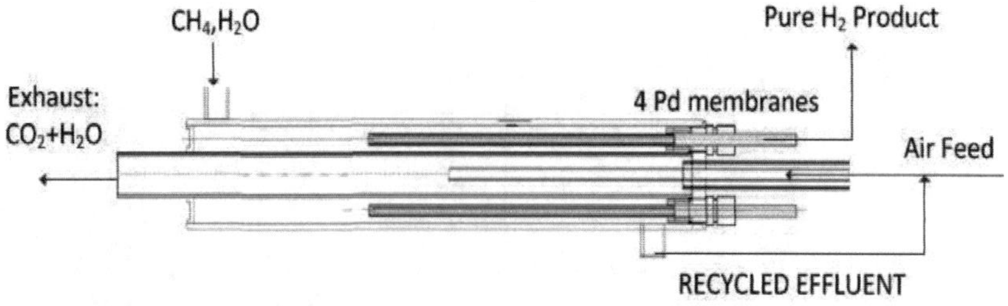

<u>Design considerations:</u>

1. Use of **air** as oxidizer.
2. **Separated compartments** for reforming and Oxidation – Avoid dilution of H_2.
3. Supply pure **H_2 at atmospheric** pressure.
4. Axially **distribute fuel** to oxidation compartment.
5. Counter current for **heat recuperation**.

IETHyCo

Asymptotic integral performance

$$efficiency:\ \eta = \frac{H_2 \text{ flow through membrane} \cdot LHV_{H_2}}{\text{total } CH_4 \text{ fed to the system} \cdot LHV_{CH_4}} = \frac{F_H^m}{\left(F_{CH_4}^{SR} + F_{CH_4}^{Ox}\right)3.3}$$

1. Equilibrium limited (very fast kinetics)
2. Maximum separation of H_2 (infinite membrane area)

$$x = f^{eq}_{(T\_work, P, P_m)}$$

3. Autothermal Adiabatic

$$0 = \left(T_{exit} - T_{feed}\right) C_p^{Ox} F$$
$$+ F_{H_2O}^{SR} \Delta H_{vap} + x \cdot F_C$$

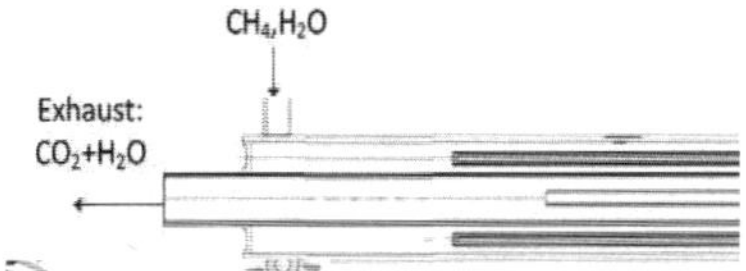

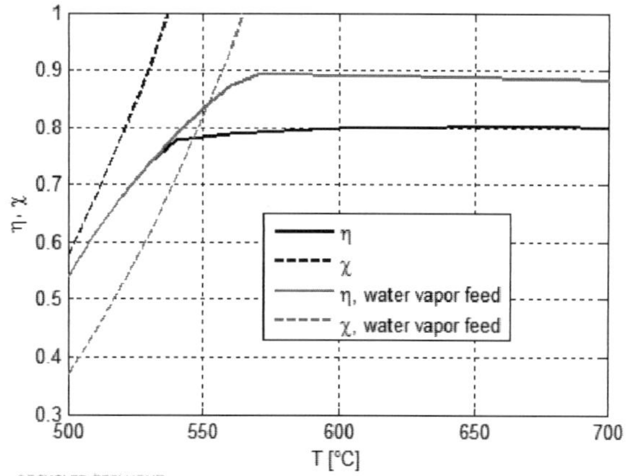

RECYCLED EFFLUENT

T_{exit}=350°C

1D pseudo-homogeneous model

changing # of recycled feed ports

$$Pe_m = \frac{\text{feed molar flow}}{\max H_2 \text{ molar flow through membrane}} = \frac{F_0}{S_m J_m \sqrt{P}}$$

$$\eta = \frac{H_2 \text{ flow through membrane} \cdot LHV_{H_2}}{\text{total } CH_4 \text{ fed to the system} \cdot LHV_{CH_4}} = \frac{F_H^m}{\left(F_{CH_4}^{SR} + F_{CH_4}^{Ox}\right)3.32}$$

$$Power\ [kW] = F_H^m \cdot 237 \cdot 0.6$$

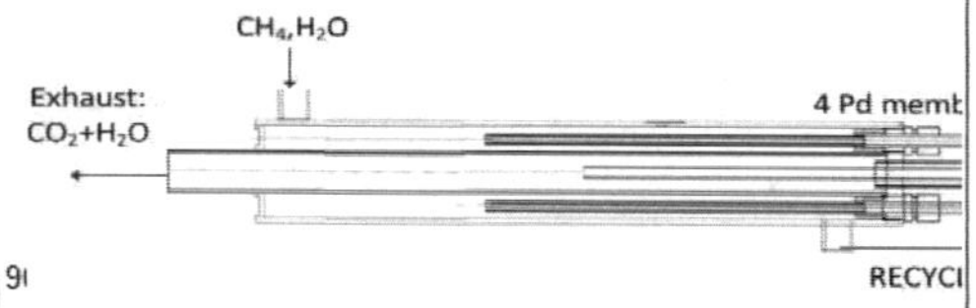

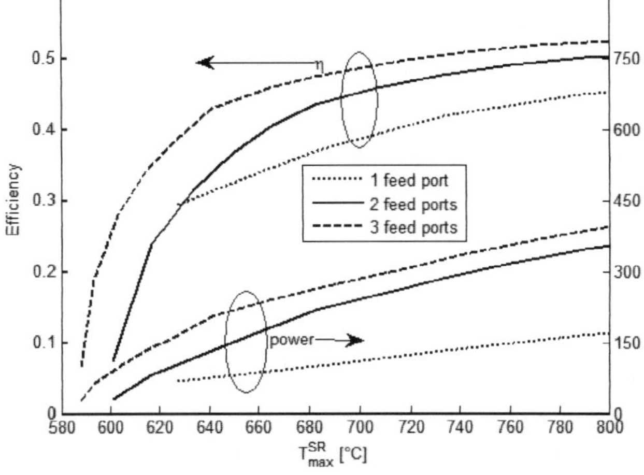

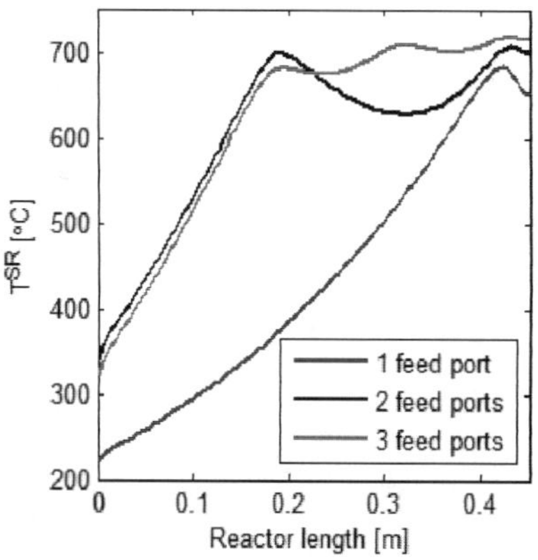

Experimental results – 2 feeds

TECHNION
Israel Institute
of Technology

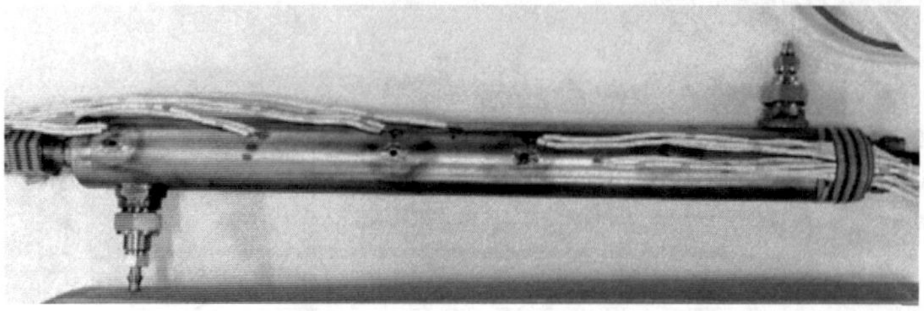

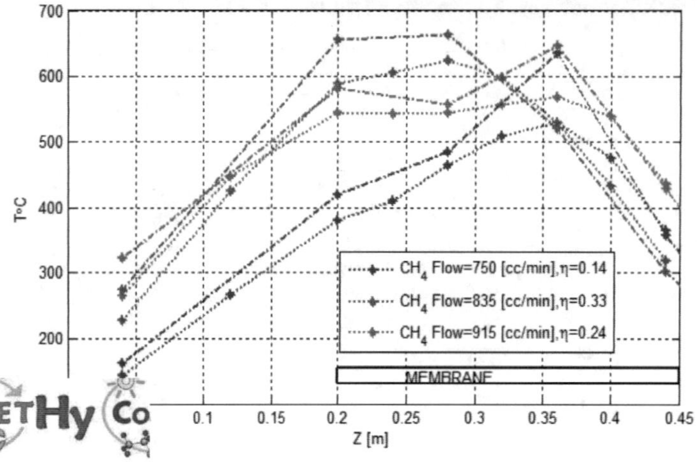

1. Up to 33% efficiency measured in lab scale system.
2. Only recycled effluent sustains the system.
3. Pure atmospheric pressure H_2 product.
4. No need for vaporizer.

Conclusions and future work

TECHNION
Israel Institute
of Technology

- Adsorption and possibly surface reactions of MSR components (specifically of CH_4) should be quantitatively studied.
- Consider using water vapor as sweep in on-site application (and possibly in on-board as well).
- What to do with unconverted methane?
- In on-board system recycle of unconverted methane is enough to supply the necessary heat and sustain the system.
- Almost isothermal operation can be obtained by distribution of recycle.
- Increase of membrane area relative to outer surface area may improve efficiency (lower relative heat losses).

Acknowledgements:

Work conducted within the framework of the CoMETHy (Compact Multifuel-Energy to Hydrogen converter) project, funded by European Union's Seventh Framework Programme (FP7/2007-2013) for the Fuel Cells and Hydrogen Joint Technology Initiative under grant agreement n. 279075. The experimental equipment was partially supported by Grand Technion Energy Program (GTEP).

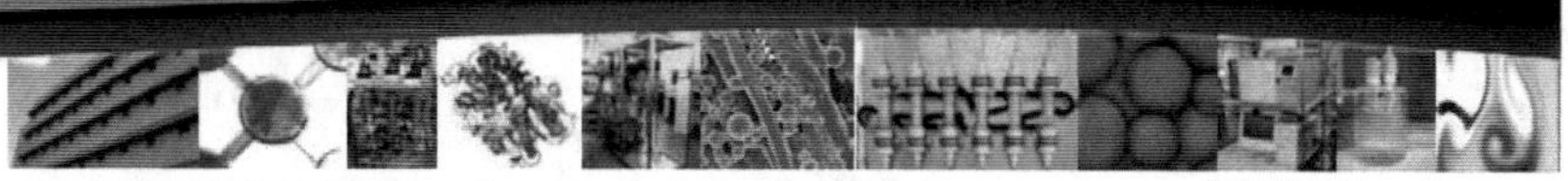

FEUP FACULDADE DE ENGENHARIA UNIVERSIDADE DO PORTO
Departamento de Engenharia Química
Rua Dr. Roberto Frias, S/N | 4200-465 Porto| Portugal
http://lsre.fe.up.pt
lsre@fe.up.pt

Synthesis of oxygenated compounds using the simulated moving bed reactor

Synthesis of the acetal 1,1-diethoxybutane (DEB).

2014 AIChE Annual Meeting
Atlanta

Nuno Graça, Mehabub Rahaman, Carla Pereira, Alírio Rodrigues

November 19, 2014

Outline

- Introduction
  - Reaction
  - Simulated Moving Bed Reactor Concept
  - Methodology
- The Batch Reactor: Thermodynamics and Kinetics
- The Fixed Bed Adsorptive Reactor
- The Simulated Moving Bed Adsorptive Reactor
- Conclusions

Reaction

- 1,1-Diethoxybutane Synthesis Reaction

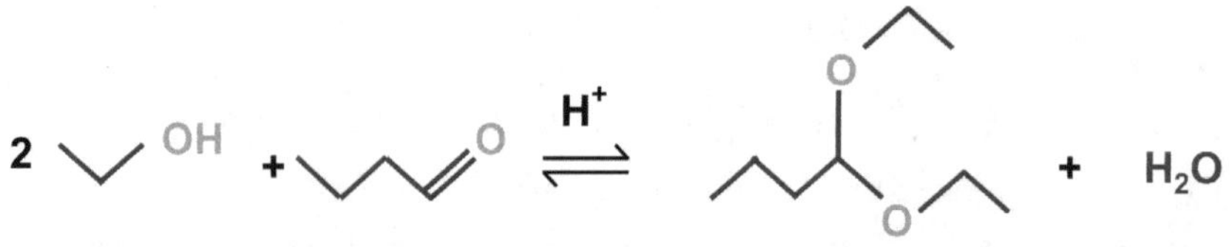

Ethanol   Butyraldehyde   1,1-Diethoxybutane   Water

Equilibrium conversion (T=20°C) ≈ 50%

Catalysed by the ion-exchange resin Amberlyst-47®

Process Intensification

Equilibrium limited reactions

$$A + B \Leftrightarrow C + D$$

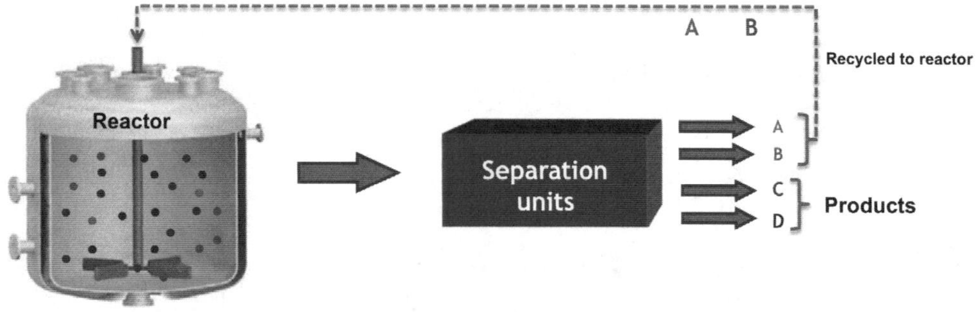

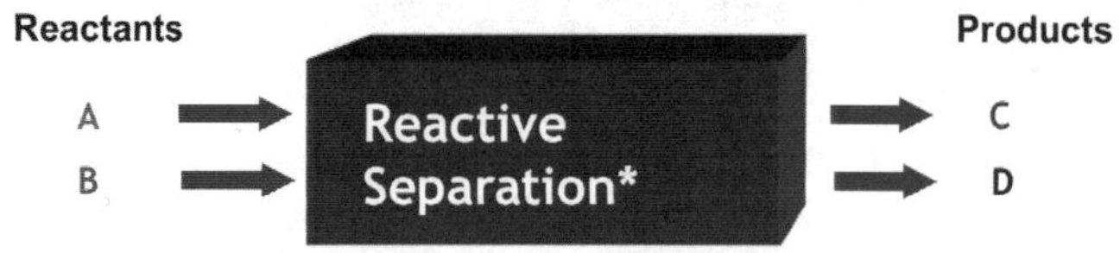

*Better energy efficiency; Conversion beyond the equilibrium value; Productivity improvement; Lower solvent consumption; Smaller production plants.

TMB and SMB concepts

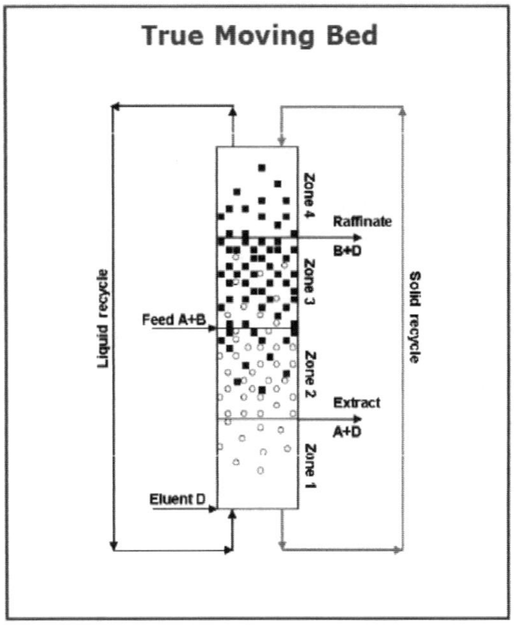

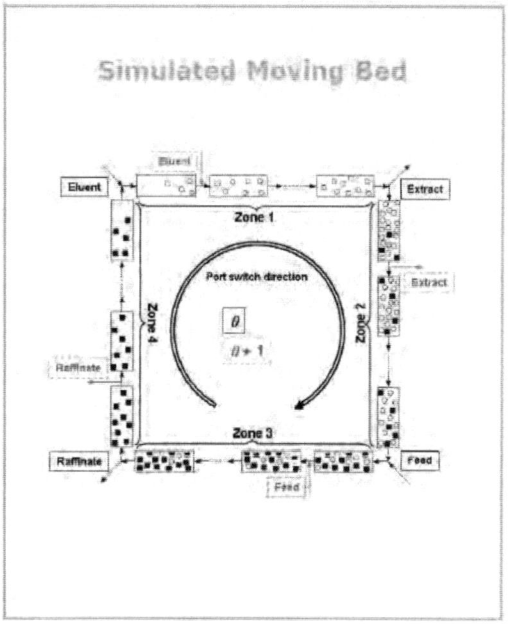

o - More retained component ■ - Less retained component

TMB and SMB equivalence

- The equivalence between the TMB and the SMB model can be made by constant the liquid velocity relative to solid velocity.

$$v_j^{TMB} = v_j^{SMB} - u_s$$

- The solid velocity must be evaluated from the value of switch time interval.

$$u_s = \frac{L_C}{t^*}$$

Simulated Moving Bed Reactor

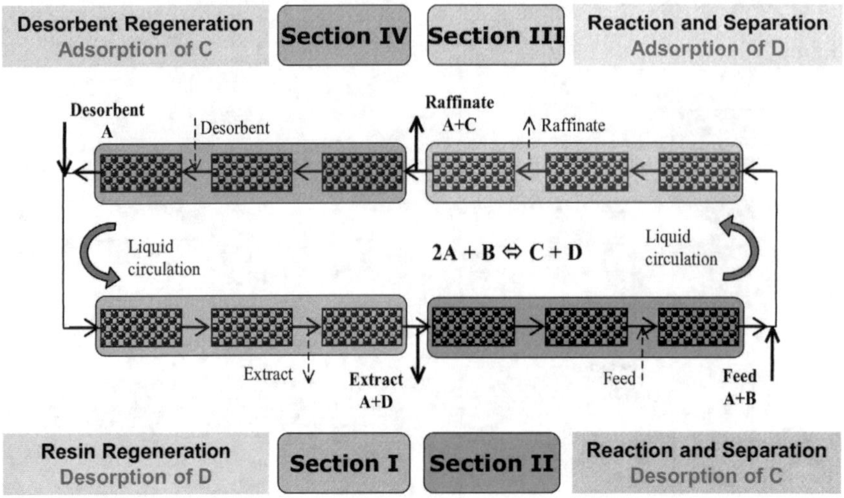

Simulated Moving Bed Reactor

— Reactant

— More retained product

— Less retained product

$$A \rightleftharpoons B + C$$

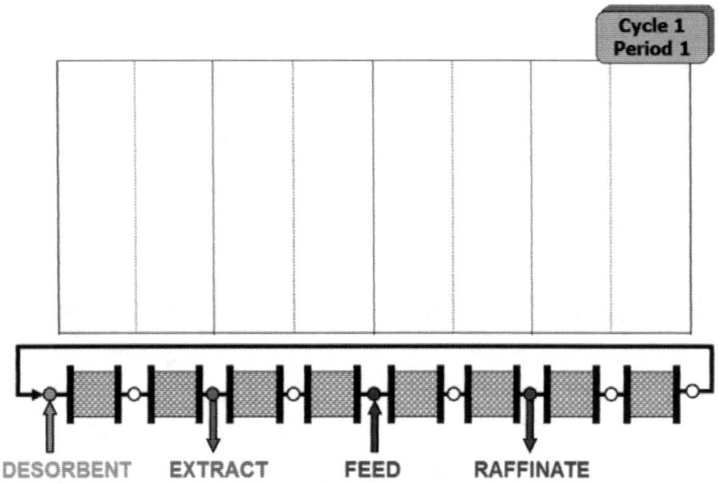

Methodology

Batch Reactor → **Thermodynamic and kinetics**

Fixed-Bed Reactor
- Non reactive pairs → **Adsorption Parameters**
- Reactive mixture → **Parameters and model validation**

Simulated Moving Bed Reactor → **Study of the SMBR operation experimentally and by simulation**

Batch Reactor

Batch Reactor

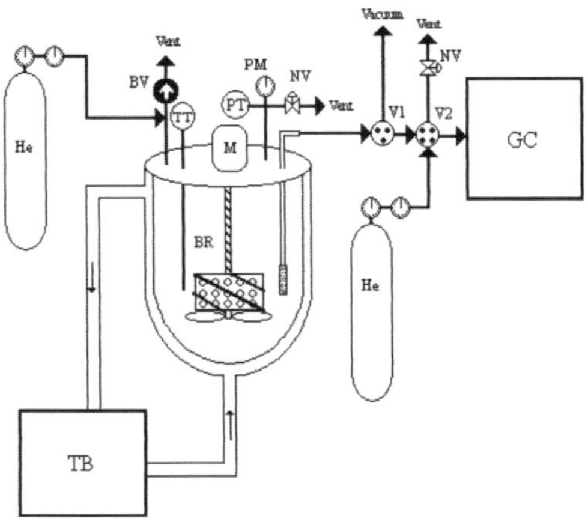

Operating Conditions:

Catalyst: Amberlyst 47

Temperature Range: 283-323 K

Mass of Dry Catalyst: 1.8-3 g

Reaction Volume: ≈ 530 ml

Initial Molar Ratio of Reactants: 1.5-3

Kinetic Model:

$$r = k_c \frac{a_A\, a_B - \dfrac{a_C\, a_D}{K_{eq}\, a_A}}{\left(1 + K_D\, a_D\right)^2}$$

Experimental Results

Effect of Temperature:

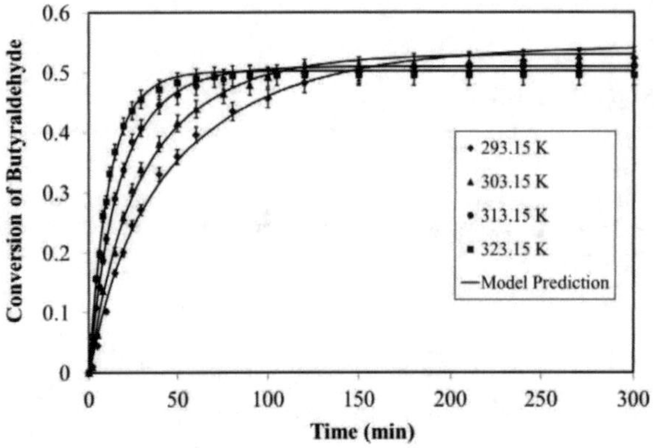

Influence of temperature on the conversion of acetaldehyde history.

$0.5<d_p<0.6$ mm

P = 6 atm

$r_{A/B}$ = 2.2

w_{cat} = 1.8 g

V = 530 ml

Equilibrium Constant

Equilibrium Constant

- **Experimental equilibrium composition were determined;**
- **The activity coefficients were evaluated by the UNIFAC method, for each equilibrium composition.**
- **The equilibrium constant was calculated according to:**

Experimental conditions

Pressure: 1.0 MPa

Initial molar ratio of reactants: 2.2

Equilibrium composition measured experimentally | **Activity coefficients computed by UNIFAC**

$$K_{eq} = \frac{a_{C_e}\, a_{D_e}}{a_{A_e}\, a_{B_e}} = \frac{x_{C_e}\, x_{D_e}}{x_{A_e}\, x_{B_e}} \cdot \frac{\gamma_C\, \gamma_D}{\gamma_A\, \gamma_B} = K_X\, K_\gamma$$

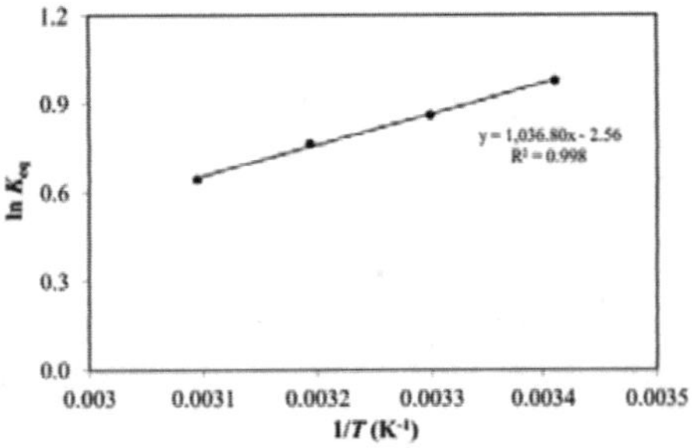

- **The dependence of the equilibrium constant on temperature was found by the Van't Hoff Equation:**

$$\ln K_{eq} = \frac{\Delta S^o}{R} - \frac{\Delta H^o}{RT} \Rightarrow \ln K_{eq} = -2.56 + \frac{1036.8}{T}$$

$\Delta H^o = -8619.7\, J\, mol^{-1}$

Exothermic

Model Parameters

Reactor Model

Mass balance in the bulk fluid:

$$\frac{dC_{\mathrm{b},j}}{dt} = -\frac{A_{\mathrm{p}}}{V_{\mathrm{liq}}} D_{e,j} \left.\frac{\partial C_{\mathrm{p},j}}{\partial r}\right|_{r=r_{\mathrm{p}}} \qquad (j = \mathrm{A}-\mathrm{D})$$

Mass balance in the particle:

$$\varepsilon_{\mathrm{p}} \frac{\partial C_{\mathrm{p},j}}{\partial t} = \frac{1}{r^2}\frac{\partial}{\partial r}\left[D_{e,j} r^2 \frac{\partial C_{\mathrm{p},j}}{\partial r}\right] + (1-\varepsilon_{\mathrm{p}})\nu_j \rho_{\mathrm{p}} \Re_{\mathrm{p}}$$

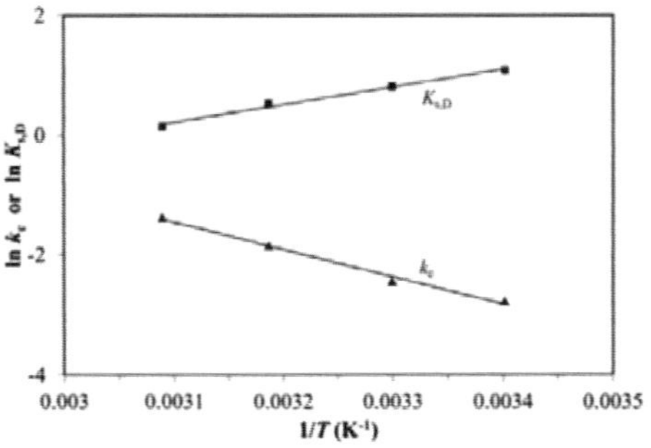

Kinetic and water adsorption constants.

T (°C)	k_C (mol/g.min)	$K_{s,D}$ (___)
20	0.061	2.939
30	0.086	2.238
40	0.156	1.699
50	0.251	1.157

Arrhenius equation

$$k_C = k_{0,C}\exp\left(-\frac{E_a}{RT}\right) \quad \Rightarrow \quad E_a = 37.92\ \mathrm{kJ\ mol^{-1}}$$

Reaction Activation Energy

Van't Hoff Equation

$$K_{s,i} = K_{0s,i}\exp\left(-\frac{\Delta H_{s,i}}{RT}\right) \quad \Rightarrow \quad \Delta H_s = -24.28\ \mathrm{kJ\ mol^{-1}}$$

Adsorption Enthalpy

Fixed-Bed Reactor

Fixed-Bed Reactor

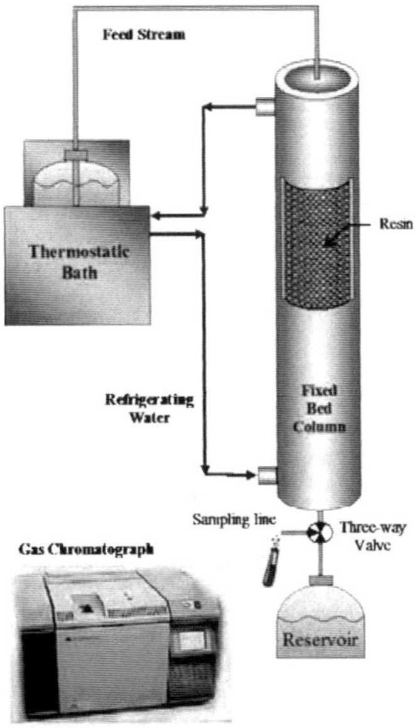

Characteristics of the fixed bed column.

Solid weight	28.4 g
Length of the bed	11.1 cm
Internal diameter	2.6 cm
Bed porosity	0.36
Bulk density	482 kg/m^3

Experimental and Simulated Breakthrough Curves

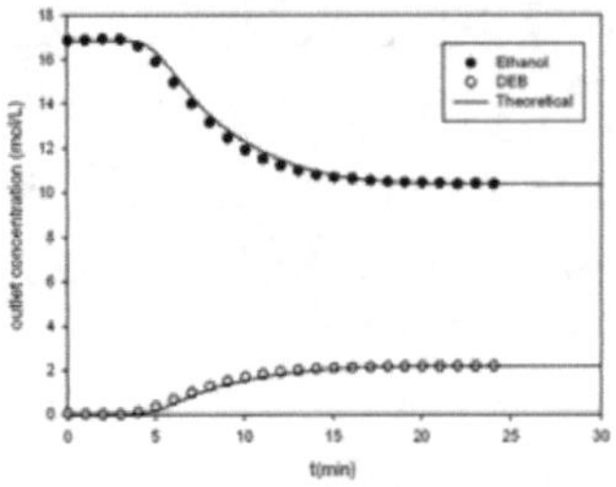

DEB displacing Ethanol, T=30°C.

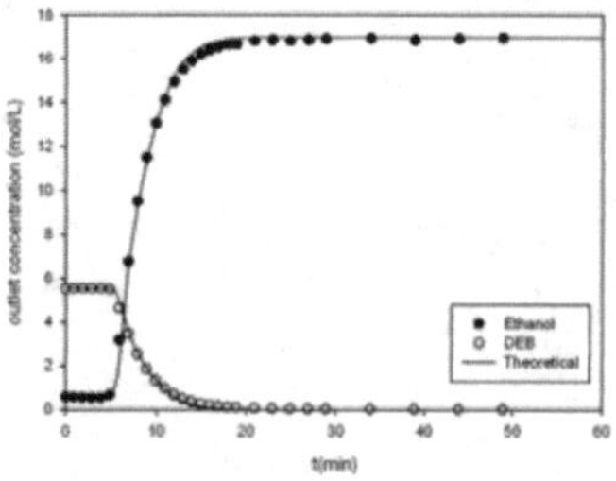

Ethanol displacing DEB, T=30°C.

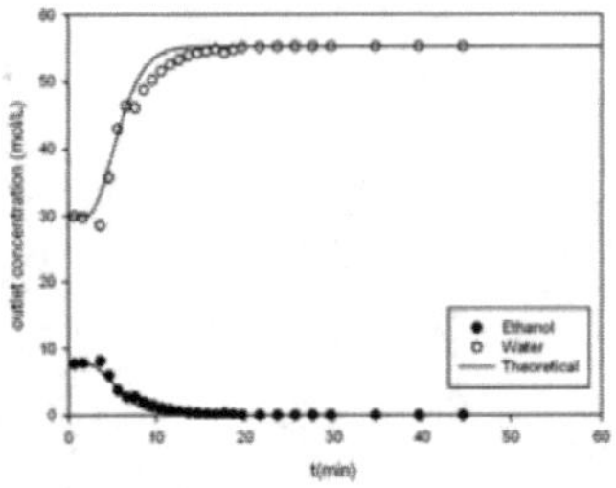

Water displacing Etahnol, T=30°C.

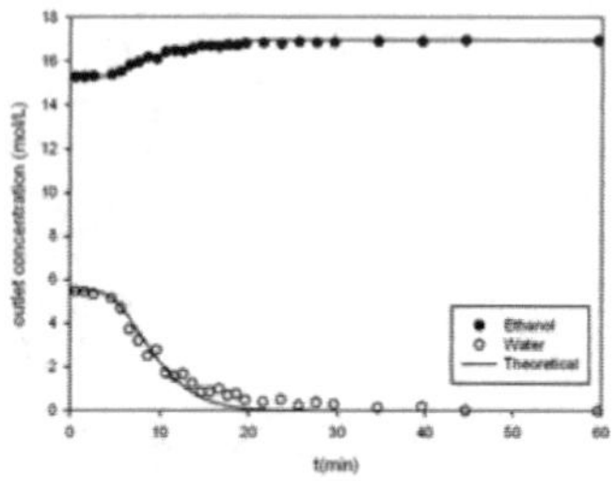

Ethanol displacing Water, T=30°C.

Adsorption Parameters

Langmuir Model:

$$\overline{q}_i = \frac{Q_i K_i \overline{C}_{p,i}}{1 + \sum_{j=1}^{NC} K_j \overline{C}_{p,j}}$$

Amount adsorbed:

$$n_{\exp}^{ads} = Q \int_0^{\infty} \left[C_f - C_{out}(t) \right] dt$$

$$n_{theo}^{ads} = \left[\varepsilon + (1-\varepsilon)\varepsilon_p \right] \left(C_f - C_0 \right) V_c + (1-\varepsilon)(1-\varepsilon_p) \left[q(C_f) - q(C_0) \right] V_c$$

Component	*Q (mol/ l)*	*K (mol/ l)*
Ethanol	10.58	7.24
Butyraldehyde	3.08	0.12
DEB	3.59	0.06
Water	34.47	8.18

Langmuir isotherm parameters

Fixed-Bed Reactor

Reaction Kinetic data

Adsorption equilibrium data

Mathematical model

Production step

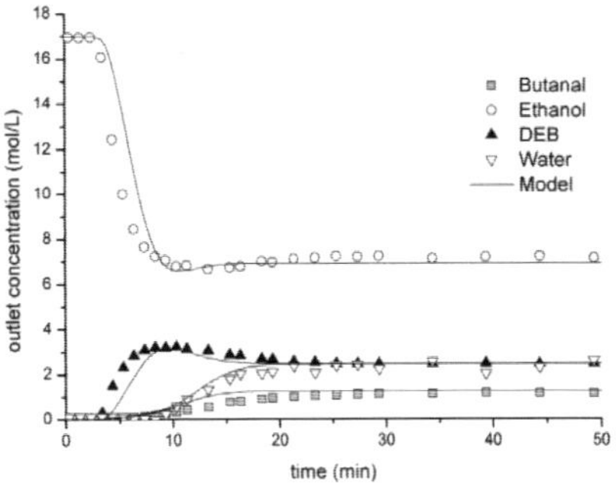

Regeneration step

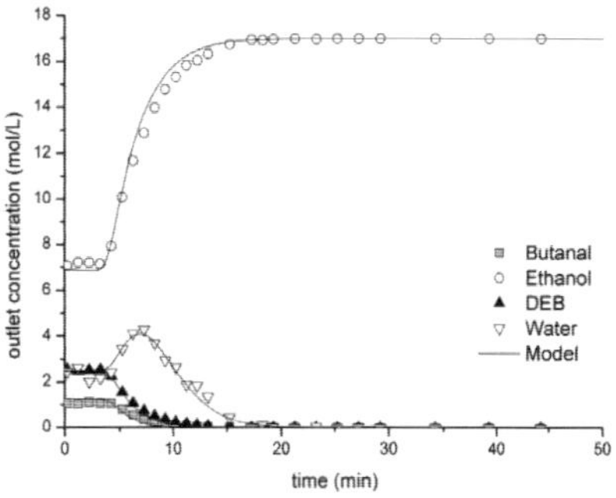

Reactive Separation Regions

Feasible regions that define the conditions to operate the SMBR to obtain specific purity and conversion requirements.

Design parameter

$$\gamma_i = \frac{u_i}{U_S}$$

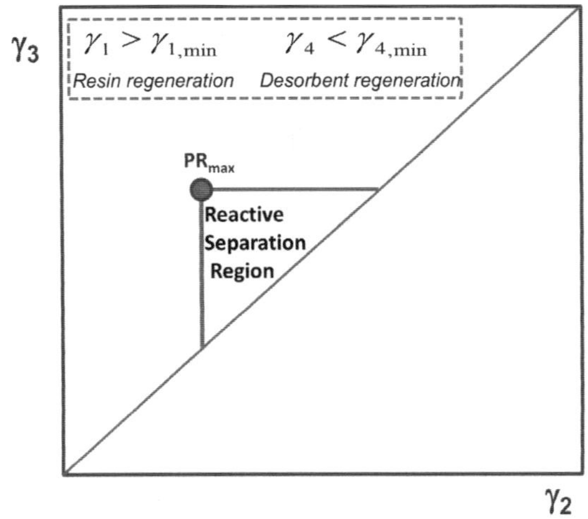

Performance

Extract purity

Raffinate Purity

Conversion

Productivity

Desorbent Consumption

Simulate Moving Bed

Critical values

$$\gamma_{1,min} = 2.827$$

$$\gamma_{4,max} = 0.553$$

Reaction/Separation Region

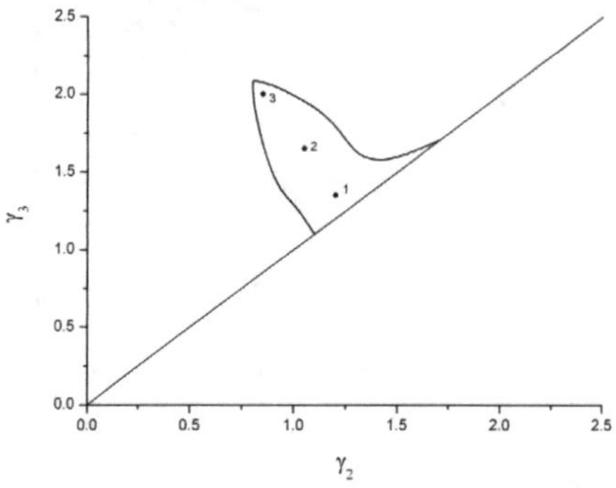

PUX, PUR, CONV >97%

SMBR operation conditions

Configuration	3-3-3-3
Temperature (ºC)	30
Catalyst	Amberlyst 47
Switching time (min)	4.4
Column Length (cm)	23
Column Diameter (cm)	2.6
Bed porosity	0.42
Peclet number	64.4

Simulate Moving Bed

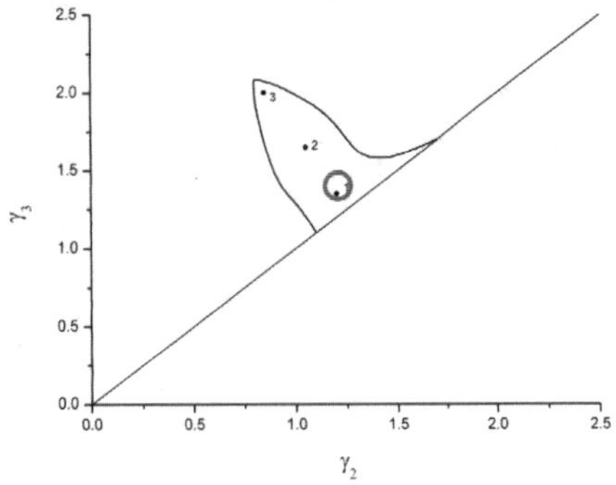

SMBR internal concentration profile

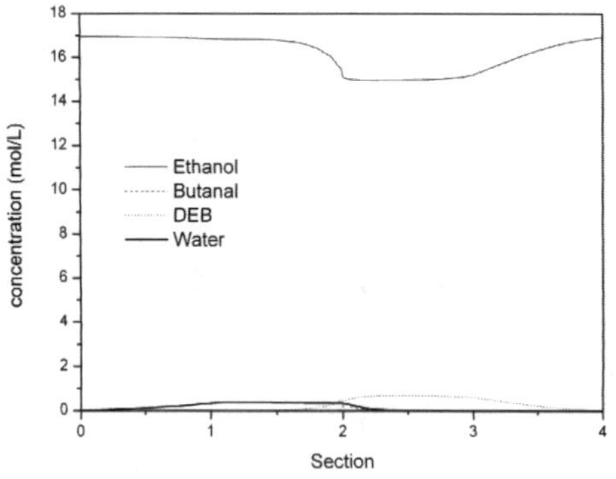

SMBR Performance

PUX(%)	98.7
PUR (%)	97.9
CONV (%)	99.8
PR $(kg_{DEB}/(L_{resin}\ day))$	1.23
DC(L_{etOH}/kg_{DEB})	30.16

Simulate Moving Bed

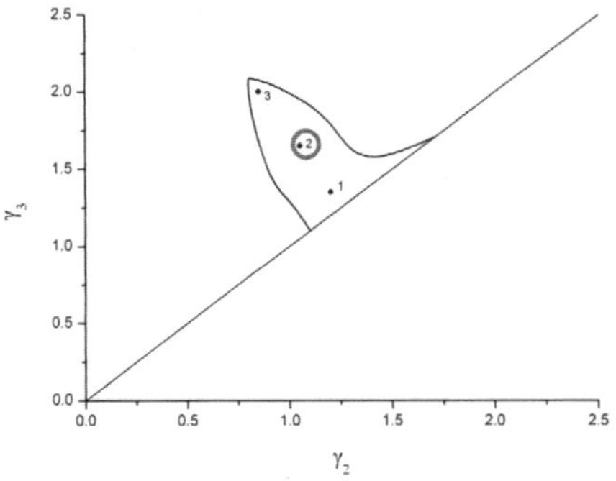

SMBR internal concentration profile

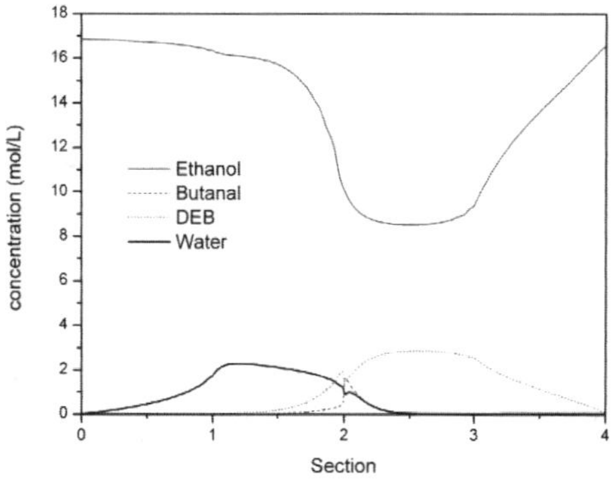

SMBR Performance

PUX(%)	98.3
PUR (%)	98.3
CONV (%)	99.6
PR $(kg_{DEB}/(L_{resin}\ day))$	5.13
DC(L_{etOH}/kg_{DEB})	7.99

Simulate Moving Bed

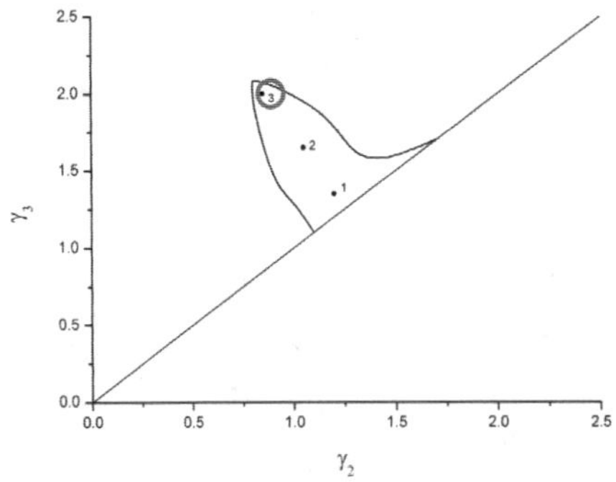

SMBR internal concentration profile

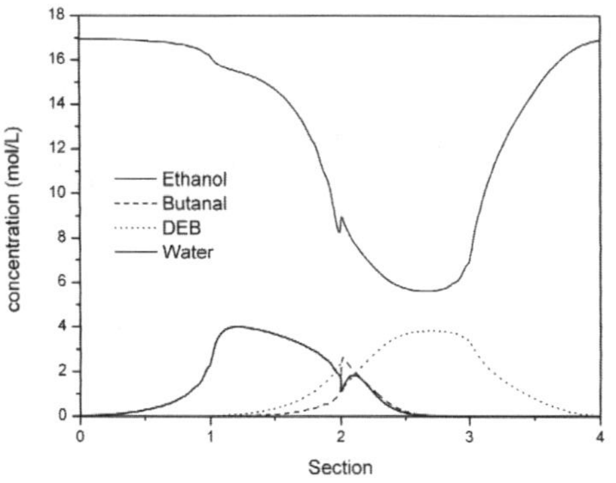

SMBR Performance

PUX(%)	97.6
PUR (%)	98.1
CONV (%)	99.2
PR $(kg_{DEB}/(L_{resin}\ day))$	10.04
DC(L_{etOH}/kg_{DEB})	4.42

Conclusions

- This work present the implementation of the synthesis of DBE in the integrated separation/ reaction process SMBR.
- The effect of different conditions on the reaction, thermodynamic, kinetic and adsorption data were obtained with the present work. This data is fundamental in further developments on the process intensification area.
- The synthesis of DBE in SMBR was by simulation. The data obtained with batch and fixed bed reactors was used to develop a mathematical model of SMBR. Simulated results showed that high purities and conversions can be obtained with the SMBR.

Thank you for your attention!

Acknowledgements

- FEDER - COMPETE – FCT - PEst-C/EQB/LA0020/2013
- QREN - ON2 - FEDER - NORTE-07-0124-FEDER-0000007

FCT Fundação para a Ciência e a Tecnologia
MINISTÉRIO DA CIÊNCIA, TECNOLOGIA E ENSINO SUPERIOR Portugal

24

362191 Systematic Process Design and Operation of Intensified Processes

Abstract:

Process Intensification (PI) is one of the many options to match current and future challenges of the (bio)chemical industry and it has the potential to improve existing processes as well as new designs of processes to achieve a more sustainable and economic production [1]. However, to date only a few intensified technologies have been implemented in the industry [2, 3] as many issues related to implementation and operations are still not clear. One such issue is the question of controllability and operability of intensified processes. A new approach is to tackle process intensification and controllability issues simultaneously, in the early stages of process design. This simultaneous synthesis approach provides optimal operation and more efficient control of complex intensified systems. Most importantly, it identifies and eliminates potentially promising alternatives but that may have controllability problems later. It may also suggest innovative process solutions which are more economically and environmentally efficient and agile. Note however, due to the integration of functions/operations into one system the controllability region of intensified equipment may become smaller.

In this work, integrated design and operation of intensified operations are considered systematically through a computer-aided framework. To assure that design decisions give the optimum operational and economic performance, operability and controllability issues are considered simultaneously with the process design issues. Operability issues are addressed to ensure a stable and reliable process design at pre-defined operational conditions whereas controllability is considered to maintain desired operating points of the process at any kind of imposed disturbance under normal operating conditions. To demonstrate the application of the computer-aided framework, production of methyl-tert-butyl-ether (MTBE) from methanol and isobutene using a reactive distillation column (RDC) is selected. First, a set of simple design methods for RDC that are easy to use and similar in concept to nonreactive distillation design methods are used. The methods are based on the element concept which is used to translate a ternary system of compounds (methanol, isobutene and MTBE) to a binary system of elements (elements A and B) [4]. For a binary element system, a simple reactive McCabe-Thiele-type method (to determine the number of reactive stages) has been used. The reactive equilibrium curve is constructed through sequential calculation of reactive bubble points. For an energy-efficient design, the driving-force approach (to determine the optimal feed location) for a reactive system has been employed [5]. For both the reactive McCabe-Thiele and driving force method, vapor-liquid equilibrium data are based on elements. ICAS-PDS has been used to compute the reactive vapor-liquid equilibrium data set. It has been shown previously that designing a reactive distillation column at the maximum driving force will result in the minimum energy consumption. Note, that the same principles that apply to a binary compound system are valid also for a binary element system [6]. Therefore, it is advantageous to employ the element based method for multicomponent reacting systems.

The operation of the RDC is investigated at the highest driving force (point 1) and two other candidate points (points 2 and 3). It is shown analytically and through rigorous dynamic process simulation (using ICAS process simulation software) that the sensitivity of the system to the disturbances in the feed at the highest driving force (point 1) is less than the two other candidate points. By application of this method, the intensified design option is verified to have the inherent ability to reject disturbances resulting in the most self-regulating design. The sustainability and economic metrics are also calculated at points 1, 2 and 3 and it is shown that the operation of RDC at point 1 not only results in an optimal operation, but also in a more economic and environmental friendly design option.

References:

[1] P. Lutze, D.K. Babi, J.M. Woodley, R. Gani, 2013. *Ind. Eng. Chem. Res.*, 52, 7127.

[2] P. Lutze, A. Román-Martinez, J.M. Woodley, R. Gani, 2012. *Compt. Chem. Eng.*, 36, 189.

[3] S.S. Mansouri, M.I. Ismail, D.K. Babi, L. Simasatitkul, J.K. Huusom, R. Gani, 2013. *Processes*, 1, 168.

[4] E.S. Pérez-Cisneros, R. Gani, M.L. Mikkelsen, 1997. *Chem. Eng. Sci.*, 52, 527.

[5] O.S. Daza, E.S. Pérez-Cisneros, E. Bek-Pedersen, R.Gani, 2003. *AIChE J.*, 49, 2822.

[6] R. Gani, E. Bek-Pedersen, 2000. *AIChE J.*, 46, 1271.

DESIGN OF A THERMALLY COUPLED REACTIVE DISTILLATION SEQUENCE FOR TRIETHYL CITRATE PRODUCTION.

Miguel Santaella, Gerardo Rodríguez, Alvaro Orjuela

Universidad Nacional de Colombia, Sede Bogotá. Grupo de Investigación Procesos Químicos y Bioquímicos.

Recent trends in the plasticizers market have shown a shift towards the consumption of renewable and bioderived ones. Although not all bioderived plasticizers can substitute petroleum derived plasticizers in all applications, biobased are gaining acceptance because their low toxicity, biocompatibility and biodegradability. Triethyl citrate (TEC) is a well-known, Generally Recognized as Safe (GRAS), bioderived plasticizer, which has been used as a substitute for various phthalic acid esters. TEC has found its main market in application such as medical devices and children toys mainly due to the safety concerns of using phthalates. Nevertheless, its competitiveness in a broader market is limited because its high production costs. These high production costs are related to the excess of ethanol required in the process, and the energy required for its recovery and recycling, as well as the TEC separation and purification. In this sense, the economics improvement by using process intensification is desirable to increase TEC's competitiveness and its market share.

Esterification of citric acid to produce TEC involves three reversible reactions with the consecutive formation Monoethyl Citrate (MEC), Diethyl Citrate (DEC) and TEC respectively. Continual removal of produced water displaces the equilibrium towards product formation and makes this process suitable to implement in a Reactive Distillation (RD) system. Because of the presence of an azeotrope, water removal also carries away some ethanol within the distillate. In the other hand, maintaining a low temperature in the bottoms to avoid product decomposition requires to remove some ethanol from bottoms of the column. These characteristics make necessary the use of an excess of ethanol in the reaction.

Ethanol recovery and recycle is a major cost in TEC production. To enhance energy efficiency of the system further intensification of the RD system is possible via thermal coupling. This would allow separation and purification of the water-ethanol mixture inside the RD column. Then, it is the aim of this article to study TEC production using the so called reactive dividing wall column (RDWC).

Process configurations were modeled and simulated using commercial software (Aspen One 7.2). A validated NRTL-HOC thermodynamic model was used to describe phase equilibria, while kinetic models for Amberlyst 15 and methane sulfonic acid catalyzed reaction was used. Energy savings for reactive distillation are predicted between 15-80% over classical configurations of a reactor followed by separation units while RDWC predict 30-50% energy reduction over simple RD.

364276 The Virtual Product-Process Design Lab

Abstract:

The Virtual Product-Process Design Laboratory (VPPD-Lab) has been developed as a computer aided tool for product-process development. Rather than design chemicals based products through an experimental-based trial and error approach, the powerful methodologies and tools in VPPD-Lab help to replace some of the experimental steps by performing virtual experiments using model-based techniques. VPPD-Lab offers a quick search of active ingredients, solvents, and additives to determine a set of promising formulations that match the desired (target) product properties (functions). This is followed by experimental-based validation, and evaluation of the most promising candidates for eventual selection as the final product, reducing thereby, valuable time and resources at the preliminary design stage.

The earlier version of VPPD-Lab consisted of templates for the design of microcapsules, pesticide uptake, and homogeneous liquid formulated products such as paint formulation, sunscreen lotion, and insect repellent lotion. The algorithms in VPPD-Lab can perform ingredients selection and testing, solubility tests, and property measurements using its available library of property models and chemicals databases. VPPD-Lab is also able to directly access ICAS (Integrated Computer Aided System) as a support tool to (i) consult several other chemicals databases, (ii) predict property values which are missing in the databases predicted through the ICAS property prediction tool (ProPred), and (iii) perform process design using process models available in ICAS MOT (Modeling testbed) to design the processes. However, the main limitations of property models available in VPPD-Lab is that the application range of the models is not large enough to include a wider spectrum of chemicals based products, such as, emulsified formations, blended synthetic fuels and lubricants, and many more. Covering a wider spectrum of products requires the use and management of a wide range of property models, calculation procedures and a very large amount of data. A systematic framework is needed to manage the data and models are necessary as well as the modules for a better communication between VPPD-Lab and support tools.

This work proposes a systematic framework to improve and extend the application range of VPPD-Lab. New property prediction models have been developed and integrated to the different steps of the product design algorithms. The chemicals databases have been extended and integrated, and property models needed for each type of products are managed through a knowledge base representation system (ontology) enabling a flexible and efficient communication between the product-design templates and supporting tools (databases, models, and connected software). The new version of VPPD-Lab has included new templates for the design of blended and emulsified products with the new user interface of a generic workflow to design products and processes. Therefore, VPPD-Lab is now generic, user friendly, and widely applicable with options for a wide range of chemicals based liquid products.

Two case studies dealing with tailor-made design of jet-fuels and an emulsified formulation product is highlighted to illustrate the new features and options available in VPPD-Lab.

368170 Oscillatory Amplitudes Regulation Method in 1,3-PD Production By Klebsiella Pneumoniae

Abstract:

Oscillatory amplitudes regulation method in 1,3-PD production by *Klebsiella pneumoniae*

Hangzhou Wang[a], Tong Qiu[b], Jinsong Zhao[b],Bingzhen Chen[b*]

[a]Memorial University, St. John's, N.L., Canada

[b]Tsinghua University, Beijing, China

1,3-propanediol (1,3-PD) is an important chemical raw material, and it can be used to produce polyesters, polyether and polyurethanes with excellent properties. Microbial fermentation production has many advantages [1, 2] over chemical synthesis of 1,3-PD. But oscillatory behavior was observed in both experiments and mathematical simulation during the continuous fermentation of *Klebsiella pneumoniae*, in which biomass, product and substrate oscillations exist under certain operating conditions. It was reported that these oscillations were caused by the existence of Hopf bifurcations in the system[3]. To design a more stable process and achieve high product quality, these operating condition points corresponding to the Hopf singularity ones should be avoided. And in some particular situations, the singularity operation point is located exactly at the maximum concentration point; the oscillation is unavoidable or is prone to generate. In these situations, it is a better way to regulate the oscillatory phenomenon, that is, to reduce the oscillatory amplitude so as to increase process stability to some extent. In this paper, a regulation method of oscillatory amplitudes is proposed so that the oscillatory amplitude can be attenuated when the oscillation itself is unavoidable. The detailed steps for this regulation method of oscillatory amplitudes as follows:

1. Calculate all steady state solutions for the dynamic system[4].
2. Identify all Hopf singularity points in the steady state curve[5, 6].
3. Simulate the oscillatory phenomenon and find the amplitudes under different operation conditions.
4. Establish the relationship model between the amplitudes and operation conditions(such as feed dilution rate, initial concentration of substrate).
5. Use the relationship model to regulate oscillatory amplitudes in process design.

According to the proposed method, with the results of oscillation simulation around the singularity operation condition points, the product concentration oscillatory amplitudes under different operation conditions can be obtained. These calculated amplitude data can be regressed to a smooth amplitude curve. The curve fits the varying tendency of the process, and can represent the relationship between amplitude and operation conditions. As a result, the characteristics of the oscillatory behavior in dynamic process can be studied quantitatively The oscillation behavior can be predicted and regulated so that a process with better stability can be designed.

Keywords: 1,3-PD, Hopf Singularity points, Oscillatory Amplitudes, Process Design and Regulations

References

1. Wong, C.-L.; Huang, C.-C.; Chen, W.-M.; Chang, J.-S., Converting crude glycerol to 1,3-propandiol using resting and immobilized Klebsiella sp. HE-2 cells. *Biochemical Engineering Journal* **2011,**58–59, (0), 177-183.

2. Menzel, K.; Zeng, A. P.; Deckwer, W. D., High concentration and productivity of 1,3-propanediol from continuous fermentation of glycerol by Klebsiella pneumoniae. *Enzyme and Microbial Technology* **1997,**20, (2), 82-86.

3. Sridhar, L. N., Elimination of Oscillations in Fermentation Processes. *AIChE Journal* **2011,**57, (9), 2397-2405.

4. Wang, H.; Chen, B.; He, X.; Zhao, J.; Qiu, T., Numerical Analysis Tool for Obtaining Steady-State Solutions and Analyzing Their Stability Characteristics for Nonlinear Dynamic Systems. *Journal of Chemical Engineering of Japan* **2010,**43, (4), 394-400.

5. Wang, H.; Zhang, N.; Qiu, T.; Zhao, J.; He, X.; Chen, B., Analysis of Hopf Points for a Zymomonas mobilis Continuous Fermentation Process Producing Ethanol. *Industrial & Engineering Chemistry Research* **2012,**52, (4), 1645-1655.

6. Wang, H.; Zhang, N.; Qiu, T.; Zhao, J.; He, X.; Chen, B., A process design framework for considering the stability of steady state operating points and Hopf singularity points in chemical processes. *Chemical Engineering Science* **2013,**99, (0), 252-264.

---

* Corresponding author. Tel.: +86 10 62781499;fax: +86 10 62770304.

E-mail address: dcecbz@tsinghua.edu.cn (Bingzhen Chen)

388102 Molecular Design of Optimum CO2 Capture Solvents: From Conceptual Screening to SAFT-Based Validation

Abstract:

The wide adoption of chemical absorption/desorption systems for post-combustion CO_2 capture in industry is currently challenged by the high energy penalty in solvent regeneration and the environmental impacts associated with solvents and their derivatives. Intense research efforts reported in recent years are predominantly based on lab and pilot-scale experiments to select solvents which may potentially improve the overall performance of absorption/desorption CO_2 capture. However, this is very challenging due to a) the highly non-ideal solvent-CO_2-water chemical interactions, b) the countless combinations of potential capture solvent and blend candidates and c) the need for combined consideration of numerous thermodynamic, kinetic and sustainability properties as performance criteria prior to selecting solvents with optimum capture features. Computer-aided molecular design methods (CAMD) can help address these challenges and have been successful in supporting the synthesis of molecules with desired physical, chemical and environmental properties in non-CO_2 separations [1]. Despite extensive developments in CAMD methods, few recent works have reported their utilization in the design of CO_2 capture solvents or mixtures for chemical and physical absorption using the Statistical Associating Fluid Theory with potentials of Variable Range (SAFT-VR) [2, 3] and for physical absorption using the Perturbed Chain Polar Statistical Associating Fluid Theory (PCP-SAFT) [4]. While these approaches enable an accurate and reliable determination of solvent-process vapour-liquid equilibria, the set of few solvents screened to date will be further expanded as research efforts extend the rigorous predictive capabilities of SAFT-based models towards additional molecular structures. On the other hand, few CAMD approaches have also been reported [5, 6, 7] that use less accurate, but well-established group contribution methods to screen a much wider set of molecular structures by approximating as solvent performance criteria mainly properties relevant to thermodynamics and toxicity.

In our current work, we propose the use of an optimization-based CAMD method to select post-combustion CO_2 capture solvents of optimum performance in molecular and mixture properties associated with thermodynamics, kinetics and sustainability. The problem is first approached in a fast screening stage where solvent structures are evaluated based on the simultaneous consideration of important pure component properties. For the first time numerous properties are considered as performance criteria reflecting solvent characteristics based on thermodynamic (e.g., vapour pressure, CO_2 solubility etc.), kinetic (solvent basicity, steric hindrance etc.) and sustainability (e.g., health and safety hazard, life cycle assessment etc.) behaviour. The simultaneous consideration of properties selected to capture the molecular chemistry effects on the absorption/desorption process compensates for the initial utilization of simpler models and ensures the selection of fewer but more effective solvents. A few highly-performing solvents are further evaluated using the SAFT-VR and SAFT-γ equations of state to predict accurately the very non-ideal solvent-water-CO_2 mixture vapour-liquid equilibrium behaviour. Different functionalities of the employed CAMD method are used both to design optimum, novel molecular structures and to screen a dataset of commercially available amine solvents suitable for CO_2 capture. The obtained results reveal interesting structure-property trade-offs and point to commercial molecules, which have very recently been considered or have yet to be employed as capture solvents.

Acknowledgements

Funding from the European Commission under grant FP7-ENERGY-2011-1-282789-CAPSOL is gratefully acknowledged.

Cited References

[1] Adjiman, C.S., Galindo, A., 2010, Process Systems Engineering: Volume 6: Molecular Systems Engineering, Pistikopoulos, E.N., Georgiadis, M.C., Dua, V. (Eds).

[2] Mac Dowell N., Pereira F.E., Llovell F., Blas F.J., Adjiman C.S., Jackson G., Galindo A., 2011, Transferable SAFT-VR models for the calculation of the fluid phase equilibria in reactive mixtures of carbon dioxide, water, and n-alkylamines in the context of carbon capture, Journal of Physical Chemistry B, 115, 8155-8168.

[3] Pereira, F.E., Keskes, E., Galindo, A., Jackson, G., Adjiman, C.S., 2011, Integrated solvent and process design using a SAFT-VR thermodynamic description: High-pressure separation of carbon dioxide and methane, Computers and Chemical Engineering, 35 (3), 474-491.

[4] Bardow, A., Steur, K., Gross, J., 2010. Continuous-molecular targeting for integrated solvent and process design. Ind. Eng. Chem. Res. 49, 2834–2840.

[5] Juan Salazar, Urmila Diwekar, Kevin Joback, Adam H. Berger, Abhoyjit S. Bhown, Solvent Selection for Post-Combustion CO2 Capture, Energy Procedia, Volume 37, 2013, Pages 257-264

[6] Bommareddy, S., Chemmangattuvalappil, N.G., Solvason, C.C., Eden, M.R. Simultaneous solution of process and molecular design problems using an algebraic approach (2010) Computers and Chemical Engineering, 34 (9), pp. 1481-1486.

[7] Chemmangattuvalappil, N.G., Eden, M.R. A novel methodology for property-based molecular design using multiple topological indices (2013) Industrial and Engineering Chemistry Research, 52 (22), pp. 7090-7103.

389735 Structure Generation of Optimal Reactants and Products Using Signature Descriptors

Abstract:

Selection of optimal chemicals and maximization of process efficiency has successfully been carried out by utilizing process design techniques that integrate product design considerations in a reverse problem formulation. The combined process-product design problem is solved for design targets in terms of properties that are later used to design optimal products. However, most of these techniques have been implemented in non-reactive systems. Different efforts to address the paucity of techniques for reactive systems have been restricted to single unknown reactants. Also, utilization of linear property models and solution schemes that do not have a provision for treating property constraints are among some of the other restrictions. In this work an algorithm has been developed for the design of optimal reactants and products that is not restricted by the number of unknown reactants and products. Nonlinear property models can be incorporated in the algorithm and the solution scheme considers property constraints. The products of the reaction(s) are constrained by property bounds arising from the process to ensure the optimality of the process. An optimization problem is formulated for each of the products in order to select the best chemical structures that satisfy the property bounds and have the best dominant property. The recently developed molecular signature descriptors have been utilized to perform molecular design. The signature descriptor is a systematic codification system over an alphabet of atoms describing the extended valence of the atoms of a molecule. The signature of a molecule can be expressed as a linear combination of its atomic signatures. Topological indices of molecules that form various property models can be linearly represented in terms of molecular signature descriptors [1]. Different quantitative structure activity/property relationships (QSARs/QSPRs) have been applied in the developed algorithm to estimate properties from molecular structures. QSARs/QSPRs can be expressed in terms of different topological indices. Thus the target properties can be expressed in terms of signatures. In the developed method, the problem is formulated in terms of the number of occurrences of each signature, i.e. the number of times each signature appears in the molecular structure. The dominant property to be optimized and the property constraints can thus be ultimately expressed in terms of the number of occurrences of the signatures in the product molecule. After identifying the occurrence values from each of the formulated problems, the molecular structure of products is generated by using an algorithm developed by Chemmangattuvalappil et al. [2]. Next, using the chemical equation to track the migration of groups from reactants to products, the structure of the reactants is determined by considering the appropriate addition and/or removal and/or rearrangement of groups. Thus, candidate reactant structures can be generated irrespective of the type and number of reactions as the introduced algorithm focuses on the design of products first and then uses the chemical equations to identify the reactant structures. Graph theory principles have been utilized to track the signatures and to avoid the generation of infeasible molecular structures. To solve the resulting nonlinear integer optimization problem(s), MI-LXPM, a real coded genetic algorithm (RCGA) developed by Deep et al. [3] has been incorporated in the solution scheme. RCGAs are general purpose population based search techniques that mimic the principles of natural selection and natural genetics laid down by Charles Darwin and encode the solution (represented as chromosomes) in real numbers. Property functions that may be discontinuous, non-differentiable and/or highly non-linear can be handled by genetic algorithms (GAs). Real coding is generally more efficent than binary and gray coding and it helps avoid the "Hamming-Cliff" problem. MI-LXPM is a robust and efficient GA for solving nonlinear integer programming problems. It utilizes the recently developed and self adaptive Laplace crossover (LX) operator and a tunable power mutation (PM) operator [4]. A special truncation procedure is utilized to handle integer restrictions on variables instead of rounding off carried out in previous algorithms. One may not begin with an array of feasible points since a parameter free constraint handling procedure dependent on tournament selection is used. Thus search of feasible points is not a constraining factor. This contribution will illustrate the developed methods and highlight their use through a case study.

[1] J.L. Faulon, D.P. Visco Jr., R.S. Pophale, 2003, Journal of Chemical Information and Computer Sciences, 43, 707-720

[2] N.G. Chemmangattuvalappil, M.R. Eden, 2013, Industrial & Engineering Chemistry Research, 52, 7090-7103

[3] K. Deep, K.P. Singh, M.L. Kansal, C. Mohan, 2009, Applied Mathematics and Computation, 212, 505-518

[4] K. Deep, M. Thakur, 2007, Applied Mathematics and Computing, 193, 211-230

390512 An Application of Computer-Aided Molecular Design (CAMD) Using the Signature Molecular Descriptor: Designing New Water-Reducing Admixtures for Cement through Cement Paste Rheology Evaluation

Abstract:

Concrete is a material that forms the literal foundation of our modern society. Concrete is no longer simply cement, water and aggregates, but rather, it also typically contains a complex array of chemical admixtures that modify its many properties to produce a cost effective, durable, formable and mechanically sound material. Water-reducing admixtures (WRAs) have been developed to reduce the required water content of concrete for a given workability or increase the workability for a given water content.

The development of new chemical admixtures for concrete is normally an experimental approach wherein one uses prior knowledge to suggest potential effective compounds. This approach is time-consuming, incremental and typically expensive. To accelerate the process, we propose a computer aided molecular design (CAMD) approch using the Signature molecular descriptor as a powerful alternative. CAMD is the application of computer-implemented algorithms that are utilized to identify and design new molecules with optimally predicted properties such that they can be tested and evaluated for efficacy. In this work, we are interested in the flow properties of cement paste, described by its rheological properties. Accordingly, a parallel-plate rheometer was used to measure the effect of eighteen low molecular weight organic chemicals belonging to different classes (sugars, sugar alcohols, and sugar acids) on rheology by directly obtaining yield stress and viscosity data. Workability, measured by a simple industry standard flow (mini-slump cone) test, was also obtained. Setting time, the time required for cement paste to stiffen, was measured since a higher workability is normally accompanied by an extended setting time, which is an undesired side-effect in some applications. After creating and refining four different quantitative structure–property relationship (QSPR) models for our four properties of interest (yield stress, plastic viscosity, workability, and setting time), a structure enumeration algorithm wass employed to generate structures outside of the original training set that have optimally predicted properties such as high workability, low yield stress and viscosity, and short setting time. Over 670 structures were newly generated and screened against a non-commercial chemical database, ZINC, to identify commercially available compounds. As a result, 6 compounds were purchased and tested for their effect on the flowability (i.e. workability, yield stress, and viscosity) and the setting time of cement paste for evaluation. Experimental results validated the previously developed QSPR models. Results confirm that the CAMD strategy is a powerful tool for generating novel and non-intuitive structures with the ability to increase flowability of cement paste with controlled setting time.

Graph Based Database Development for Gasoil Reactive Systems

AIChE Abstract
Charles Solvason and Jorge Martinis

Sessions:

- 10A03 Product and Molecular Design

In the petroleum refining industry, the accurate design and optimization of hydrocarbon conversion processes for gasoil feedstocks demand a growing effort in developing rigorous and reliable kinetic models (Van Geem *et al.*, 2002). Existing techniques rely on the approximation of gasoil stream compositions using a set of pseudo-components called "oil cuts" characterized by a set of lumped kinetic models. Although the lumped models are easy to construct and implement, they lack molecular information which makes them feedstock dependent and of limited use in process simulation and design. Developments by Neurock *et al.* (1994), Trauth *et al.* (1994), and Stark *et al.* (1995) address this limitation using molecular reconstruction techniques to approximate the individual components of the gasoil streams via stochastic optimization methods coupled with Monte Carlo kinetic models. The application of stochastic methods is, however, limited by the computational intensity imposed by the large sampling frequency needed to achieve statistical significance (Van Geem 2007). For specific classes of gasoils, Hudebine and Verstraete (2011) utilize a hybrid stochastic method that maximizes an entropy function with Lagrangian parameters associated with analytical constraints in order to reduce the computational burden. However, the method strongly depends on the initial set of molecules which must be chosen and built in relation to the type of gasoil studied.

An alternative, deterministic approach to molecular reconstruction has been developed by Bryan Research & Engineering (BRE) for the process simulator, ProMax®. The approach replaces the stochastic methods with a parallelized SQL call to a molecular database that contains all chemical constituents important to gasoil products and streams. The method is dependent on two competing factors, the completeness of the database and the size of the database. Unfortunately, most gasoil reconstructions require representations of chemical constituents of up to 40 atoms in a hydrogen suppressed, molecular graph. Enumerating all chemical constituents up to this size is cumbersome for even state-of-the-art graph generation algorithms and would result in a database too large to be efficiently searched (Peironcely *et al.*, 2012).

To address these limitations, a hybrid method is proposed that limits the size of the graph database using a set of constraints provided by molecular group and point group theory. In the first step of the method, a modified set of molecular functional groups described by Joback and Reid (1987) are used as building blocks to construct a complete set of vertex subgraphs that contain all double bonds, aromatic ring structures, and non-aromatic ring structures necessary to model gasoil systems. Next, these vertices are perturbed using point group symmetry operations to create a set of isomeric vertex graphs. Of these graphs, constitutional isomeric vertices are selected and separated into terminal and internal vertex sets. The internal vertices are constructed using graph generation algorithms and the terminal vertices are added at the final step to create complete molecular graphs. By utilizing coarse-grained sub-graphs of vertices, specific isomer classes common to gasoil systems can be created, without overloading the database, thus ensuring the capability of the SQL call to search the database in a timely manner. In addition, the database generation routing is constructed such that it is expandable by toggling various stereoisomer point groups and vertex rotations. The molecular library will be utilized to illustrate the database generation method as it is applied to a diesel hydrotreater case study.

Van Geem, K.M., Hudebine D., Reyniers M.F., Wahl F., Verstraete J.J., Marin, G.B. (2007): "Molecular reconstruction of naphtha steam cracking feedstocks based on commercial indices." *Computers & Chemical Engineering,* 31, pp. 1020-1034.

Hudebine D. and Verstraete J.J. (2011): "Reconstruction of Petroleum Feedstocks by Entropy Maximization. Application to FCC Gasolines." *Chemical Reaction Modelling of Refining Processes,* 66(3), pp. 461-477.

Joback K.G. and Reid R.C. (1987): "Estimation of pure-component properties from group contributions." *Chemical Engineering Communications*, 57(16), pp. 233-243.

Peironcely, J.E., Rojas-Chertó M., Fichera D., Reijmers T., Coulier L., Faulon, J. and Hankemeier T. (2012): "OMG: Open Molecule Generator." *Journal of Cheminformatics* 4(21), pp. 1-13.

Neurock, M., Nigam, A., Trauth, D., and Klein, M.T. (1994): "Molecular representation of complex hydrocarbon feedstocks through efficient characterization and stochastic algorithms." *Chemical Engineering Science*, 49, pp. 4153-4177.

Stark S.M., Neurock M., Klein M.T. (1995): "Comparison of MIMD and SIMD Strategies for Monte Carlo Modeling of kinetically coupled reactions." *Computers& Chemical Engineering*, 19(6), pp. 719 -742.

Trauth D.M., Stark S.M., Petti T.F., Neurock M., and Klein, M.T. (1994): "Representation of the molecular structure of petroleum resid through characterization and Monte Carlo modeling. *Energy & Fuels*, 8(3), pp. 576-580.

386157 Determining Optimal Groups for Group Contribution Methods

Abstract:

Given the sheer number of chemical compounds in use today, determining a new or untested compound's properties is a problem beyond current laboratory resources. This necessitates the need for quantitative structure activity relationships (QSARs), which are models used to predict certain properties based solely on the chemical structure of a compound. Often, QSARs are in the form of group contribution methods, which assume that a property of a compound can be modelled based on the number of occurrences of molecular subunits, or "groups". Many group contribution methods are derived from regression over large datasets, but many of these regression problems require first specifying the identity of these groups, a requirement which introduces some bias in the model building. While many group contribution methods have seen a substantial degree of success [1], many new methods are still based to some degree on the original UNIFAC groups [2]. Some recent approaches have applied best subset selection methods to alleviate this bias [3,4], but these methods still must specify a larger set from which to select the best components. We propose a method that aims to investigate the entire space of subgroups and select the set of groups which provides the best predictive accuracy from a given dataset. This is essentially optimal group selection for a given model. This method has promise in improving group contribution methods for properties that have historically been difficult to predict and design for as well as determining important substructures for more complicated properties like toxicity or bioactivity.

[1] Group-contribution based estimation of pure component properties. Jorge Marrero and Rafiqul Gani.Fluid Phase Equilibria. 2001. 183, 183-208.
[2] Group-contribution estimation of activity coefficients in nonideal liquid mixtures. Aage Fredenslund, Russel L. Jones, and John M. Prausnitz. AIChE Journal. 1975. 21(6), 1086-99.
[3] Choosing Feature Selection and Learning Algorithms in QSAR. Martin Eklund, Ulf Norinder, Scott Boyer, and Lars Carlsson. Journal of Chemical Information and Modeling 2014 54 (3), 837-843
[4] Towards Optimal Descriptor Subset Selection with Support Vector Machines in Classification and Regression. Holger Fröhlich, Jörg K. Wegner and Andreas Zell. QSAR & Combinatorial Science. 2004. 23(5), 311-318

Short-cut statistical models for estimating the consumption of energy utilities in chemical batch plants

Cecilia Pereira, Ioannis Kourlimpinis, Stavros Papadokonstantakis, Konrad Hungerbühler

Institute of Chemical- and Bioengineering, Swiss Federal Institute of Technology Zurich (ETHZ), Switzerland

The minimization of energy consumption in the chemical industry is one of the key principles of green chemistry. This has led to the development of evaluation tools, which include energy use as a metric, not only in academia but also in the industry. The use of these evaluation tools requires process specific data of energy consumption, which is usually scarce particularly in multiproduct and multipurpose batch plants.

In this study we focus on shortcut models for estimating the consumption of energy utilities (steam, cooling water, brine, etc.) in chemical batch plants. For this purpose, we developed statistical models, namely probability density functions (PDF) and classification trees on the basis of real production data. These models take the form of generic intervals defined as interquartile ranges derived from the PDF parameters, and as classes derived from the classification trees. The use of the models is possible at different levels of process design, the minimal required information being the reaction type. The PDF models were assessed with diverse metrics for the goodness-of-fit and the classification trees by means of cross-validation. Additionally, the transparency of the classification trees facilitated the analysis of the effect of the predictor variables (i.e., reaction type and operational parameters) upon the consumption of the energy utilities, resulting in a set of dominating, descriptive classification rules.

The prediction performance of both types of models was further evaluated in case studies. The validation results show that the statistical models proposed in this work provide satisfactory interval estimations for energy consumption, for instance similar to the expected accuracy of cost estimations in early process design phases (i.e., 30-40% modeling estimation error). Moreover, we have applied the PDF models to a case study for the production of the intermediate substance 4-(2-methoxyethyl)-phenol, which can be produced from seven different synthesis routes. In a previous work, a decision-making framework, which considers environmental and economic proxy indicators for screening of chemical batch process alternatives during early phases of process design (Albrecht et al., 2010), was applied to the same case study. Both methodologies allow a ranking of the different synthesis routes. It is demonstrated that the results depicted in this case study show the applicability of the PDF models – considering median values or intervals – for a fast ranking of process alternatives with respect to energy consumption, when only the reaction synthesis routes are known at early stages of process design. The PDF models provided a similar ranking compared to more complex indicators (i.e., mass loss index, energy loss index, productivity loss index) which require more detailed chemical and process information.

These new shortcut models developed in this work, especially the classification trees, represent a potentially useful tool for estimating energy utility consumption of production processes, when limited process information is available or when overwhelming processes have to be screened in short time (e.g., early design phase screening of process alternatives, life cycle assessment studies, etc.). Even though the PDF models also allow reasonable predictions of energy utility consumption, their most interesting potential applications will be for providing a benchmark framework for labelling chemical reaction types and rigorous uncertainty analysis.

Finally, the short-cut models are put into perspective with respect to a set of previously developed methodologies for estimation of energy consumption developed in the Safety and Environmental Technology Group of ETH (Szijjarto et al., 2008, Wernet et al., 2009, Rerat et al., 2012, Pereira et al., 2013) based on standard process documentation, rules of thumb, expert knowledge, and thermodynamic principles. Depending on the application target, the proper use of these models is summarized, mentioning the modeling challenges for further research.

References

A. Szijjarto, S. Papadokonstantakis, U. Fischer, K. Hungerbühler, 2008. Bottom-up modeling of the steam consumption in multipurpose chemical batch plants focusing on identification of the optimization potential, *Industrial and Engineering Chemistry Research,* 47, 7323-7334.

G. Wernet, S. Papadokonstantakis, S. Hellweg, K. Hungerbühler, 2009. Bridging data gaps in environmental assessments: Modeling impacts of fine and basic chemical production, *Green Chemistry*, 11, 1826-1831.

T. Albrecht, S. Papadokonstantakis, H. Sugiyama, K. Hungerbühler, 2010. Demonstrating multi-objective screening of chemical batch process alternatives during early design phases, *Chemical Engineering Research and Design*, 88, 529-550.

C. Rerat, S. Papadokonstantakis, K. Hungerbühler, 2012. Estimation and analysis of energy utilities consumption in batch chemical industry through thermal losses modeling, *Industrial and Engineering Chemistry Research,* 51, 10416-10432.

C. Pereira, S. Papadokonstantakis, C. Rerat, K. Hungerbühler, 2013. Industrial documentation-based approach for modeling the process steam consumption in chemical batch plants, *Industrial and Engineering Chemistry Research,* 52, 15635-15647.

387167 Ring-Closing Metathesis (RCM) – a New Route for Producing Nylon 11, 12 and 13 Precursors from Oleic Acid

Abstract:

The production and use of biomass-derived renewable materials, such as polymers, has been a topic of intense research in recent years, as alternatives to petroleum-derived products are increasingly sought. Nylons 11, 12 and 13 are high-strength materials that have use across many industries. The amino-acid precursor of Nylon 12 is conventionally manufactured from petroleum resources in 6 steps. Syntheses of Nylon 11 and 12 precursors from plant-derived fatty acids have also been reported, but these too involve multiple steps (4–6 steps) and the feed material used is restricted to ricinoleic acid, a fatty acid available only from castor beans. Our program focuses on development of new and simple synthetic strategies towards these nylon precursors from oleic acid, an abundant natural fatty acid available from common plant sources as well as oleaginous microorganisms including microalgae.

Our strategy to Nylons 11–13 involves first converting the oleic acid to alkenyl amides by reaction with allyl, homoallyl, and bis-homoallyl amines, respectively. Each amide is then subjected to "ring closing metathesis" generating an unsaturated ene-lactam, which is then hydrogenated to a lactam – the usual precursor for Nylon 12 and Nylon 13 synthesis. Production of Nylon 11 is from a linear amino ester, which can be readily accessed by hydrolysis of the lactam.

The first step, amide formation and the third hydrogenation step occurred smoothly under conventional conditions giving the desired products in excellent yields. Thus, our focus has been the optimization of ring-closing metathesis steps. When homoallyl oleate or bis-homoallyl oleate were subjected to ring-closing metathesis conditions, the desired lactams were obtained in good to excellent yield after tuning of several reaction parameters (temperature, solvent, reaction time, concentration, and mode of catalyst addition). However, allyl oleate only gave a very low conversion and produced significant amount of oligomers. This issue was resolved by the attachment of a protecting group to the amide. Upon systematic screening, we found the best protecting group that reverses the selectivity of oligomerization to macro-cyclization, which may have general value for other macro-lactam cyclization reactions.

Our approach to nylon precursors from oleic acid represents one of the shortest approaches to date. This strategy features better selectivity than other cross coupling methods and allows use of stoichiometric amount of coupling partners (vs excess of materials), leading to less complex product mixtures. Full details of our optimization efforts, including screening of reaction parameters and amide protecting groups, will be presented.

Process development for the production of a novel protein source

Background

The world population will grow towards 2050 and onwards [1]. One of the great challenges coming with this growth in population is the production of nutritious food in sufficient quantities. The nutritional value of food depends for a great deal on the protein content, hence a local, sustainable protein source, applicable in the food chain, will contribute positively to the world wide food security.

The protein

The protein from leaves, mainly the enzyme RuBisCo (ribulose-1,5-bisphosphate carboxylase/oxygenase), is a very promising, locally produced protein source. Most leaves are nowadays discarded on the field when crops are harvested, emitting CO_2 and NO_x in the rotting process that follows. In order to demonstrate the potential of this sustainable protein source TNO has worked towards a protein production process. This demonstration process is based on sugar beet leaves and realizes the production of a leaf protein concentrate that has a good nutritional value (AAS: 87%) [2], is highly soluble, has good functional properties [3] and possesses no known allergenicity [4]. These properties together make this protein very interesting for food applications.

The process

This proof of principle is based on sugar beet leaves, which contain on average 1% RuBisCo on wet weight bases [5]. The leaves are harvested with a machine that can match the capacity of 1 hectare per hour of a beet combine harvester. The leaves are collected without stems or dirt and transported to a processing facility. At this location the leaves are pressed with a conical screw press in order to release the water-soluble protein in the green juice fraction of the press. The green fragments in the juice are subsequently coagulated by heat treatment (50°C, 20 minutes) using plate-type heat exchangers and removed by centrifugation in a disc stack centrifuge. The remaining protein solution is concentrated and diafiltrated over a 100 kDa regenerated cellulose ultrafiltration membrane to remove salts, water and phenolic components from the cell walls. The last purification step is performed by hydrophobic adsorption using an XAD-resin to remove residual chlorophyll and phenolic components and ensure the production of a white and highly purified dry protein product after spray drying. This whole process has been executed at a scale of 1 metric ton wet leaves per hour input stream.

The product

The produced protein has been extensively tested on its functional properties, as this greatly determines the value of the protein for food manufacturers. The gelling capacity was found to be the most promising property of the protein, being more than four times better than whey protein and ovalbumin. This will allow applications in for example meat replacers, in which both gelling and nutritional value are of key importance as well as having a sustainable, plant-based origin.

Outlook

The coming years, the protein isolate will be tested by food producers in various products. Approximately 100 kg of protein isolate will be produced from sugar beet leaves by TNO in 2014 in order to facilitate these tests. Production on this scale will allow us to generate accurate numbers regarding sizing of equipment and detailed cost estimates for a full scale production facility.

References

1. UN. 2012. http://esa.un.org/wpp/Documentation/pdf/WPP2012_HIGHLIGHTS.pdf.
2. Barbeau, W. E. 1990. Functional properties of leaf proteins: criteria required in food applications. Italian journal of food science 4:213-225.
3. Barbeau, W. E. and J. E. Kinsella. 1988. Ribulose bisphosphate carboxylase/oxygenase (rubisco) from green leaves - potential as food protein. Food Rev. Int. 4:93-127.
4. Leduc, V., de Laval, A.D., Ledent, C., Mairesse, M. Respiratory allergy to leaf proteins involvement of a new allergen. Revue française d'allergologie et d'immunologie Clinique 2008, 48, 521-525
5. FAO. 2011. www.fao.org/DOCREP/005/AC854T/AC854T00.htm#TOCOnline.

357375 One-Step Process for the Production of 1,1,3,3-Tetrachloropropene

Abstract:

Hydroclurocarbon (HFC) products are widely utilized in many applications, including refrigeration, foam expansion, and as propellants for aerosol products. HFC is being phased out with Hydrofluoroolefins (HFO) such as 1,3,3,3-tetrafluororopene (HFO-1234ze) to meet the expected higher standard environmental impact. This HFO has properties that meet European requirements of low global warming potential and no impact on the ozone layer. HFO-1234ze can be produced from 1,1,3,3-tetrachloropropane (TCPe). In this work, a one-step process is proposed to produce TCPe from readily available trichloroethylene and methylene chloride.

382196 Alternative Routes for Production of Propylene and BTX through Catalytic Cracking of Naphtha

Abstract:

Abstract: Demand of light olefins like propylene and ethylene is increasing day by day in petrochemical industries. Propylene is commonly produced as a byproduct of steam cracking of naphtha. The steam cracking units are more energy intensive as well as its propylene to ethylene weight ratio is also low (less than 1). Furthermore, much of the new steam cracking capacity is based on ethane feed, which produces little propylene, creating gap between future propylene supply and demand. Therefore, there is an increasing need for producing propylene employing alternative method utilizing surplus naphtha, being widely available in refineries. The other important raw materials for polymer and other petrochemical synthesis are benzene, toluene and xylene (BTX). The worldwide demand for BTX has also been increasing continuously. Conventionally, in refineries, BTX is produced from catalytic reforming process from naphtha of specified quality. However, a refiner can increase the refinery margin by producing both light olefins and BTX from undesirable naphtha streams by alternate routes.

IndianOil R&D Centre offers a suite of naphtha cracking technologies that can help to meet the propylene and BTX demand, depending on the feed type, as follows.

Technology	Feed	Ethylene + Propylene, wt.% of fresh feed	BTX, wt.% of MS
INDALIN	Olefinic naphtha	40-45	40-45
INDALINPLUS	Any type of naphtha including paraffinic naphtha, condensates	2-3	55-65

The present paper discusses the features and status of each of these technologies.

Keywords: Naphtha, Cracking, INDALIN, INDALINPLUS

375672 Studying Inhibitory Effects of Lignocellulosic Hydrolysate on Fungal Lipid Accumulation

Abstract:

Microbial lipid accumulated from lignocellulose has received growing attention as a precursor for advanced biofuels and functional oils. Despite its promise, the full potential of lignocellulose has yet to be unlocked partly because of the major challenges associated with its bioconversion such as liquid-solid separation, large water usage, detoxification, and low xylan utilization. Therefore, it is critical to develop new methods to tailor lignocellulose processing for enhanced lipid fermentation. The novel co- and combined hydrolysis process we developed could remove steps of detoxification, liquid-solid separation and washing as well as improve the efficiency of chemical usage during bioprocessing of lignocellulose. The oleaginous fungus *M. isabellina* cell mass and lipid fermentation on hydrolysate was compared with that from synthetic ones, indicating the integration of co- and combined-hydrolysis with oleaginous fungal lipid fermentation is a technically feasible way to lignocellulose-based biodiesel production. However, the inhibitors present in thermal-chemically pretreated hydrolysates showed negative effect on the growth and lipid formation of microorganisms. Understanding their inhibitory mechanism on oleaginous microorganism could help improving the efficiency of lignocellulose hydrolysis and lipid fermentation. Four lignocellulosic byproducts in lignocellulosic hydrolysates, including furfural, HMF, ferulic acid and coumaric acid were systematically investigated to elucidate their effects on the cell growth and lipid accumulation of *M. isabellina*. The results showed that their inhibitory effect on cell growth was more serious than on the lipid synthesis.

377452 Production of Butanol and Hydrogen from Agriculture Residues By Using Clostridium Sp. G117

Abstract:

Abstract

Butanol and hydrogen are recognised as alternative energy to fossil fuels. They can be simultaneously produced during anaerobic fermentation of starchy materials (maize, wheat, millet, rye, etc) and sugars (glucose, xylose and molasses) by using *Clostridia*. Herein, we report a new isolate *Clostridium* strain G117 which is able to utilize agricultural residues including rice straw, corn cob, rice bran and sugarcane bagasse in reinforced clostridial medium (RCM) for butanol and hydrogen production. Among the selected substrates, rice bran is found to support high level of butanol (3.48±0.12g/l) and hydrogen (1.98±0.10l/l) production. Rice bran is a good carbon source for culturing strain G117 and butanol production, because it contains a high percentage of hemicellulose (37-40%) and cellulose (27-30%). Because *Clostridium* also requires nitrogen sources and metal ions for cellular functions and synthesis of biomass, oil cakes from caster, groundnut and sesame were added to culture medium as low-cost nitrogen and mineral sources. Among them, sesame oil cake gives the highest level of butanol (4.71±0.15g/l) and hydrogen (3.73±0.05l/l). Consequently, rice bran (20-60g/l) and sesame oilcake (20-40g/l) were used together to enhance butanol and hydrogen production. Using response surface methodology (RSM), we obtain an optimized medium which contains 48.5g/L of rice bran and 32.8g/L of sesame oilcake. In this medium, 13.7±0.20g/l of butanol and 4.23±0.10l/l of hydrogen can be produced simultaneously. Based on RSM analysis, rice bran has higher impact on butanol and hydrogen production than sesame oilcake. This is the first report dealing with direct fermentation of agriculture residues for butanol and hydrogen production. This bioprocess would be beneficial to industrial-scale production of hydrogen and butanol.

High cumulative amount of ethanol fermentation in continuous and closed-circulating fermentation process with pervaporation membrane bioreactor

Paper number: 380382

Senqing Fan, Zeyi Xiao, Yan Zhang, Qing Deng, Peina Yao
School of Chemical Engineering, Sichuan University, 610065, Chengdu, China

Introduction

With increasing gap between the energy requirement of the industrialized world and inability to replenish such needs from the limited sources of fossil fuels, an ever increasing level of greenhouse pollution from the consumption of fossil fuels in turn aggravated the perils of global warming and energy crisis (Balat & Balat, 2009). Carbon neutral bioethanol, derived from a renewable biomass, has emerged as one of the most strategically important sustainable fuel sources and is considered an important way of progress for limiting greenhouse gas emissions, improving air quality and finding new energetic resources (Nigam & Singh, 2011). As a biofuel, ethanol is by far the most widely used in the worldwide for the transportation sector, since it has a higher octane number, broader flammability limits, higher flame speeds and higher heats of vaporization (Amillastre et al., 2012; Balat & Balat, 2009). Therefore, the development of cost-effective technologies for bioethanol production is a priority for many research centers, universities and private firms, and even for different governments(Cardona & Sanchez, 2007).

During the traditional batch fermentation process, the accumulation of ethanol in the broth inhibits the cell growth and ethanol fermentation, which leads to low cell density in the broth, low ethanol productivity, low production and great waste water processed. In the subsequent process, a lot of energy is consumed for ethanol recovery from the fermentation broth. Separation technologies, such as gas stripping, extraction and pervaporation, have been developed to remove ethanol *in situ* from the fermentation broth and meet the requirement of reducing the cost of ethanol fermentation (Gyamerah & Glover, 1996; Taylor et al., 1995; Taylor et al., 1998). Pervaporation is one of the most promising approaches for the removal of ethanol from fermentation broth because it does not cause any harmful effect on the microorganisms and has nice separation performance (Lee et al., 2012). During the pervaporation process, the fermentation broth is forced flowing on one side of the membranes with a gaseous permeate released and a vacuum pump or sweep gas less common is exerted at the other side of the membranes to supply a driven force between the two streams (Gaykawad et al., 2013). Coupling the pervaporation with ethanol fermentation, a smaller effect of products inhibition could increase the ethanol production and improve the equipment utilization with less amount waste water processed (Cardona & Sanchez, 2007; Ding et al., 2011; O'Brien et al., 2004).

In our previous works, the conception on a continuous and closed-circulating fermentation (CCCF) process with a pervaporation membrane bioreactor was proposed, and a lot of experimental evidences has been disclosed (Chen et al., 2012; Ding et al., 2011; Ding et al., 2012). In this process, ethanol inhibition could be eliminated owing to simultaneous ethanol removal by the membrane pervaporation, so that long fermentation period could be maintained and then high amount of ethanol production could be accomplished. The aim of the study was aim to explore the ethanol production in the CCCF process with three operation mode: coupling pervaporation intermittent with fermentation, coupling pervaporation continuous with fermentation and coupling pervaporation continuous with fermentation at higher broth pH during the later fermentation period.

Methods

Membranes and modules

The composite polydimethylsiloxane (PDMS) membrane was self-prepared. It was nonporous and hydrophobic with good stability and nice tolerance to the organic solvent. The operating temperature of the PDMS material was in the rage of 20 ℃ and 90 ℃ (Vane, 2005). A quadrate flat-plate membrane module with a piece of membrane filled providing the effective area of 0.08 m^2 was used in the intermittent mode and a quadrate flat-plate membrane module with two piece of membrane filled in series providing the effective area of 0.16 m^2 was used in the continuous modes. The preparation procedure of the membrane and the modules had been fully described in the previous work (Li et al., 2004).

Microorganism and medium

The thermostable alcohol-active dry yeast (ADY), *Saccharomyces cerevisiae*, purchased from Angel Yeast Corporation (Hubei, China) was employed in the CCCF system. Prior to inoculation, 9 g of the yeast was rehydrated for 20 min in 250 ml glucose (15 gL^{-1}) solution at 38℃ followed the revival operation for 1.5 h at the temperature of 34℃.

The synthetic medium (in 1 L) used in the CCCF system contained: glucose 100 g, yeast extract 8 g, $(NH_4)_2SO_4$ 5 g, KH_2PO_4 1.5 g, $MgSO_4{\cdot}7H_2O$ 0.55 g, $CaCl_2$ 0.15 g. All the materials in the medium were purchased from Kelong Chemical Reagent Corporation (Chengdu, China) and sterilized in an autoclave at 121℃ for 20 min.

Fermentation

The schematic diagram of the CCCF process with a pervaporation membrane bioreactor for ethanol fermentation was shown in Fig.1. The continuous and closed-circulating fermentation process was conducted after inoculation of 6 L of fermentation medium with the rehydrated seed. Three runs of ethanol fermentation were carried out in the CCCF process. During the first 4 h after inoculation in the first run, the filtered air was sparged into the broth by a compressor for the aerobic fermentation, and then aeration was stopped for the anaerobic fermentation during the remaining period. When the ethanol concentration in the broth reached about 70 gL^{-1} after 24 h, membrane pervaporation was started and the broth was circulated through the membrane module and fermentor at a flow rate of 100 L h^{-1} by the circulating pump. During the fermentation process, the broth was kept at 35±1℃ by the thermostat. The pH was adjusted 4.0±0.5 with the ammonia water of 25 wt% and the pervaporation was intermittently coupled with fermentation.

Another two runs of ethanol fermentation were carried out in the CCCF process with pervaporation and fermentation continuously coupled. The second run was lasting for 264 h and the fermentation was operated without membrane separation for the product ethanol during the initial 24 h (similar to general batch fermentation process). The continuous pervaporation was started at the time point of the 24th h for ethanol removal with the broth pH kept at 4.0±0.5. The thirs run was lasting for 360 h and the pervaporation and fermentation were coupled entirely at the whole course. During the early 72 h of the process, the broth pH was kept in 4.0±0.5 and then gradually increased to 6.5 so as to relieve the inhibition of the organic acids accumulated in the broth. The detailed procedure of the CCCF process with the pervaporation membrane bioreactor had been fully described in our previous work (Fan et al., 2014b). The second run of fermentation was terminated when the viable cell ratio decreased to about 8% and the average ethanol productivity was about 0.85 $gL^{-1} h^{-1}$ during final 12 h. The third run experiment was terminated when the viable cell ratio was about 5% and the average ethanol productivity was about 0.56 $gL^{-1} h^{-1}$ during final 12 h.

Glucose and water were added to maintain the glucose concentration in the range of 20 gL^{-1} and 50 gL^{-1} and broth volume of 6 L. At the downstream of the membranes, some of the vapor was cooled into the liquid by the condenser with a spiral coil pipe and the running water as coolant. At the bottom of the condenser was the catch pot A for collecting the condensate. The remaining vapor released from the condenser was sucked and transported to the atmospheric environment by the vacuum pump, and condensed into a liquid in the second condenser by air at the room temperature. The fermentation coupling with pervaporation was continually running through day and night, without any discontinuity till the experiment ended. The absolute pressure at membranes downstream was kept for 35 mmHg by the vacuum pump.

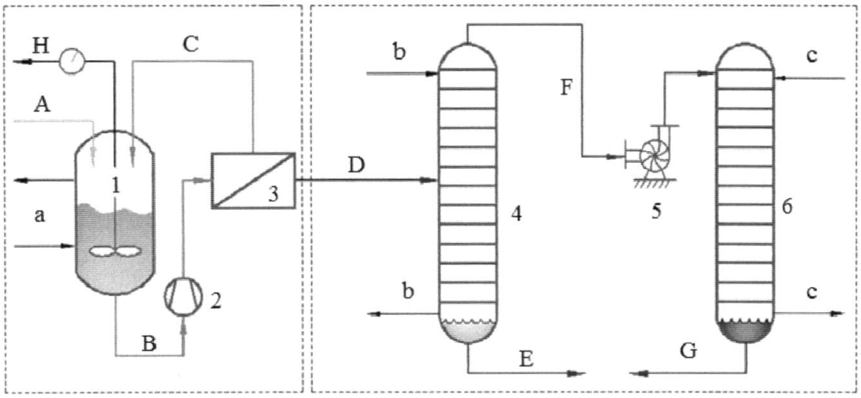

Fig.1 Schematic diagram of the experimental set-up for the designed pervaporation membrane bioreactor. 1, fermentor; 2, circular pump; 3, membrane module; 4, condenser C1; 5, vacuum pump; 6 condenser C2; A, feed; B, fermentation broth or model solution; C, retentate; D, permeate; E and G, condensates; F, non-condensed permeate vapor; H, vent; a, thermostat; b, running water; c, air

Assay

The ethanol concentration was determined with a density meter DMA 4500 (Anton Paar, Austria). The secondary metabolites of ethanol fermentation (acetic acid, lactic acid, propionic acid, citric acid, succinic acid, glycerol) in the broth of the pervaporation membrane bioreactor were determined with a HPLC system equipped with a Agilent Hi-Plex H (300×7.7 mm, 8μm). The mobile phase was 5mM degassed H_2SO_4 at a flow rate of 0.8 mL min^{-1} and a column temperature of 35 ℃. A UV detector was used and the wavelength was 210 nm (Chen et al., 2013).

Results and discussion

During traditional ethanol batch fermentation processes, the cumulative production amount is low, due to the inhibition of ethanol to the microorganism. Coupling pervaporation with ethanol fermentation, ethanol is *in situ* removed from the broth by the pervaporation membrane and the inhibition is released or eliminated. Therefore high cumulative production with long fermentation period would be achieved. A continuous and closed-circulating fermentation (CCCF) process with a pervaporation membrane bioreactor had been proposed for ethanol fermentation. Three runs of fermentation were carried out in CCCF process. Fermentation performance of the CCCF system for ethanol production was shown in Table 1. It could be seen that the accumulative ethanol fermentation production exceeded 600 gL^{-1} in the process with ethanol removed *in situ* from the fermentation broth. The yeast cell used in the CCCF usually experience strong ethanol inhibition above 6 wt% and thus higher cell density could be achieved when ethanol was removed by the pervaporation membranes during the fermentation period, as shown in Table 1. Increasing the concentration of viable cells in the fermentor would increase the ethanol productivity for a long time and thus ethanol production was improved. The time course profile of ethanol productivity in the CCCF process was shown in Fig.1 and it could be seen that the ethanol productivity when pervaporation was intermittently coupled with fermentation was much lower than that in continuous coupled and obvious oscillations of the productivity was performed owing to ethanol inhibition when pervaporation separation was not operated. Therefore, ethanol production in the intermittent case could not increase although the fermentation period lasted for 500 h. The fermentation process could be improved when the pervaporation was continuously coupled with fermentation. 610 gL^{-1} of ethanol production and 26.3 gL^{-1} h^{-1} of cell density could be obtained with the process lasting for 264 h.

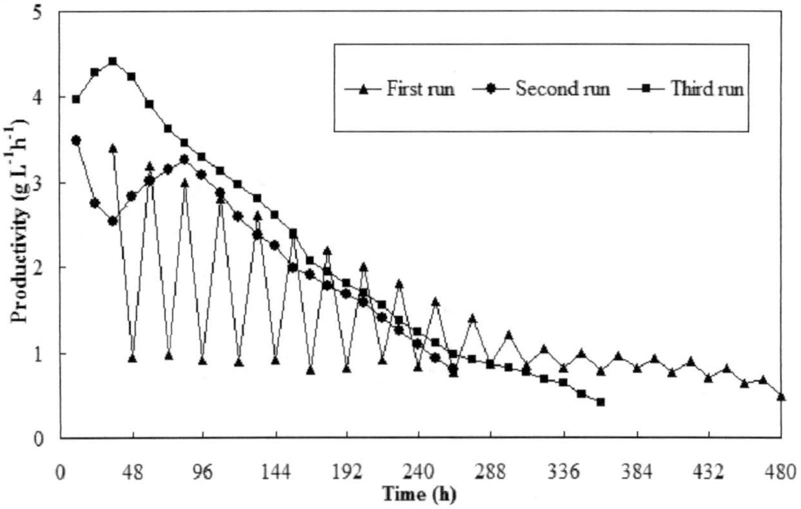

Fig.1 Time course profile of ethanol productivity in CCCF process

During ethanol fermentation from the glucose, complex metabolic pathways exist in the yeast cells and ethanol is the end product of the metabolisms. Therefore, several by-products or secondary metabolites would be produced during the process of ethanol fermentation. In the CCCF system for ethanol fermentation trace of the secondary metabolites could accumulated to a point where they could inhibit the cell growth and ethanol fermentation (Chen et al., 2012; Fan et al., 2014a; Maiorella et al., 1983). In our previous study, it was found that only 0.58 g of acetic acid, 1.97 g of lactic acid, 0.45 g of propionic acid, 0.21 g of citric acid, 0.41 g of succinic acid and 8.91 g of glycerol could be accumulated in the broth of the pervaporation membrane bioreactor along with per 100 g of ethanol produced in the CCCF system. When these compounds concentrations in the broth accumulated to

3.52 g L^{-1} of acetic acid, 14.87 g L^{-1} of lactic acid, 3.05 g L^{-1} of propionic acid, 1.49 g L^{-1} of citric acid, 2.48 g L^{-1} of succinic acid and 57.18 g L^{-1} of glycerol with fermentation proceeding, the cell could hardly grow and no ethanol could be produced (Fan et al., 2014a). During ethanol fermentation with the secondary metabolites in the medium, the inhibitory effect is related to the concentration of the undissociated forms of the weak acids, which could diffuse through the cell membrane and acidify the cytoplasm, decreasing the intracellular pH. In according to others' studies, it has been found that the concentrations of the undissociated weak acids were dependent on the pH of the fermentation medium and a higher pH would decrease the undissociated weak acids. The results indicated that the inhibition effect of the secondary metabolites accumulated in a pervaporation membrane bioreactor for ethanol fermentation would be less pronounced at a higher pH, since the weak acids dissociate in the medium and cannot permeate the cell membrane. In third run experiment, the pH of the medium was gradually increased from 4.0-6.5 after 72 h fermentation. In this case the inhibition of the organic acids was released and enhanced fermentation could be achieved with higher cell density, longer fermentation period and hence higher cumulative ethanol production amount.

During batch fermentation the cell concentration was about 9 g L^{-1} with ethanol production of about 70 g L^{-1} obtained, which indicated that less 8 g of ethanol was produced by 1 g of yeast cell in this mode. It could be easily deduced that in the CCCF process for ethanol fermentation, 32.5 g and 23.1 g and 19.6 g of ethanol could be produced by 1 g of cell in the first run, in the second run and in the third run, respectively. This implied that ethanol yield against cell (ratio of ethanol production to cell mass) in the CCCF process would be much superior to that in a conventional batch fermentation process, since the ethanol inhibition was relieved.

Table 1 Ethanol fermentation performance in CCCF system

Fermentation performance	Intermittent PV	Continuous PV	Continuous PV+ higher pH
Lasting time (h)	500	264	360
Ethanol Production (gL^{-1})	645	610	750
Cell concentration (gL^{-1})	19.8	26.3	38.1
Ethanol productivity (gL^{-1} h^{-1})	1.29	2.31	2.08

Conclusions

Three runs of ethanol fermentation were carried out in a continuous and closed-circulating fermentation (CCCF) process with a pervaporation membrane bioreactor. High ethanol production could be achieved in the process with ethanol removal in situ from the broth. When the pervaporation had been intermittently coupled with fermentation, the fermentation process had been lasting for 500 h and the cumulative ethanol production was about 670 gL^{-1}. The fermentation process could be improved when the pervaporation was continuously coupled with fermentation. An experiment in the CCCF process lasting for 264 h had been carried out with cumulative ethanol production of 610

gL^{-1} obtained. During the later period of the CCCF process, the fermentation was inhibited by the accumulative organic acids in the broth. Increasing the pH of the broth in the range of 6.0 and 8.0 at the later period of the process, the inhibition of the organic acids was released and enhanced fermentation could be achieved with higher cell density, longer fermentation period and hence higher accumulative ethanol production amount.

References

1.Amillastre, E., Aceves-Lara, C.A., Uribelarrea, J.L., Alfenore, S., Guillouet, S.E. 2012. Dynamic model of temperature impact on cell viability and major product formation during fed-batch and continuous ethanolic fermentation in Saccharomyces cerevisiae. Bioresour. Technol., 117, 242-50.

2.Balat, M., Balat, H. 2009. Recent trends in global production and utilization of bio-ethanol fuel. Appl. Energ., 86, 2273-2282.

3.Cardona, C.A., Sanchez, O.J. 2007. Fuel ethanol production: process design trends and integration opportunities. Bioresour. Technol., 98, 2415-57.

4.Chen, C., Tang, X., Xiao, Z., Zhou, Y., Jiang, Y., Fu, S. 2012. Ethanol fermentation kinetics in a continuous and closed-circulating fermentation system with a pervaporation membrane bioreactor. Bioresour. Technol., 114, 707-710.

5.Chen, C., Xiao, Z., Tang, X., Cui, H., Zhang, J., Li, W., Ying, C. 2013. Acetone-butanol-ethanol fermentation in a continuous and closed-circulating fermentation system with PDMS membrane bioreactor. Bioresour. Technol., 128, 246-251.

6.Ding, W.-W., Wu, Y.-T., Tang, X.-Y., Yuan, L., Xiao, Z.-Y. 2011. Continuous ethanol fermentation in a closed-circulating system using an immobilized cell coupled with PDMS membrane pervaporation. J. Chem. Technol. Biot., 86, 82-87.

7.Ding, W., Xiao, Z., Tang, X., Deng, K., Fu, S., Jiang, Y., Yuan, L. 2012. Evolutionary engineering of yeast for closed-circulating ethanol fermentation in PDMS membrane bioreactor. Biochem. Eng. J., 60, 56-61.

8.Fan, S., Xiao, Z., Tang, X., Chen, C., Zhang, Y., Deng, Q., Yao, P., Li, W. 2014a. Inhibition effect of secondary metabolites accumulated in a pervaporation membrane bioreactor on ethanol fermentation of Saccharomyces cerevisiae. Bioresour. Technol., 162, 8-13.

9.Fan, S., Xiao, Z., Zhang, Y., Tang, X., Chen, C., Li, W., Deng, Q., Yao, P. 2014b. Enhanced ethanol fermentation in a pervaporation membrane bioreactor with the convenient permeate vapor recovery. Bioresour. Technol., 155, 229-234.

10.Gaykawad, S.S., Zha, Y., Punt, P.J., Groenestijn, J.W.v., Wielen, L.A.M.v.d., Straathof, A.J.J. 2013. Pervaporation of ethanol from lignocellulosic fermentation broth. Bioresour. Technol., 129, 469-476.

11.Gyamerah, M., Glover, J. 1996. Production of Ethanol by Continuous Fermentation and Liquid-Liquid Extraction. J. Chem. Technol. Biotechnol., 66, 145~152.

12.Lee, H.J., Cho, E.J., Kim, Y.G., Choi, I.S., Bae, H.J. 2012. Pervaporative separation of bioethanol using a polydimethylsiloxane/polyetherimide composite hollow-fiber membrane. Bioresour. Technol., 109, 110-115.

13.Li, L., Xiao, Z., Tan, S., Pu, L., Zhang, Z. 2004. Composite PDMS membrane with high flux for the separation of organics from water by pervaporation. J. Membr. Sci., 243, 177-187.

14.Maiorella, B., Blanch, H.W., Wilke, C.R. 1983. By-Product Inhibition Effects on Ethanolic Fermentation by Saccharomyces cerevisiae. Biotechnol. Bioeng., 105, 103-121.

15.Nigam, P.S., Singh, A. 2011. Production of liquid biofuels from renewable resources. Prog. Energ. Combust, 37, 52-68.
16.O'Brien, D.J., Senske, G.E., Kurantz, M.J., Craig, J.C. 2004. Ethanol recovery from corn fiber hydrolysate fermentations by pervaporation. Bioresour. Technol., 92, 15-19.
17.Taylor, F., Kurantz, M.J., Goldberg, N., Jr, J.C.C. 1995. Continuous Fermentation and Stripping of Ethanol. Biotechnol. Prog., 11, 693~698.
18.Taylor, F., Kurantz, M.J., Goldberg, N., Jr, J.C.C. 1998. Kinetics of continuous fermentation and stripping of ethanol. Biotechnol. Lett., 20, 67~72.

Cell growth model in a pervaporation membrane bioreactor for continuous ethanol fermentation by *Saccharomyces cerevisiae*

Paper Number: 380575

Senqing Fan, Zeyi Xiao, Yan Zhang, Peina Yao, Qing Deng
School of Chemical Engineering, Sichuan University, 610065, Chengdu, China

Introduction

Environmental issues, the growing demand for energy, political concerns and the medium-term of petroleum has created the need for development of sustainable technologies based on renewable raw materials(Luque et al., 2008). Carbon neutral bioethanol, derived from a renewable biomass, has emerged as one of the most strategically important sustainable fuel sources and is considered an important way of progress for limiting greenhouse gas emissions, improving air quality and finding new energetic resources(Nigam & Singh, 2011).

Currently, batch fermentation is the most widely used in ethanol fermentation industry all around the world. In the batch process, the accumulation of ethanol in the broth inhibits the cell growth and leads to low ethanol productivity, low cell density and great wastewater discharge (Nguyen et al., 2011). Pervaporation is one of the most promising approaches for the removal of ethanol from fermentation broth because it does not cause any harmful effect on the microorganisms and has nice separation performance (Lee et al., 2012). Coupling the pervaporation with ethanol fermentation, a smaller effect of products inhibition could increase the ethanol production and improve the equipment utilization with less amount waste water processed (Ding et al., 2011).

During ethanol fermentation by the *Saccharomyces cerevisiae* cells, ethanol production is directly related to the cell growth and each cell is an "independent reactor" in which the substrate is converted to ethanol in the function of enzymes expressed by the cell. The cell growth model is regarded as an important step in developing an ethanol fermentation process, since models help in process control, increasing product quality and reducing process costs (Dodić et al., 2012). Once the models are defined, they can enable us to understand the fermentation process in a deep insight, design a new fermentation process with different condition such as temperature, pH, aeration and mixing (Zajšek & Goršek, 2010). They can also be used to control the fermentation process in the optimal condition. In the previous studies of ethanol fermentation, various kinetic models have been proposed to describe the cell growth, ethanol production and substrate utilization.

During continuous ethanol fermentation in a pervaporation membrane bioreactor, a continuous circulating-closed fermentation (CCCF) system could be achieved (Chen et al., 2012; Ding et al., 2011). The accumulated secondary metabolites in the broth and cell lysis along with the long fermentation time were the two focus of the CCCF system for ethanol production (Chen et al., 2012; Fan et al., 2014a). The aim of this study was to develop kinetic models to describe the cell growth in the CCCF system for long time continuous ethanol fermentation.

Methods

Membranes and module

The membranes used in the study were the composite PDMS membranes and the module was of a plate-frame structure. The detailed description has been shown in our previous work (Li et al.,

2004).

Microorganism and medium

The thermostable alcohol-active dry yeast (ADY), *Saccharomyces cerevisiae*, obtained from Angel Yeast Corporation (Hubei, China) was used in the study and the detailed composition of synthetic medium has been fully described in our previous work (Chen et al., 2012; Fan et al., 2014b).

Fermentation

Two runs of ethanol fermentation were carried our in the CCCF system with a pervaporation membrane bioreactor. The first run lasted for 264 h and the fermentation was operated in the classic batch mode during the initial 24 h. The continuous pervaporation was started at the time point of the 24th h for ethanol removal with the broth pH kept at 4.0±0.5. The second run lasted for 360 h and the pervaporation was coupled with the fermentation at the start of the process. During the first 72 h of the fermentation the broth pH was kept in 4.0±0.5 and then the pH was gradually increased to 6.5 to relieve the inhibition of the accumulated organic acids in the broth. The detailed procedure of continuous ethanol fermentation in the pervaporation membrane bioreactor had been fully described in our previous work (Fan et al., 2014b). The first run fermentation was terminated when the number of the viable cell ratio was decreased to about 8% and the average ethanol productivity during the last 12 h was about 0.85 $gL^{-1}h^{-1}$. The second experiment was terminated when the number of the viable cell ratio was about 5% and the average ethanol productivity during the last 12 h was about 0.56 $gL^{-1}h^{-1}$.

Assay

The cell concentration was determined by the dry cell mass. Cell viability was determined by direct microscopic count in a counting chamber after staining with methylene blue. The detailed measure procedures of cell had been fully described in our previous work (Chen et al., 2012; Fan et al., 2014b).

Kinetic model

During the continuous ethanol fermentation in the CCCF system with a pervaporation membrane bioreactor, ethanol removal *in situ* from fermentation broth could eliminate the ethanol inhibition (Ding et al., 2011; Vane, 2005). However, the secondary metabolites of the *Saccharomyces cerevisiae* metabolism not permeating through the membranes were accumulated in the broth of the pervaporation membrane bioreactor and these accumulative compounds could inhibit the yeast cell growth (Chen et al., 2012; Fan et al., 2014a; Fan et al., 2014b). In the previous study, it was found that the specific cell growth rate was decreased linearly with the accumulating concentration of the secondary metabolites (acetic acid, lactic acid, propionic acid, citric acid, succinic acid and glycerol) in the broth.(Fan et al., 2014a). Therefore, the cell growth model with considering these compounds inhibition in the CCCF system could be described as follows:

$$\frac{dx}{dt} = kx(1 - ai) \tag{1}$$

where, x, t, i stood for the cell concentration, fermentation time and the total concentrations of the secondary metabolites, respectively; a, k were the constants.

During ethanol fermentation complex metabolic pathways exist in the yeast cell with kinds of products produced in the process and these products follow the stoichiometric rules of the cell metabolism. During ethanol fermentation in the CCCF system with a pervaporation membrane bioreactor, the concentrations of the secondary metabolites not permeating through the membranes were found to be proportional to ethanol production with the fermentation process proceeding (Fan et al., 2014a). Therefore, Eq. (1) could be described as follows:

$$\frac{\mathrm{d}x}{\mathrm{d}t} = kx(1 - abp) \tag{2}$$

where, b was the constant representing the metabolic stoichiometry of the yeast cell in the CCCF system and p was the ethanol production with the fermentation proceeding.

It was widely accepted that during ethanol fermentation by *Saccharomyces cerevisiae*, the ethanol formation was directly related to the yeast cell growth since ethanol was the product of the yeast cell energy metabolism (Bailey & Ollis, 1994). The correlation of yeast cell growth with ethanol formation could be described as follows:

$$c\frac{\mathrm{d}x}{\mathrm{d}t} = \frac{\mathrm{d}p}{\mathrm{d}t} \tag{3}$$

where, c was a constant, representing the yeast cell metabolism characteristics.

Provided that p was equal to zero at the initial hour of fermentation and a solution on ethanol production by the integration of Eq. (3) was given as follows:

$$p = c(x - x_0) \tag{4}$$

where, x_0 was the cell concentration at the initial hour of fermentation.

Combining Eq. (4) and Eq. (2), the cell growth model in the pervaporation membrane bioreactor with considering the accumulative secondary metabolites inhibition could be described as follows:

$$\frac{\mathrm{d}x}{\mathrm{d}t} = \mu_0 x(1 - \frac{x}{x_\infty}) \tag{5}$$

where, μ_0 and x_∞ were defined as the specific cell growth rate and the maximum cell concentration including the active and dead cell as well as the broken cell in the broth, respectively. The two defined parameters could be correlated to constants k, a, b, c and x_0, as follows:

$$\mu_0 = k(1 + abcx_0) \tag{6}$$

$$x_\infty = x_0 + \frac{1}{abc} \tag{7}$$

During ethanol fermentation in the CCCF system with a pervaporation membrane bioreactor lasting hundreds of hours, the natural apoptosis and lysis of cell would lead the broth gradually deteriorating to inhibit the cell growth. Volterra had ever developed a model to predict a phase of decline in cell growth time profiles which led an integral term into the model (Bailey & Ollis, 1994). We simply borrowed Volterra model, and defined K_0 as a constant to describe effects of the cell lysis on cell growth:

$$r_\mathrm{d} = \int_0^t K_0 x(t)\mathrm{d}t \tag{8}$$

consequently, specific cell lysis rate could be defined as r_d/x.

It could be reasonable to suppose that during CCCF process the cell growth rate could be expressed as a linear superposition of Eq (5) and Eq (8):

$$\frac{dx}{dt} = \mu_0 x(1 - \frac{x}{x_\infty}) + \int_0^t K_0 x(t) dt \quad (9)$$

From Eq. (9), it could be seen that during the fermentation process the cell growth in the pervaporation membrane bioreactor was determined with the specific cell growth rate (μ_0), the maximum cell concentration without considering the cell lysis (x_∞) and constant (K_0) concerning with the other inhibition effects. The parameters in Eq. (9) were obtained by the numerical simulation method.

Results and discussion

Cell growth and viability

Two runs of ethanol fermentation were carried out in the CCCF system and the cell growth performance and viability were illustrated in Fig.1. The viable cell concentration was determined by the multiplying of the total cell concentration and the viable cell ratio. From Fig.1 it could be seen that the total cell concentration could be increased for a long time to a maximum density and then a stationary stage or decline stage could be presented. During the earlier fermentation period the viable cell growth performed almost the same as the total cell growth, since the cell viability was high. But the viable cell concentration was lower than that of the total cell with the fermentation proceeding, since the cell viability was decreased gradually. During the later fermentation period, the difference was performed more and more obvious. From Fig.1 it could be also seen that a lag stage (24 h-36 h) was performed in the first run owing to the delayed response of yeast cell to the environment change of the fermentation broth after the pervaporation was started. The cell could increase continuously to a point with the maximum viable cell concentration of 13.3 g $L^{-1}$and then to another point with the maximum total cell concentration of 26.1 g L^{-1} after going through the shorter lag stage. In the second run the cell could increase continuously without lag stage to a point with the maximum viable cell concentration of 25.5 g L^{-1} and then to another point with the maximum total cell concentration of 38.5 g L^{-1}, since pervaporation was coupled at the start of the fermentation. Fermentation at higher pH would relieve the inhibition of the accumulated secondary metabolites (main the organic acids) and therefore higher cell concentration could be achieved in the second run with the broth pH being increased gradual from 4.5 to 6.5 after 72 h fermentation (Chen et al., 2012; Fan et al., 2014a).

During ethanol fermentation in the CCCF system the fermentation process would operating for hundreds of hours and therefore the cell aging, death and lysis must be emerging in the broth with the fermentation proceeding. As shown in Fig.1, with the fermentation proceeding, the cell viability was less and less, which means the ratio of the aged cell to the total cell was increased gradually. Undoubtedly, the cell aging, death and lysis should be imputed to the gradually deteriorating culture environment of the broth in the CCCF system for ethanol fermentation. The deterioration factors would include the accumulative inhibitors in the broth and the harm to the cell population increase induced by the individual cell competing with each other for surviving (Atkinson & Mavituna, 1991; Bailey & Ollis, 1994). During the later fermentation period the culture deterioration became more and more serious, leading to the acceleration of the cell apoptosis and lysis and therefore the stationary stage and decline stage were presented in the cell growth profile.

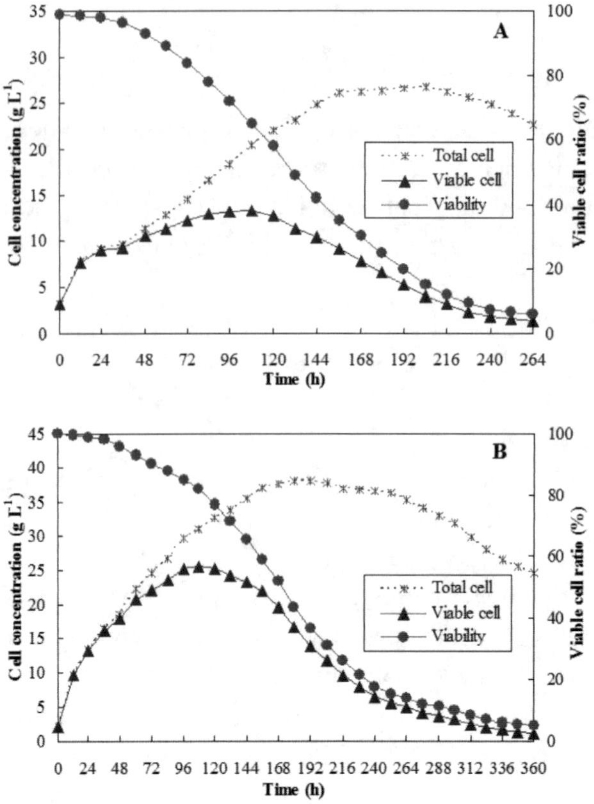

Fig.1 Time course profile of cell growth and cell viability in CCCF system for ethanol fermentation. A, the first run; B, the second run

Simulation of the cell growth

The kinetics of cell growth in the CCCF system with a pervaporation membrane bioreactor for ethanol fermentation was shown in Fig.2. The correlation coefficients of non-linear analysis by Eq. (9) were 0.995 and 0.987 for the first run and the second run respectively, which suggested that the kinetic model of Eq. (9) could be able to adequately describe the cell growth in the CCCF system for ethanol fermentation. If the cell lysis was not considered in the kinetic model, the fitting curves had appeared the shape of the Logistic curves, described as dashed lines in Fig. 2. From the dashed lines of the two runs, it was found that the maximum cell concentration in the CCCF system for ethanol fermentation had been about 40 g L^{-1} in the first run and 47 g L^{-1} in the second run respectively if there was no cell lysis in the fermentation process. Obviously, the value had not been actually obtained in the experiments due to the cell death and lysis during the long time of fermentation.

The simulating parameters of the kinetic model for cell growth in the CCCF system was shown in Table 1. As a contrast, the experimentally measured maximum cell concentration x_m was also listed in Tab. 1, which was also the peak point of the fitting curves obtained from Eq. (9). The differences between the x_∞ and the x_m were 13.7 g L^{-1} and 7.9 g L^{-1} in the first run and the second run, respectively. These would be considered as evidences of the cell death and lysis during the long time of ethanol fermentation in the CCCF system. Owing to less (x_∞-x_m) difference in the second run of fermentation, it seems that the cell death could be relieved by elevating pH value of the broth. The same result could be also obtained from the parameter K_0, since the value of K_0 in first run was higher than that in the second run. The parameter μ_0 in Table 1 represented the overall cell growth performance during the whole fermentation period and it was the combined effect of several complex factors in the CCCF system, including cell breeding, products inhibition and cell lysis. The accumulated ethanol inhibition after 24 h batch fermentation and the accumulated organic acids inhibition without increasing the medium pH in the first run led to the cell growth performance poor and therefore lower value of μ_0 was obtained.

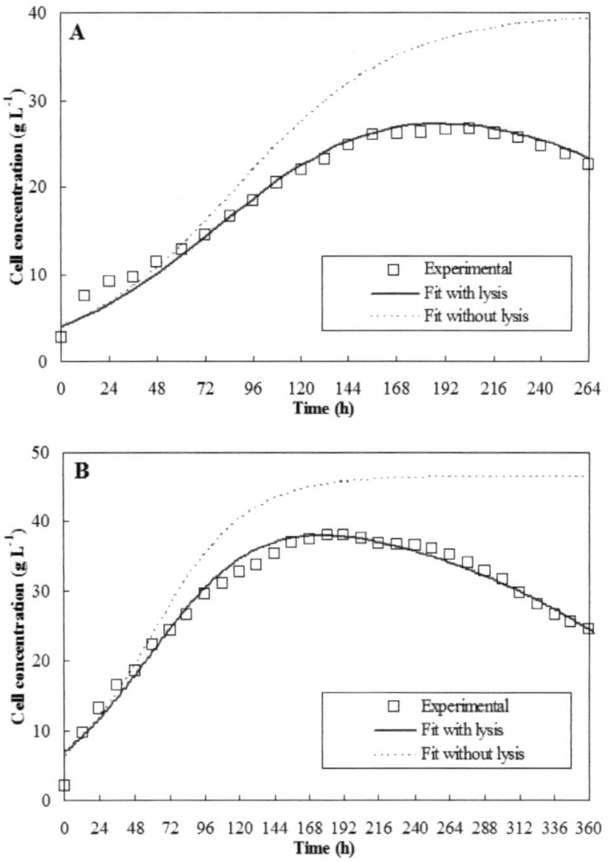

Fig.2 Simulation on the cell growth of the CCCF system for ethanol fermentation. A, the first run; B, the second run

Table 1 Calculated parameters using the kinetic model for cell growth in CCCF system

Kinetic parameters	First CCCF	Second CCCF
μ_0 (h^{-1})	0.025	0.031
x_∞ (g L^{-1})	39.8	46.4
x_m (g L^{-1})	26.1	38.5
K_0 (h^{-2})	-6.48×10^{-5}	-4.18×10^{-5}

Conclusions

A kinetic model of cell growth in the CCCF system with a pervaporation membrane bioreactor was proposed. During the process, ethanol was removed in *situ* from the broth by the membranes with ethanol inhibition being eliminated and the accumulated secondary metabolites in the broth became the main inhibitors. The cell death rate at a certain time was considered as a function of all past values of the cell population density. The model was validated by the simulation results of the dynamic model agreeing well with the experimental data.

References

1.Atkinson, B., Mavituna, F. 1991. Biochemical Engineering and Biotechnology Handbook. second ed. Stockton press, New York.

2.Bailey, J.E., Ollis, D.F. 1994. Biochemical Engineering Fundamentals. second ed. McGraw-Hill Book Company, Singapore.

3.Chen, C., Tang, X., Xiao, Z., Zhou, Y., Jiang, Y., Fu, S. 2012. Ethanol fermentation kinetics in a continuous and closed-circulating fermentation system with a pervaporation membrane bioreactor. Bioresour. Technol., 114, 707-710.

4.Ding, W.-W., Wu, Y.-T., Tang, X.-Y., Yuan, L., Xiao, Z.-Y. 2011. Continuous ethanol fermentation in a closed-circulating system using an immobilized cell coupled with PDMS membrane pervaporation. J. Chem. Technol. Biot., 86, 82-87.

5.Dodić, J.M., Vučurović, D.G., Dodić, S.N., Grahovac, J.A., Popov, S.D., Nedeljković, N.M. 2012. Kinetic modelling of batch ethanol production from sugar beet raw juice. Appl. Energ., 99, 192-197.

6.Fan, S., Xiao, Z., Tang, X., Chen, C., Zhang, Y., Deng, Q., Yao, P., Li, W. 2014a. Inhibition effect of secondary metabolites accumulated in a pervaporation membrane bioreactor on ethanol fermentation of Saccharomyces cerevisiae. Bioresour. Technol., 162, 8-13.

7.Fan, S., Xiao, Z., Zhang, Y., Tang, X., Chen, C., Li, W., Deng, Q., Yao, P. 2014b. Enhanced ethanol fermentation in a pervaporation membrane bioreactor with the convenient permeate vapor recovery. Bioresour. Technol., 155, 229-234.

8.Lee, H.J., Cho, E.J., Kim, Y.G., Choi, I.S., Bae, H.J. 2012. Pervaporative separation of bioethanol using a polydimethylsiloxane/polyetherimide composite hollow-fiber membrane. Bioresour. Technol., 109, 110-115.

9.Li, L., Xiao, Z., Tan, S., Pu, L., Zhang, Z. 2004. Composite PDMS membrane with high flux for the separation of organics from water by pervaporation. J. Membr. Sci., 243, 177-187.

10.Luque, R., Herrero-Davila, L., Campelo, J.M., Clark, J.H., Hidalgo, J.M., Luna, D., Marinas, J.M., Romero, A.A. 2008. Biofuels: a technological perspective. Energy Environ. Sci., 1, 542-564.

11.Nguyen, V.D., Auresenia, J., Kosuge, H., Tan, R.R., Brondial, Y. 2011. Vacuum fermentation integrated with separation process for ethanol production. Biochem. Eng. J., 55, 208-214.

12.Nigam, P.S., Singh, A. 2011. Production of liquid biofuels from renewable resources. Prog. Energ. Combust, 37, 52-68.

13.Zajšek, K., Goršek, A. 2010. Modelling of batch kefir fermentation kinetics for ethanol production by mixed natural microflora. Food Bioprod. Process., 88, 55-60.

385950 Process Modeling and Economic Evaluation of the Ozone and Soaking Aqueous Ammonia Pretreatment Process with Aspen Plus

Abstract:

The ozone and soaking aqueous ammonia pretreatment (OSAA) is a novel process that operates at low temperatures and ambient pressure to effectively remove lignin from wheat straw and other herbaceous biomasses. Prior work has demonstrated that the OSAA pretreatment process is able to remove up to 80% of lignin from the biomass, resulting in greater than 90% sugar yield after enzymatic hydrolysis for 72 hours using 15 FPU and 15 CBU cellulase and β-glucosidase loading, respectively. In this work, a comprehensive process model has been developed for the sequential pretreatment of wheat straw with the OSAA process. The pretreatment process has been simulated using collected data for ozone pretreatment and soaking aqueous ammonia pretreatment. In addition, a techno-economic assessment was performed to calculate the operating costs and to estimate the capital cost. The results indicate that operating at low temperatures and pressures, while requiring moderately longer residence time than higher severity pretreatments, is able to effectively pretreat lignocellulose at lower production costs.

Title: "Process development and scale-up of ionic liquid-tolerant cellulase cocktail"

F. Tachea[1], C.Li[1], L. Sandoval[1], S Hubbard[1], Q.He[1], J. Wong[1], P. Coffman[1], J. Gardner[1], J. Baez[1], and J. Gladden[2]

The availability of lignocellulose biomass makes it a promising feedstock for commercial scale production of biofuels. However, current commercial enzyme cocktails are less efficient and more expensive to make fermentable products from lignocellulose biomass. Researchers at the Joint BioEnergy Institute (JBEI) have identified a novel ionic liquid (IL) tolerant cellulase cocktail called JTherm, a mixture of recombinant β-glucosidase (BG), cellobiohydrolase (CBH) expressed in E.Coli and endoglucanase/xylanase produced from a thermphilic Bacterial Community, which can efficiently break down the IL pretreated cellulosic biomass to fermentable sugars. In collaboration with JBEI, the Advanced Biofuels Process Demonstration Unit (ABPDU) is performing process optimization and development of the JTherm enzymes at 2L scale. At the ABPDU before scaling up the JTherm cellulose enzyme cocktail, process scale up parameters' were defined at 2L scale and all the scale up issues were resolved. In addition, a fed-batch fermentation process was developed for both BG and CBH enzymes and the biomass growth and enzyme expression were investigated in complex, defined and auto-induction medium. Moreover, the effect of early and late IPTG induction on biomass growth and enzyme expression was investigated. Based on these studies, we were able to improve the BG biomass growth about 80% and employing the fed-batch fermentation improves the activity of the enzymes. More findings are underway for the 300 L scale fermentations, and the knowledge gained from these scale up studies is an essential step for demonstrating the commercial viability of the enzyme technology for biofuels production.

[1]Advanced Biofuels Process Demonstration Unit, Lawrence Berkeley National Laboratory, Emeryville, CA;

[2]Joint BioEnergy Institute, Emeryville, CA

393623 Sustainable Winemaking – Challenges and Opportunities

Abstract:

In the making of wine one can expect to generate 2 gallons to as high as 10 gallons of water for every gallon of wine produced. This water is purely for cleaning/sanitation since only the water contained in the grape can be added to wine. The variation is largely driven by the scarcity/cost of available water, which suggests conservation is intimately tied to economics and the economic conditions are changing at an accelerated rate due to the severity and suddenness of the drought in the western US. A more proactive approach is to consider the problem of water use reduction, treatment (including energy recovery) and synergistic use of the solids streams that exit a typical winery which will be the topic of this talk.

393625 The Challenges of Water Use Reductions and Food Safety in Dairy Processing

Abstract:

The UC Davis Viticulture & Enology Department is developing technologies necessary to realize a self-sustainable, zero carbon winery. Through the use of rainwater capture, filtration and recycling, green cleaning chemistries, Clean-In-Place technology, carbon capture, as well as use of non-carbon energy sources, UCD proposes to develop the world's first energy positive, water positive, and zero carbon winery. Advanced data systems will enable analysis and optimization of processing technologies. Our goal is to develop systems and processes that will enable the food and beverage industry to significantly reduce their water and energy requirements as well as show how process wastewater can be reduced to nearly zero.

THE CHALLENGES OF FOOD SAFETY AND WATER USE REDUCTION IN DAIRY PROCESSING

Montgomery Bohanan and Joe Herrud, Leprino Foods, Denver, CO

Overview

We will discuss the challenge of limiting water usage in conflict with food safety requirements in the dairy industry. As the food industry on a whole strives to deliver the safest food products possible, cleaning requirements and demands continue to increase every year - which is compounded by the fact that the majority of new equipment installations have greater volume demands for cleaning water/fluid. The ability to re-use water is dependent upon and limited by the Federal Pasteurized Milk Ordinance (PMO), and by the infrastructure in place throughout the industrial facilities. With improved equipment designs and the possibility of using enzymes or other cleaning methodologies that are much less water intensive, a significant reduction in water consumption could be made across the dairy manufacturing segment. Chemical engineers working on water reduction around sanitation practices will be vital to simultaneously achieving food safety and water consumption goals within the dairy industry.

Food safety is becoming a greater and greater concern each year as negative health events drive food growers, processors, and manufacturers to work harder to eliminate food-borne risks. The efforts usually result in more cleaning/sanitizing and those steps typically generate more water use. And at the same time, companies are also working hard on reducing the amount of water consumed in their operations to meet sustainability goals or to reduce costs for usage and/or waste water treatment. The availability of potable water in many arid regions (which is more typical in many of the higher dairy production areas of the country) coupled with the ability to treat the additional wastewater generated each day in the sanitation process becomes a balancing act for every organization regarding cleaning and sanitizing and food safety.

Historically, the variables used to achieve the desired level of cleanliness for a clean in place (CIP) system in any unit operation has consisted of:

- Time,
- Temperature,
- Concentration, and
- Mechanical action.

Each variable is dependent upon the others and also upon the soil type and soil load. Let's review the criticality of each of these variables and how they relate to sanitation/food safety.

Time

With enough time, a just about any cleaning strategy may be able to attain the desired soil removal and bacteriological kill. Increased cleaning time could lead to lower temperature requirements or lower cleaning chemical concentration needed. From a food safety standpoint, time is critical in microbial growth and the development of biofilms - so that should also be taken into account. And to maximize capital utilization, the amount of time the equipment is down has to be limited - because time is money. Uptime is vital. Additionally, in most dairy manufacturing processes, sanitation must begin after no longer than every 24 hours of continuous operations (as required in the PMO). Therefore, the ability to sanitize in an efficient amount time will dictate the daily throughput of each manufacturing facility. Typically cleaning time is minimized rather than maximized.

Temperature

Temperature ranges to inhibit bacteriological growth when storing food are typically below 40 degrees F or higher than 140 degrees F. Also, to achieve the necessary kill level as required during the pasteurization step in dairy operations, temperature is time dependent. As an example, raw milk must be held to 161 degrees F for a minimum of 15 seconds for proper pasteurization according to the PMO. So the higher the temperature, the shorter the amount of time needed to achieve desired bacteriological kills. Still, in the thermal process the food flow must pass through the food safety danger zone (between 40 - 140 F), typically twice. These thermal processes for storing, pasteurizing, and otherwise handling food products effect the soil load on equipment surfaces and therefore effect the cleaning and sanitizing process. Soil residues in the food safety danger zone are especially worrisome as they can become microbiological harborages. Typically, cleaning a process requires a temperature minimally 10 F higher than the processing temperature for hot processes. Additionally, the reactivity of cleaning solutions is dependent upon the energy supplied by the temperature of the solution. Temperature can also be dictated by the nature of the soil to be removed; for example in the dairy industry butterfat typically melts at 128 F so wash solutions would exceed that temperature to ease removal of the soil, and usually exceed 140 F. And on the other hand excessively high temperatures can denature or burn on soils, thus making them even harder to remove.

As stated above in the section discussing time needed for the CIP process, there is a balance between the amount of energy utilized to attain desired temperatures and the amount of time the system can be down for sanitation. Typically, cleaning temperatures are limited by energy requirements and equipment tolerance.

Concentration

Water may be the "universal" solvent, but not all soils are water soluble. To improve the efficacy of cleaning and shorten the time required to remove soils, chemicals are used in cleaning solutions dependent upon the soil to be removed and the equipment or surface to be cleaned. Alkaline and/or chlorine solutions have typically been used to remove organic soils, while acid solutions have been used to remove mineral soils by helping to dissolve or emulsify the soil into the cleaning solution.

In the dairy industry, the organic soil load typically consists of butterfat, proteins, and carbohydrates; while the inorganic soils are minerals found in milk such as calcium or minerals deposited from the water source or alkaline cleaners. The amount of cleaning time needed can be shortened with higher concentrations of caustics or acids, but there is a point of diminishing return - when the costs of the higher concentrations outweigh the time saved, and when the higher chemical concentrations begin to damage equipment or prove too costly in the waste water treatment plant.

Sanitizers used in providing bacteriological kill after cleaning are EPA registered pesticides and must be used following label instructions. The efficacy of sanitizing solutions is dependent upon time, temperature and concentration. In a CIP application, the sanitizing solution needs to make contact with the surfaces with enough exposure time to achieve bacteriological kills, and enough volume to ensure complete coverage. For no-rinse sanitizing solutions, a specific concentration range must be maintained to ensure enough chemical to effectively kill microorganisms but still not exceed the maximum level allowed for safe human consumption/exposure. In some instances, this can require significant amounts of potable water to ensure coverage and exposure time on large pieces of equipment or long or complicated line circuits with multiple valve clusters. While critical to ensuring food safety, most sanitizer chemicals prove detrimental to the waste water treatment flora, which can add significant cost to the overall operation.

Typically, chemical concentrations are limited due to usage costs and to waste water treatment and handling costs (COD).

Mechanical Action

The sanitation fluid used in the CIP process must have enough energy to make it to all the surface areas of the equipment, the ability to penetrate biomass or "soil" to be able to chemically dislodge, and be able to scour/remove all residues. The minimum velocity desired in piping systems is typically 5 to 7 feet per second to attain the turbulence needed. Laminar flow will not ensure all soil films are removed from the surfaces. As far as face velocities on other unit operations, the numbers can vary significantly. The flow of the cleaning solution in these applications typically delivers mechanical action through impingement provided by spray devices. Traditionally, supplying improved mechanical action through increased flow rates and thus increased water usage has been the accepted practice. Additionally, as companies have tried to optimize production, the size of equipment and especially piping has increased significantly. The large diameter piping requires large flow rates to fill the entire pipe to ensure coverage of all surfaces plus turbulent flow - when the needed action and contact is only on the inner surface area of the pipe. So, now with cost optimization and water conservation efforts becoming more prevalent, new and improved methods for delivering mechanical action need to be explored.

Additionally, CIP cleaning requires at least two and sometimes several water rinse steps. The initial rinse step in a CIP program is designed to remove as much of the gross soils (and easily removed soils) from the system as possible (targeting 90 - 95% of the soil load). This allows the chemicals in the cleaning solution to work off the hard the remove soils - the films and hardened soils, and can require a large amount of water if a large amount of product is left in the system. Also, there are rinses after each wash step to assist in remove the dissolved soils and to ensure there are no chemical residues left behind. Sometimes these rinses are necessary to ensure incompatible chemical solutions do not inadvertently mix and potentially damage surfaces or equipment. Also, the rinse prior to sanitizing must not only ensure removal of all soil and chemical residues so that microorganisms are not protected from exposure to the sanitizer, but the rinse also is necessary at times to ensure the temperature of the surfaces are appropriate for the sanitizer application.

Even if these variables were always accepted as key to achieving proper sanitation, they are typically incorporated into the design as an afterthought following equipment efficiency, throughput, and capital costs. Cleaning of equipment and processes is often an afterthought to operations, and are therefore difficult to cost or performance optimize. So to achieve sanitation targets in such designed systems and processes, it has been necessary to design equipment to include:

- Higher flow rates and volumes (to increase mechanical action),
- More CIP nozzles and segments (to achieve the necessary coverage),
- Higher temperatures (to maximize solution reactivity), and/or
- Higher concentrations or more effective chemicals (to increase solubility using wetting agents, surfactants, and detergency).

All of which can have a negative impact on water usage, energy usage, and waste water treatment – ultimately the cost of production.

Design for Success

If we were truly optimizing the sanitation process during the design phase of equipment or processes, the following would be considered:

- Minimize residue/biofilm create or left on equipment during operations
  - Keep material deposits to a minimum
    - Eliminate hot spots favorable to biological growth or soil degradation
    - Self cleaning velocities during processing
    - Correct surface finishes or improved surface finished to enhance soil removal
- Make equipment easier to clean
    - Don't have places for bacteria to hide
    - Surface finishes
      No dead spots
    - No ledges
    - Easily and completely drains
    - Accessibility for cleaning
- Ability to remove soils without water (Potential yield improvement)

    - Dry clean
    - Line "pigs" or other air purging devices
    - New methods for purging the system or equipment
- Easy teardown to allow manual removal instead of using water
- Correct style, number, and location of CIP nozzles
    - Design equipment around a spray pattern to minimize wash solution needs instead of adding spray balls to create coverage in an existing piece of equipment
    - Utilize nozzles that maximize coverage and impingement for the volume of water used.
    - Develop new CIP impingement devises.
    - Treat CIP systems and solutions as a limited utility resource and design equipment around those limitations and availability.

Sanitation with Minimal Water or Chemicals

Just pumping more water and additional sanitation chemicals can't always be the answer for proper cleaning. What other options should be investigated to make step changes in the amount of water used?

- Alternative dry cleaning, such as:
    - Dry ice/CO2 residue removal/pretreatment
    - Air pulse or turbulent air blow
- Enzymatic cleaners to break down soils into more readily soluble forms
    - Develop/utilize enzymes that are recoverable and reusable
    - Potential for high concentrations that will not damage equipment while minimizing water usage
- Cleaning chemical recovery and re-use, such as
    - Membrane recovery of caustic/acid
    - Mineral precipitation from acid or of caustic
    - Phase separation from acid or caustic
    - Removal of sanitizing agents from water prior to traditional waste water treatment
    - Associated water recovery from the process listed above
- Different chemical combinations, such as
    - solvent based cleaners
    - acid override washes
    - nanotechology or encapsulation
- Different methods for delivering/applying cleaning chemicals
    - Automated foam application, followed by minimal water impingement
    - Automated and/or robotic brushes and/or spray devices
    - Nanotechnology or encapsulation

The majority of the cleaning techniques/strategies we've discussed are pertaining specifically to processing equipment and CIP systems. There are many other examples of cleaning requirements that each facility must tackle that deal directly with environmental conditions and water consumption at the facility. What are the other practices that can be put into place to address:

- COP (Clean-Out-of-Place) tanks
- Manually cleaning (Foam, Boosted Pressure) and rinsing
- Proper hand cleanliness
- Foot traffic exposure
- Forklift contamination
- Tool contact and cleanliness

Conclusion

Balancing food safety with water reduction is a challenging task within the dairy industry. There are existing best practices we can put in place today to help achieve reduction goals while not sacrificing food safety requirements. But to make a step change for the future, new designs and "out of the box" thinking will be needed. Who will discover the disruptive technology that will allow a reduction in water consumption while maintaining - or improving - plant hygiene and food safety? Who are the creators of "what could be"? You!

393627 Maximizing Water and Energy Efficiency in Food Processing While Minimizing Negative Environmental Impact

Abstract:

The long term health of the food processing industry depends on development of technologies and practices that maximize efficiency of water and energy use while minimizing negative environmental impact. Realizing sustainable food processing hinges on understanding how water and energy use are intertwined throughout food processing facilities down to the level of individual unit operations. The water-energy nexus (WEN) assessment is a framework for auditing all supply and demand points for water and energy within a food processing facility. The WEN analysis provides processors with a complete breakdown of how water and energy use are distributed throughout their processing pipeline and identifies targets for improving efficiency (e.g., by identifying inefficient equipment or opportunities for water and heat recovery). To demonstrate, the results of a WEN analysis from a tomato processing facility are provided. Additionally, opportunities for improving management strategies for food processing solid and liquid wastes are discussed. These include advancing technologies for biofuel production and emerging technologies for agricultural application of food processing organic residues and microbial desalination of wastewater streams.

385686 Open-Cell Foam Supports for Structured Catalytic Reactors

Abstract:

Structuring concepts for catalytic reactors such as the use of monolithic honeycombs and open-cell foams are promising for achieving process intensification in the chemical industries. In particular, these structured catalyst supports have the potential to eliminate the classical drawbacks of conventional randomly packed fixed-bed reactors, i.e. high pressure drop and hotspots.

The most prominent and successful examples of structured reactors are honeycomb-shaped monolithic catalysts. Owing to their excellent properties such as, e.g., exceptionally low ratio of pressure drop to geometric specific surface area, they have become the standard catalyst shape in many environmental applications. However, monolithic honeycombs lack some important properties such as, e.g., tortuosity of the flow and radial mixing since they consist of a large number of parallel straight channels with no interconnectivity.

In contrast to this, open cellular materials combine the advantages of packed beds (i.e., radial mixing and tortuosity of the flow) and honeycombs (i.e., high geometric specific surface area and low pressure drop) owing to their high porosities and their characteristic 3D cellular architecture.

In our recent research, we have investigated open cellular materials with both periodic and random cell geometries regarding their use as catalyst support. These structures were characterized with respect to their morphological/geometrical properties, fluid dynamic behavior and mass and heat transport properties. Models for the estimation of properties important for reactor design such as, e.g., specific surface area and pressure drop of such cellular materials were developed and validated.

Detailed numerical studies of such geometries and the 3D flow field along with mass and heat transport processes were carried out in order to analyze the influence of the local foam geometry on the macro transport characteristics. This allows for a more fundamental understanding of the relationship between the support structure, the local flow field, the local and global transport characteristics, and finally the reactor performance.

In our contribution, we present a comprehensive overview on our recent research on cellular materials.

Process intensification of electrochemical processes by enhanced mass transfer using the rotor-stator spinning disc reactor

P. Granados Mendoza[1], T. de Groot[2], J.T.F. Keurentjes[1], J.C. Schouten[1] and J. van der Schaaf[1,*]

[1] *Eindhoven University of Technology, Den Dolech 2, P.O. box 513, 5600 MB Eindhoven, The Netherlands*
[2] *Akzo Nobel Industrial Chemicals B.V., Stationsstraat 77 P.O. box 247 3800 AE Amersfoort, The Netherlands*

**Corresponding author: j.vanderschaaf@tue.nl*

Electrochemical processes play an important role in the chemical industry. Nowadays many chemicals are produced using an electrochemical route, for example: chlorine, sodium hydroxide, sodium chlorate, hydrogen, oxygen, aluminum, copper, magnesium, zinc, and adiponitrile [1]. In an electrochemical process, the process is limited by the rate of mass transfer at high current densities. The largest rate of the electrochemical reaction is reached at the limiting current density [2]. It is therefore desired to intensify the mass transfer rate in electrochemical reactors. An option for process intensification is the use of high shear forces that promote a rapid mixing of fluids and a high surface renewal rate. A type of rotating equipment that uses these principles is the rotor-stator spinning disc reactor (RS-SDR) [3]. The RS-SDR consists of a rotating disc in a cylindrical housing, with a typical gap distance between the rotor and the stator of 1 mm. The results reported by Meeuwse et al. [4] for gas-liquid and liquid-solid applications make this reactor a promising process intensification technology for single and multiphase electrochemical processes. In gas producing electrochemical reactions, an additional benefit of the spinning disc reactor is the high centrifugal force. This force rapidly removes gas bubbles from the electrodes which would become insulated by a gas layer under gravitational conditions. Several authors have studied the mass transfer to a rotating disc, in a rotor-stator configuration [4], or in a free disc configuration [5]. This work compares the mass transfer coefficient from the liquid to the rotor and to the stator of a RS-SDR, and evaluates the effect of electrolytic gas bubbles on the mass transfer in RS-SDR. The mass transfer coefficient at RS-SDR is determined at different rotation speeds using the limiting current density method. Measurements of the limiting current density are taken and used to calculate the liquid solid mass transfer coefficient as follows: $j_L/nF=k_{LS}C^{bulk}$. The reaction studied is the reduction of ferricyanide from an electrolyte mixture of ferro-ferricyanide ions in KNO_3, which is pumped around with 7 – 42 mL/s. The RS-SDR consists of a transparent PMMA with a rotating disc (13 cm diameter, rotational speeds from 20 to 1300 rpm) located at a small gap (1 mm) from the stator. The reactor configuration used in this study allows placing the working electrode on the rotor and on the stator respectively. Both working and counter electrodes are made of pure nickel. An Autolab PGSTAT30 is used for the potentiostatic control of the reactor. Results of the liquid-solid mass transfer without the influence of gas evolution showed that the mass transfer at the rotor disc surface is higher than the mass transfer at the stator. These results show that the mass transfer to the rotor and to the stator are enhanced by a factor 7 and 2 respectively (4.5 average) when compared to a non-rotating and a free rotating disc. Results of the influence of gas electrolytic bubbles on the mass transfer will be presented. Sh-correlations are established that correlate mass transfer to disc radius, rotation speed and gap distance. This work presents the opportunities of process intensification of electrochemical processes using the rotor-stator spinning disc reactor.

References

[1] M. Grotheer, R. Alkire, R. Varjian, Electrochem. Soc. Interf., Spring 2006, 52-53.
[2] S.L. Marchiano, A.J. Arvia, in: Ismail, M.I (Ed.) Electrochemical Reactors: Their Science and Technology, Elsevier, 1989, 145-186.
[3] J. van der Schaaf, J.C. Schouten, Current Opinion Chem. Eng., 1 (2011) 84-88.
[4] M. Meeuwse, Rotor-stator spinning disc reactor, PhD-thesis, Technische Universiteit Eindhoven, 2011.
[5] B. Levich, Acta Physiochim. URSS 19, 117 (1944).

Novel stator-rotor-stator spinning disc reactor for intensified integration of reaction and heat exchange

M.M. de Beer[1], J.T.F. Keurentjes[1], J.C. Schouten[1] and J. van der Schaaf[1,*]

[1]Eindhoven University of Technology, P.O. Box 513,
5600 MB Eindhoven, The Netherlands
*Corresponding author: J.Vanderschaaf@tue.nl

The stator-rotor-stator spinning disc reactor (SDR) is a novel continuous flow reactor particularly aiming at intensified heat transfer rates for fast, highly exothermic reactions. For the "conventional" rotor-stator SDR (consisting of a rotating disc enclosed by a stationary cylinder, separated by an axial clearance in the range of a few millimetres) high rates of gas-liquid [1], liquid-liquid [2] and liquid-solid [3] mass transfer have been reported. These high mass transfer rates, caused by the high shear force acting on the fluid between the rotor and the stator, enable increased production rates of mass transfer limited processes. In case of highly exo- or endothermic reactions however, intensified heat transfer is necessary to accommodate these increased production rates. In the rotor-stator SDR heat is withdrawn from the reactor fluid through cooling channels inside the stator walls. Fluid-stator heat transfer coefficients increase with rotational velocity. However, effective overall heat transfer in the rotor-stator SDR is limited by stator wall conduction and cooling channel convection; a maximum overall heat transfer coefficient of U_{ov} = 1990 $Wm^{-2}K^{-1}$ is obtained for a single water phase, at a rotational velocity of $\omega = 40$ rad s^{-1}.

The stator-rotor-stator spinning disc reactor addresses this issue. It consists of a stationary disc (the first stator), enclosed by a rotating cylinder (the rotor), which is again enclosed by a stationary cylinder (the second stator). In this way two rotor-stator cavities are created, separated by the common rotor. The axial clearance between the rotor and both stators is small, comparable to the rotor-stator SDR. In the stator-rotor-stator SDR, transfer of heat between two fluid streams is thus realised with a rotating heat transfer surface.

For a single water phase in both rotor-stator cavities the overall heat transfer coefficient increases with rotational velocity, from U_{ov} = 550 $Wm^{-2}K^{-1}$ at $\omega = 0$ rad s^{-1} up to U_{ov} = 7000 $Wm^{-2}K^{-1}$ at $\omega = 100$ rad s^{-1}. Similar to results obtained in the rotor-stator SDR, no influence of throughflow velocity on the heat transfer is observed; the heat transfer rates are determined by rotation. The fluid-rotor heat transfer coefficients obtained in the current work are in agreement with the values of fluid-stator heat transfer coefficients obtained in the rotor-stator SDR.

It can be concluded that in terms of heat transfer performance the stator-rotor-stator SDR is a major improvement compared to the rotor-stator SDR. Since the rotor-stator cavities are similar for both types of reactors, it is expected that mass transfer rates in multiphase reactions are equally high in the stator-rotor-stator SDR. This makes the stator-rotor-stator SDR a versatile multiphase reactor, where exo- and endothermic processes can potentially be integrated into a single reactor and where rotation simultaneously enhances both mass and heat transfer. A comparison is made with other compact heat exchanger/reactor concepts, such as microstructured and plate reactors, for the application of both single and multi-phase processes.

References

[1] M. Meeuwse, J. van der Schaaf, B. F. M. Kuster, J. C. Schouten, Gas-liquid mass transfer in a rotor-stator spinning disc reactor, Chem. Eng. Sci. 65 (2010) 466-471

[2] F. Visscher, J. van der Schaaf, M. H. J. M. de Croon, J. C. Schouten, Liquid-liquid mass transfer in a rotor-stator spinning disc reactor, Chem. Eng. J. 185-186 (2012) 267-273

[3] M. Meeuwse, S. Lempers, J. van der Schaaf, J. C. Schouten, Liquid-solid mass transfer and reaction in a rotor-stator spinning disc reactor, Ind. Eng. Chem. Res. 49 (2010) 10751-10757

367699 Enhanced Heat Transfer in the Helical Baffled Heat Exchanger with Optimal Design of Flow Stream

Abstract:

The helical baffled heat exchanger is the upgraded shell and tube heat exchanger, except supporting the pipe bundle, the main function of the baffles is to create the flow channel for the shell side fluid. In this paper, the flow stream analyze in the helical baffled heat exchanger were investigated. Compared the stream dividing results in the segmental baffled heat exchanger, the flow stream in the helical baffled heat exchanger was divide into 5 parts, and the factors causes the leakage stream flow *F* through the axis of heat exchanger was present via analytical calculation. By means of numerical calculation for helical baffled heat exchanger with clearances between tubes and baffled as well as the clearances between baffled and shells, following conclusions are obtained: first, with the increase of Reynolds number, both the ratio of main flow stream *B* and the ratio of the leakage stream flow *F* fall; second, as the helical angle raises, the ratio of B and the ratio of *F* increases and decreases, respectively; third, the correlations for the ratios of all flow streams under different Reynolds numbers and helical angles were present. With the optimal design of flow stream for a heat exchanger in industrial scale, the heat transfer rate increased 32% and the pressure drop dropped 23% respectively.

387608 Modeling, Simulation and Scale-up Study of Biomass Drying in a Vortex Chamber

Abstract:

High-G fluidized beds can be generated in vortex chambers. Different lab- and pilot-scale studies show the potential process intensification that can be achieved [1]. Scale-up of the technology is, however, not straightforward. Detailed simulations can help to deal with the scale-up. The present paper addresses the modeling and simulation of vortex chamber biomass dryers. Data for the modeling and the estimation of the model parameters are obtained from batch experiments with woody biomass. The biomass moisture content and the air in- and outlet temperature and humidity are measured as a function of time. The non-stationary nature of the batch experiments allows obtaining different data points from a single experiment, as such minimizing the amount of experiments required for the modeling. The model is validated using available data on continuous woody biomass drying in a vortex chamber. The retained model accounts for interfacial mass and heat transfer limitations, a non-uniform distribution of the moisture in the particles and intra-particle diffusion limitations. Scale-up is then studied focusing on continuous drying. Optimization of the product uniformity and of the air consumption and utilization are studied, simulating different possible configurations.

References:

[1] Eliaers P., De Wilde J., Drying Technology, 31:2, p. 236-245, 2013.

391639 Land Use Emissions and Environmental Impacts of Agricultural Systems

Abstract:

This poster presents results related to direct land use emissions and fluxes associated with agricultural activities. These emissions are further translated into environmental and climate impacts. The presented results relate to the case study of San Luis Valley in south central Colorado. The emissions (e.g. CO2, N2O etc.) are estimated using DNDC model and impact assessment methods are used to translate them into environmental impacts. We present results related different scenarios such as varying crops, cropping patterns, agricultural practices etc. We also discuss the relevance of these results in the context of regional sustainability.

388174 Development of a Scalable Friedel-Crafts Acylation: A Mechanistic Approach

Abstract:

The commercial route of a late-stage asset under development was initially performed via a Friedel-Crafts reaction with chloroacetyl chloride followed by an alkylation. Due to the projected demand and manufacturing cost, the development of an efficient, scalable route was essential to the project. The first generation process was successful in laboratory scale reactions; however, steady erosion in yield with increasing batch size was observed in production scale campaigns. This scale dependence was found to result from polymerization of the input material. Mechanistic studies of the chemistry, aided by in-line FTIR, identified the unfavorable equilibrium between intermediates.

Efforts to understand and mitigate the risk of polymerization lead the team to an alternate approach. A second generation process was subsequently developed to replace the single lower yielding step with two robust high yielding reactions. This new process was found to provide a higher overall yield to the isolated product, as demonstrated at the vendor. This presentation will provide a case study on the challenges of the scale dependent first generation process and the mechanistic approach taken to identify and design the advantages of the second generation process.

Multiphase Transport Modeling for Vacuum Drying of Pharmaceutical Products

Aditya G. Dodda[a], Kostas Saranteas[b] and Michael A. Henson[a]

[a]Department of Chemical Engineering, University of Massachusetts, Amherst, MA 01003
[b]Process Chemistry and Engineering, Sunovion Pharmaceuticals, Marlborough, MA 01752

Abstract

Downstream processing of Active Pharmaceutical Ingredients (APIs) typically involves extraction, crystallization, filtration and drying. In addition to being the most energy intensive unit operation, drying is often a major manufacturing bottleneck due to its relatively long batch processing times. Furthermore, drying conditions must be carefully selected and maintained to avoid problems such as API degradation. As a result, prediction and optimization of drying dynamic are needed to ensure product quality and minimize operating time. In principle, drying dynamics can be accurately predicted by detailed models that capture the relevant physics. Traditional models used to understand drying of APIs are based on simple solvent diffusion and penetration mechanisms that fail to capture important heat and mass transport effects. More sophisticated transport models are invariably based on the assumption that all three phases (solid, liquid, vapor) are at the same temperature. We have found that these models can be parameterized to satisfactorily predict solvent concentration dynamics, but they lack the necessary detail to capture temperature dynamics within the drying cake.

In this presentation, we discuss the theoretical development and experimental validation of a new multiphase model for API vacuum drying. Drying experiments were performed over a range of temperatures and pressures using glass beads of different sizes and acetone as the solvent. An on-line mass spectrometer was used to measure vapor composition in the head space and thermocouples were used to measure temperature at different locations in the drying cake. We developed a two-phase drying model with a separate liquid phase and a combined solid–vapor phase to capture the relevant heat and mass transport dynamics. We show that the multiphase model qualitatively captures the complex drying dynamics revealed from our experiments. By using the measured vapor phase composition and cake temperature profiles to estimate unknown transport parameters, we show that the multiphase model is capable of producing quantitatively accurate predictions over the examined range of bead sizes, temperatures and pressures.

383740 Die Face Pelletizing of Sticky HME Formulations

Abstract:

Hot melt extrusion is a continuous process with growing importance for the pharmaceutical industry [1]. An extruder processes raw materials into homogeneous strands of molten material and offers, in that way, robust production opportunities for solid dispersions. Solid dispersions are required to face current drug development challenges such as poor water solubility or modified release of the active pharmaceutical ingredient (API).

The molten material is shaped in the downstream process into the product of the extrusion line. Often, pellets are needed as intermediates for tablet compaction, capsule filling or injection molding. Pellets can be produced by die face pelletizing or strand pelletizing. The pelletizing methods distinguish themselves in terms of material temperature where the cutting takes place. In strand pelletizing the emerging material is drawn as cylindrical strands from the extrusion die. It runs through a cooling section and is chilled near or below the softening point and cut by a knife roller. Thus, solid bodies are cut resulting in cylindrical shaped pellets. In contrast, die face pelletizing is executed with a rotating knife pressed on the extrusion die plate. Thus, the material is in a viscous state during cutting and can deform due to surface tension into spherical pellets [2], [3]. Spherical pellets lead to better flowability and enhance dosing accuracy in subsequent handling steps. In addition, the pellets are conveyed with a cooling air stream to the product container where they are separated from the air stream, e.g. with an impact separator. This arrangement results in a simple self-contained construction and offers high spatial flexibility. However, air cooled die face pelletizing was limited to few formulations because of stickiness [4], although the plastics industry uses a similar principle to process sticky materials for several decades. The plastics industry uses underwater pelletizing systems, where the melt stickiness is prevented by higher cooling intensities. However, water-cooled systems are not suitable in the pharmaceutical field due to water solubility of the API and high purity requirements.

The aim of the presented study was a comprehensive understanding of polymer melt stickiness and its avoidance during die face pelletizing. The influence of heat transfer on adhesion properties has been investigated and a hypothesis on melt stickiness proposed. The hypothesis relates heat transfer properties to the materials phase transition temperature. The proposed hypothesis is supported with probe tack investigations at defined heat transfer conditions. This realization triggered a novel die plate design, which thermally decouples the die face from the melt flow channel. Thus, the surface temperatures of the pelletizer are low as compared to underwater pelletizing so that stickiness is suppressed. The novel die plate's heat balance has been calculated and optimized in-silico and subsequently built and tested as a real prototype.

The prototype was incorporated into an extrusion line consisting of an 18 mm twin-screw extruder (Coperion ZSK 18) and a die face pelletizer (Automatik Plastics Machinery GmbH, Sphero®-THA). The setup has successfully been applied to manufacture pellets from different sticky pharmaceutical polymers (e.g. copovidone). The obtained pellets are almost spherical and in the size range of 1 mm. Thus, it has been proven that the new design enables processing of sticky materials so that a broader range of formulations can benefit from the advantages of die face pelletizing.

References:

[1] M. A. Repka, N. Langley, and J. DiNunzio, Eds., *Melt Extrusion - Materials, Technology and Drug Product Design*. New York, Heidelberg, Dordrecht, London: Springer, 2013.

[2] E. Roblegg, E. Jäger, A. Hodzic, G. Koscher, S. Mohr, A. Zimmer, and J. Khinast, "Development of sustained-release lipophilic calcium stearate pellets via hot melt extrusion.," *Eur. J. Pharm. Biopharm.*, vol. 79, no. 3, pp. 635–645, Nov. 2011.

[3] S. Bialleck, "Herstellung von Polysaccharidpellets mittels Schmelzextrusion," Rheinischen Friedrich-Wilhelms-Universität Bonn, 2011.

[4] D. Treffer, P. Wahl, D. Markl, G. Koscher, E. Roblegg, and J. Khinast, "Hot Melt Extrusion as a Continuous Pharmaceutical Manufacturing Process," in *Melt Extrusion: Equipment and Pharmaceutical Applications*, M. Repka, Ed. Springer Publishers, 2013.

372442 Continuous Purification of Artemisinin and Artesunate

Abstract:

Continuous purification of artemisinin and artesunate

Andreas Seidel Morgenstern,[1,2] Ju Weon Lee,[1] Zoltan Horvath,[1] Elena Horosanskaia,[1]

Heike Lorenz,[1] Kerry Gilmore,[3] Daniel Kopetzki,[3] D. Tyler McQuade,[3] Peter H. Seeberger[3,4]

1) Max Planck Institute for Dynamics of Complex Technical Systems, Sandtorstrasse 1, 39106 Magdeburg, Germany

2) Otto-von-Guericke-University, Chair for Chemical Process Technology, Universitaetsplatz 2, 39106 Magdeburg, Germany

3) Max-Planck-Institute of Colloids and Interfaces, Department of Biomolecular Systems, Am Mühlenberg 1, 14476 Potsdam, Germany

4) Freie Universität Berlin, Institute of Chemistry and Biochemistry, Arnimallee 22, 14195 Berlin

Germany

Artemisinin, contained in sweet wormwood, and its derivatives β-artemether, β-arteether and α-artesunate are currently the basis for the most effective ways to cure the malaria disease [1]. Continuous photocatalytical synthesis of artemisisin from the waste product dihydroartemisinic acid was recently reported to be capable to produce in a laboratory flow reactor up to 3.5 kg/L/day with approx. 70 wt. % yield [2, 3]. The possibility of further synthesizing the mentioned derivatives in a continuous way could be demonstrated just recently for artesunate [4].

For a model target compound it was shown, that the outlet of a continuously operated flow reactor could be successfully fed directly into a simulated moving bed (SMB) chromatography unit in order to carry out both processes continuously [5, 6].

We will present results for various scenarios of the continuous isolation of artemisinin and artesunate from the effluent of continuously operated reactors. One of the investigated options is based on applying a 3-zone open-loop SMB process with an additional regeneration zone. The potential of combinations of chromatography and crystallization processes [7, 8] will be also discussed.

References

[1] World Malaria Report, World Health Organization, Geneva, 2011.

[2] F. Lévesque, P.H. Seeberger,

Angew. Chem. Int. Ed., 2012, 51, 1706 – 1709.

[3] D. Kopetzki, F. Lévesque, P.H. Seeberger,

Chem. Eur. J., 2013, 19, 5450 – 5456.

[4] K. Gilmore, D. Kopetzki, J.W. Lee, Z. Horvath, D. T. McQuade, A. Seidel-Morgenstern,

P.H. Seeberger, manuscript submitted

[5] A.G. O'Brien, Z. Horváth, F. Lévesque, J.W. Lee, A. Seidel-Morgenstern, P.H. Seeberger,

Angew. Chem. Int. Ed., 2012, 51, 7028 – 7030.

[6] J. W. Lee, Z. Horváth, A.G. O'Brien, P.H. Seeberger, A. Seidel-Morgenstern,

Chem. Eng. J., 2014, in print

[7] E. Horosanskaia, A. Seidel-Morgenstern, H. Lorenz,

Themochimica Acta, 2014, 578, 74 - 81.

[8] H. Lorenz, A. Seidel-Morgenstern,

Angew. Chem. Int. Ed., 2014, 53, 1218 – 1250.

392616 Safe Scale-up of Energetic Processes: Recent Case Studies

Abstract:

The safe scale-up of energetic materials requires a rigorous approach to reaction hazard characterization and criticality assessment. Several recent cases where the thermal hazard criticality assessment (based on reaction calorimetry, ARC, ARSST data and resulting heat flow, MTSR,and TD24) required a process redesign or modification to safely scale-up. Our approach to process safety and details related to redesign and safe scale-up of these examples will be discussed.

388819 Technology Transfer: Success Is in the Details

Abstract:

Successfully scaling technology safely from the laboratory bench to the pilot plant and from the pilot plant to a commercial plant, either owned or contracted, requires attention to detail --- not just technical detail, but planning detail and support detail.

This presentation discusses the application of dimensional analysis and the concept of the rate limiting regime to scaling technology from the labortatory to the commercial plant. This presentation also discusses the lessons learned from thirty years of scaling technology.

370715 Use and Abuse of Crystallization for Process Separations

Abstract:

INTRODUCTION:

Often chemical processes require an intermediate refining step to separate reaction products and the physical properties of the components exclude typical separation unit operations such as distillation or extraction. Every effort is made to exclude crystallization because of uncertainties in the performance and/or reliability of the methodology. This presentation will address the challenges (and solutions) of integrating crystallization techniques into processes where component yield and purity are the driving forces and crystal size and shape are generally secondary concerns.

PRELIMINARY EVALUATION:

Careful evaluation of the physical and chemical properties of the proposed feed stream from which a desired component is to be recovered is essential. Reliable solubility data for two components or a phase diagram (for more than two components) is necessary for preliminary process evaluation and is best determined experimentally. MATRIC has a variety of analytical equipment and experience in obtaining and evaluating solid/liquid phase data. For a two component system MATRIC computes a chart which combines product yield and total solids in suspension, as a function of temperature and initial feed concentration. The chart basically organizes the overall material balance for the system and serves as a basis for preliminary evaluation of product yield as a function of process variables. Development and use of this chart narrows the choices available for zones of probable operation and also provides guidance regarding necessary staging of the process. A typical recovery chart will be presented and discussed.

LABORATORY TESTING:

Bench scale testing is appropriate at this stage in the process development to evaluate the crystallization behavior of the desired product (solute). A variety of techniques is available to determine the so called meta-stable or labile portion of the phase diagram. Some systems sustain limited super saturation and others may require seeding to initiate nucleation. MATRIC has equipment and experienced personnel to investigate the need for a seeding policy. A chart describing a situation where high supercooling and careful seeding are required is included. At this stage of investigation it is advisable to conduct small scale batch crystallization tests to obtain information regarding particle shape (habit). As the preliminary testing proceeds it is advisable to document the crystal habit with photomicrographs to detect any dramatic change as the process development (and impurities) change. Preliminary information regarding particle size distribution is also desirable at this stage. Generally product purity studies are a little premature at this stage; however, product recovery via filtration techniques followed by washing will give a reasonable idea of the ease or difficulty in achieving desired purity.

PILOTING:

Piloting of the proposed process serves several purposes. If the proposed process operation is batch the techniques for crystallizing, separating and washing the product can be verified on a small scale and in similar equipment to a full scale operation. Sufficient material can be produced to conduct separation and washing studies and downstream drying investigations if the product must be dried. Solid-liquid separation schemes should be considered and can be evaluated at MATRIC facilities. At this stage of process development continuous operation may be of interest and process equipment can be designed to test this concept. Often product testing by the client for acceptability is also a goal during this phase.

If particle size distribution is specified in some way studies can be conducted in a continuous mode to develop the kinetics of the system. MATRIC has piloted several systems and conducted kinetic studies.

NOTE: MATRIC also has laboratory equipment to investigate the feasibility of using so-called "Melt Crystallization" to separate and purify components. The product from this type of separation operation is a "melt" which is ideal if the product is not required to be in a particulate form.

Authors:

Charles Moyers is a senior engineering scientist at Mid-Atlantic Technology, Research and Innovative Center (P.O. Box 8396, So. Charleston, WV 25303). He is a Fellow of AIChE, a PE (WV), and has co-taught several AIChE education courses including Industrial Crystallization, Liquid-Solid Separation, and Drying. He holds a BS from Virginia Tech, an MS from West Virginia U. and a PhD from U. of Delaware, all in chemical engineering. (E-mail: charles.moyers@matricinnovates.com)

Brooke Albin (presenter) is a chemical engineer at Mid-Atlantic Technology, Research and Innovative Center (P.O. Box 8396, So. Charleston, WV 25303). She earned a BS in biology and BA in chemistry from Alderson-Broaddus College and an MSE in chemical engineering from West Virginia University. (Email: brooke.albin@matricinnovates.com)

AIChE
The Global Home of Chemical Engineers

國立中央大學
National Central University

Stabilization and Spheroidization of Ammonium Nitrate: Co-crystallization with Crown Ethers and Spherical Crystallization by Solvent Screening

Chem. Eng. J. **2013**, *225*, 809–817

Department of Chemical & Materials Engineering
National Central University, Taiwan (R.O.C.)

Reporter: Chih-Wei Chen
Advisor: Prof. Tu Lee
Date: 11/20/2014
E-mail: tulee@cc.ncu.edu.tw

Bio-Inspired Materials & High Throughput Screening Laboratory

Outline

- **Introduction**
  - Propellant
  - Ammonium Nitrate (AN)
- **Results and Discussion**
  - Solvent Screening — Form Space of Ammonium Nitrate Recrystallized with Benzo-18-crown-6 (B18C6)
  - Manufacturing — 1:1 Co-crystal of AN-B18C6
  - Smoke Test
  - Burning Characteristics — Gas Generator Propellant
  - Manufacturing — Spherical agglomerates
- **Conclusions**
- **Acknowledgement**

2

Propellant

- **Propellant** — pressurized gas escape through a nozzle and create thrust.

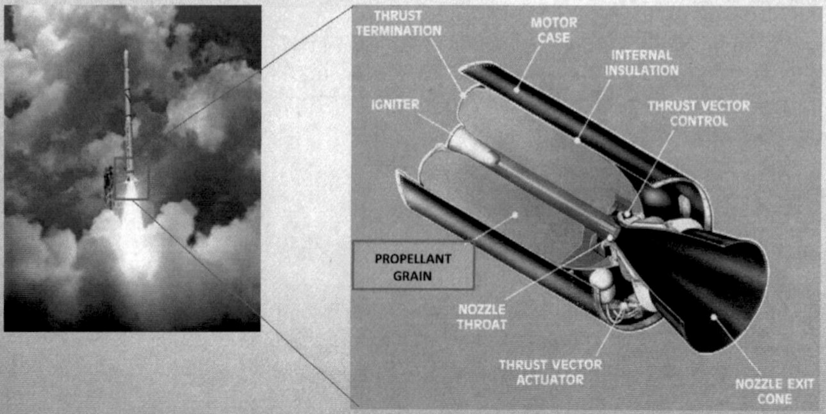

▲ Structure of rocket engine.

- **Solid-propellant gas generator** — generate electric power.

Century of Flight. http://www.century-of-flight.net

3

- **Propellant**
  - Fuel: Hydrocarbons
  - Oxidizer: Ammonium nitrate (AN) | High altitude — No oxygen
  - Binder: Polymer
  - Catalyst: Ammonium dichromate (ADC)
- **Oxygen source for burning:**

$NH_4NO_3(s) \rightarrow N_2(g) + 2H_2O(g) + \frac{1}{2}\mathbf{O_2}(g)$

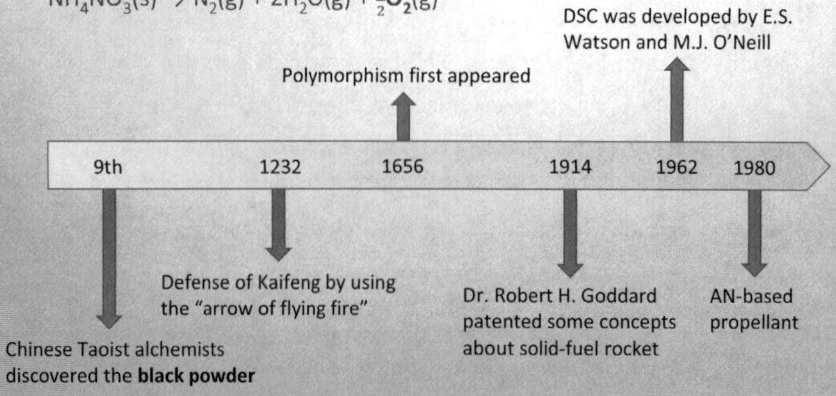

4

Ammonium nitrate

- Ammonium nitrate (NH_4NO_3):
  - ✓ **advantages**: clean burning, non-corrosive, low-hazard, low-flame temperature, inexpensive, easily available energetic material
  - ✗ **drawbacks**: (1) polymorphic transition near ambient temperature - density(volume) change
    (2) caking and smoke generation upon burning
    (3) poor flowability for packing

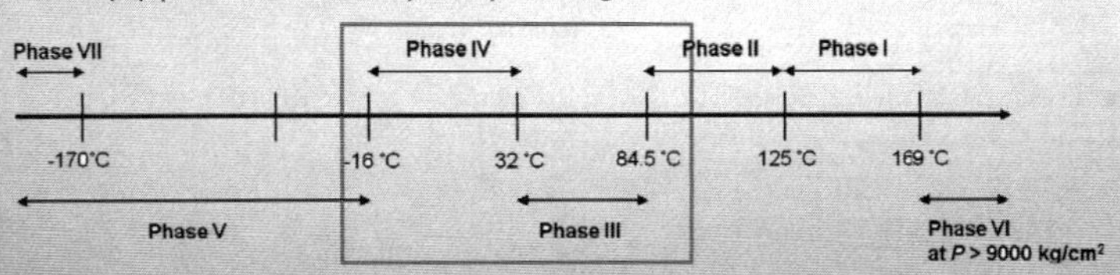

- The first two disadvantages could be minimized by forming **co-crystals**.
  The third drawback could be eliminated by forming **spherical agglomerates of co-crystals**.

Chem. Eng. J. **2013**, *225*, 809–817 5

Method

- **Co-crystallization:** chose crown ethers such as 15-crown-5 (15C5), benzo-15-crown-5 (B15C5), 18-crown-6 (18C6) and **benzo-18-crown-6 (B18C6)** as a co-former to form 1:1 co-crystals.

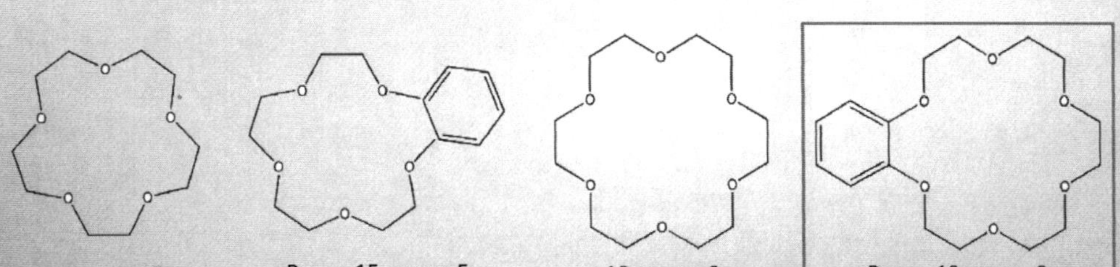

15-crown-5 Benzo-15-crown-5 18-crown-6 Benzo-18-crown-6

- **Spherical crystallization:** the formation of agglomerates was promoted by the addition of a small amount of the bridging liquid, which was immiscible with the antisolvent.

Chem. Eng. J. **2013**, *225*, 809–817 6

Use Test

Identification test

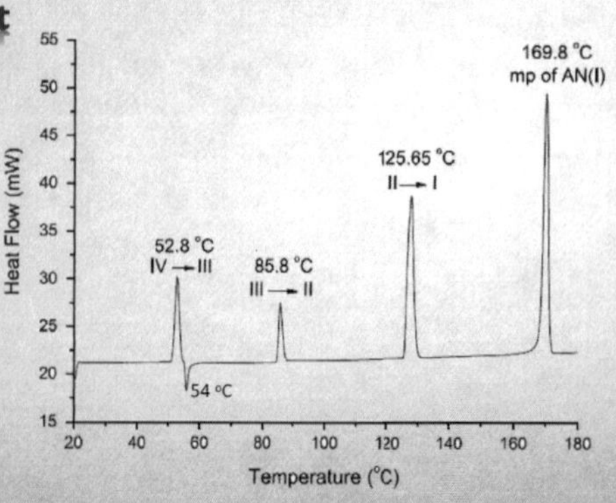

- According to differential scanning calorimetry (DSC) scan, there are **four** forms of the purchased ammonium nitrate.
- The polymorph of the purchased ammonium nitrate powders was Form IV at room temperature.

Chem. Eng. J. **2013**, *225*, 809–817 7

Solvent Screening

To construct the form space for Form IV ammonium nitrate by solvent screening

- 4 good solvents (yellow blocks): **ethanol**, **dimethyl sulfoxide (DMSO)**, **methanol** and **water**.
  [Solubility > 10mg/mL]
- 18 bad solvents (red blocks) as **antisolvent** to recrystallize: *n*-heptane, xylene, *p*-xylene, **ethyl acetate**, toluene, **methyl *tert*-butyl ether (MTBE)**, benzene, methyl ethyl ketone (MEK), chloroform, tetrahydrofuran (THF), *N,N*-dimethylaniline, acetone, 1,4-dioxane, nitrobenzene, *n*-butyl alcohol, isopropyl alcohol (IPA), benzyl alcohol and acetonitrile.

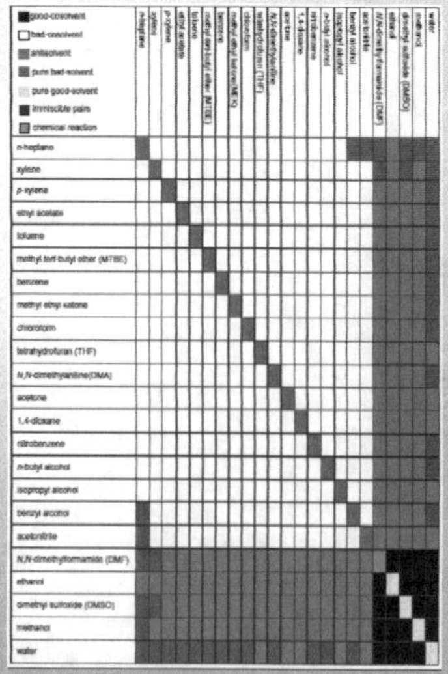

Chem. Eng. J. **2013**, *225*, 809–817 8

Procedure of Manufacturing Co-crystal

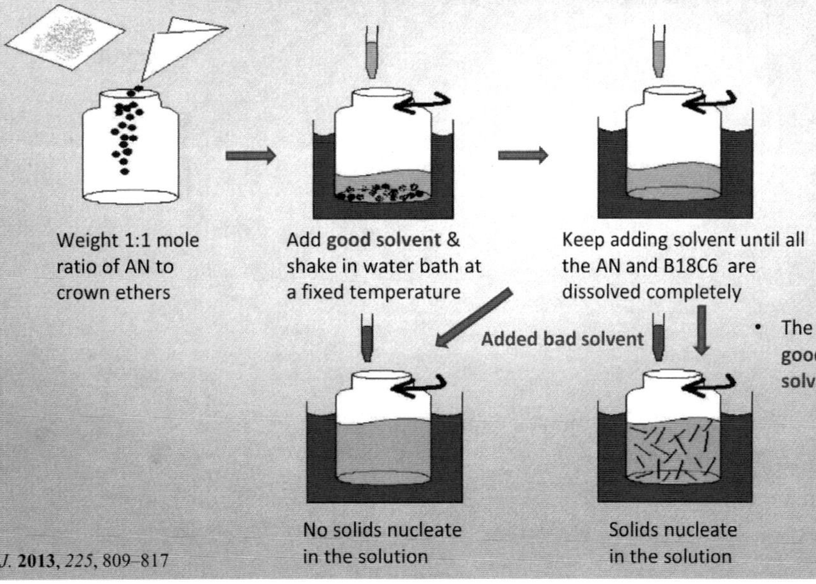

- The volume ratio of **good solvent to bad solvent is 1: 8.**

Recrystallized with B18C6

OM

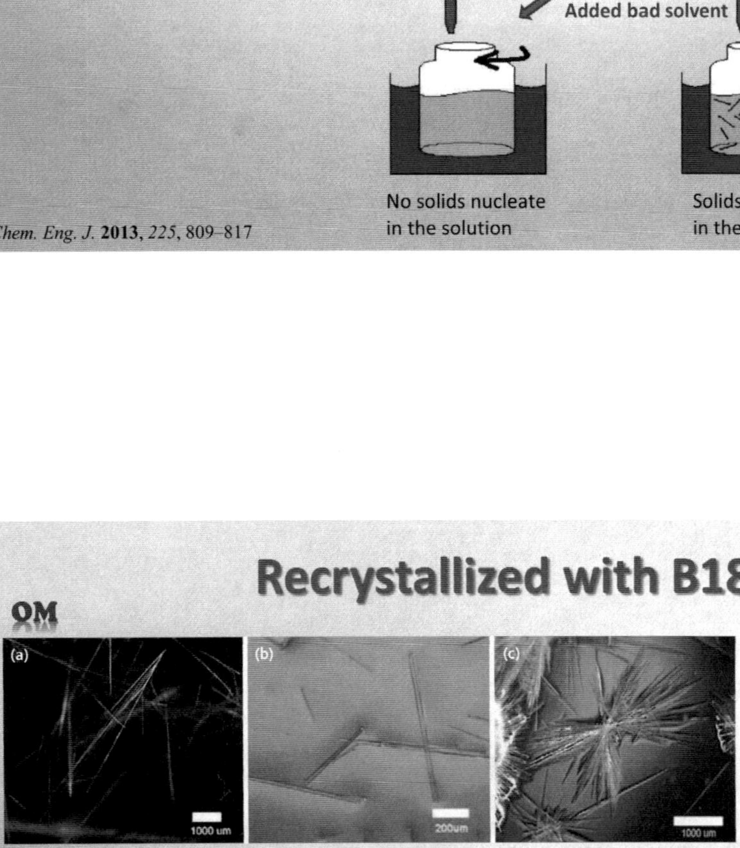

◄ Crystals of 1:1 cocrystal of AN-B18C6. (a) Ethanol-MTBE (b) Methanol-EtOAc (c) Methanol-MTBE

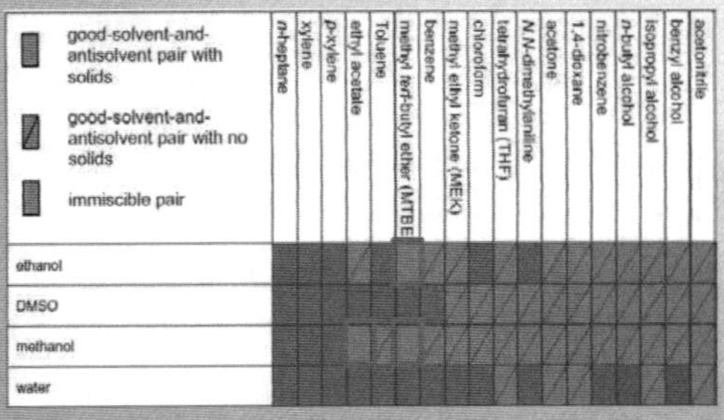

- Only **B18C6** was used, **needle-like crystals** were produced; other crown ethers such as 18C6, 15C5 and B15C5 were not.
- According to the experimental result, there are only three good-solvent-and-antisolvent systems to form co-crystals: **ethanol-MTBE**, **methanol-ethyl acetate** and **methanol-MTBE**.

Characterization of Co-crystal

DSC **PXRD**

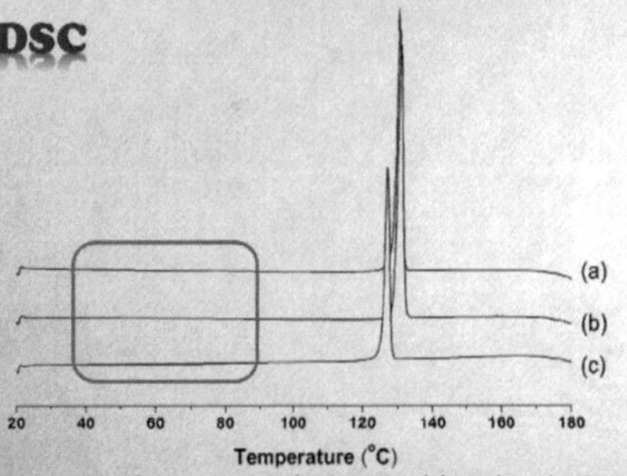

▲ Crystals of 1:1 cocrystal of AN-B18C6. (a) Methanol-MTBE (b) Methanol-EtOAc (c) Ethanol-MTBE

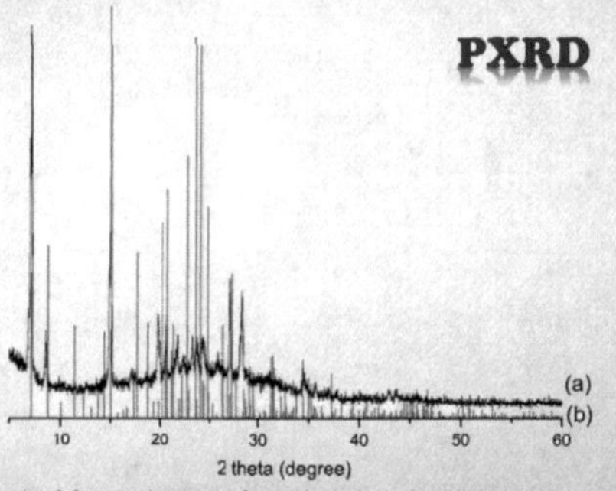

▲ (a) crystals grown from the ethanol-MTBE system (b) the calculated PXRD pattern based on SXD of crystals grown from the ethanol-MTBE system

- The DSC scans, endotherm near **125-129°C**, it illustrates that these crystals are the **same** kind; no polymorphic transition occurred between **32 to 85.5°C** .
- In addition, the good matching between the XRD diffraction diagrams of crystals validated the antisolvent precipitation method and indicated the purity of all crystals.

Chem. Eng. J. **2013**, *225*, 809–817 11

Characterization of Co-crystal

TGA

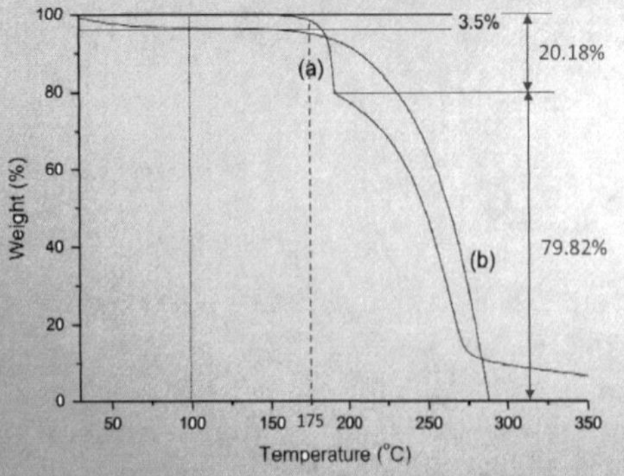

(a) crystals harvested from ethanol-MTBE system:

✓ the first weight loss at about 175°C
→ melting points = 125-129°C

✓ stoichiometric ratio:

$$\frac{\text{the first weight loss (AN)}}{\text{the second weight loss (B18C6)}} = \frac{20.18\%/80.04g\cdot mol^{-1}}{79.82\%/312.34g\cdot mol^{-1}} \approx 1$$

(b) purchased Form IV ammonium nitrate

- According to the thermogravimetric analysis (TGA) scan, there was **no significant weight loss** of 1:1 co-crystals of AN-B18C6 until 175°C, on the contrary, the purchased Form IV ammonium nitrate with **3.5 wt% weight loss from 30 to 100°C** due to dehydration. Therefore, hygroscopicity of ammonium nitrate was also minimized by co-crystallization.

Chem. Eng. J. **2013**, *225*, 809–817 12

Smoke Test

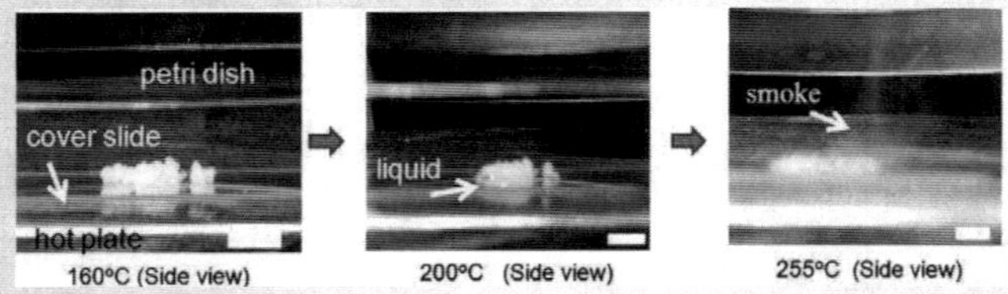

The purchased **Form IV** ammonium nitrate powders *were transformed to* **Form I** ammonium nitrate powders when heated on the hot plate in air.

- The gaseous ammonia reached cooling air and condensed to form ammonium nitrate again, which would lead to smoky appearance. $NH_4NO_3 \leftrightarrow NH_3 + HNO_3$

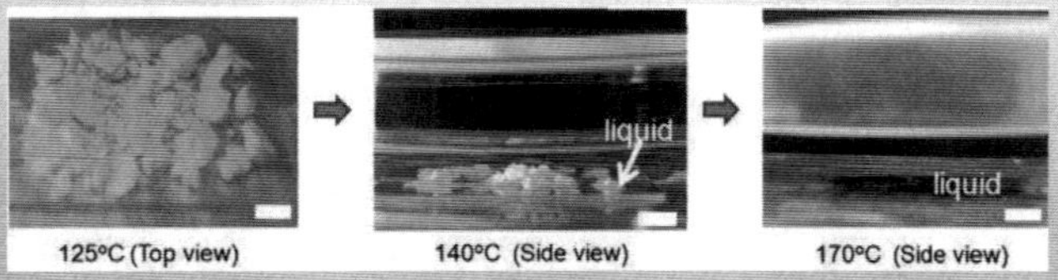

- The 1:1 co-crystals of AN-B18C6 was bonded with crown ether through **hydrogen-bonding** and was not allowed to evaporate into the gas phase to form smoke.

Chem. Eng. J. **2013**, *225*, 809–817 13

Burning Characteristics

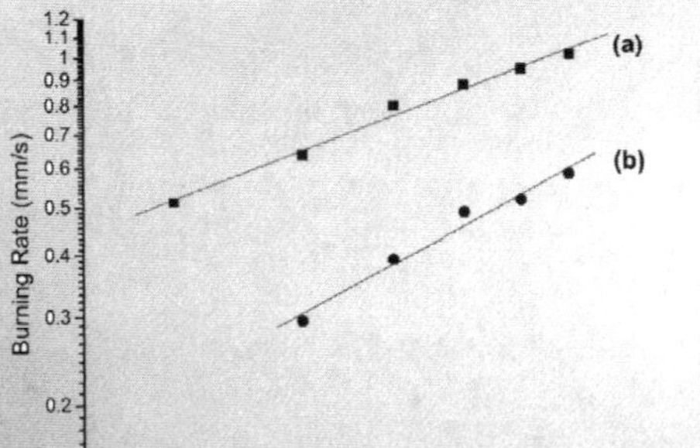

(a) AN-based propellant contained 20% **ammonium dichromate** by weight

(b) AN-based propellant contained 20% **1:1 co-crystal of AN–B18C6**

Saint-Robert's law (Veille's law)

- $r = a \cdot P^n$ (P < 2000 psia)
  - a is a constant
  - n is pressure exponent

a) $r = 0.35 \cdot P^{0.62}$

b) $r = 0.13 \cdot P^{0.8}$

➢ **Lower burning rate**

{ Lower flame temperature
Slower burning

- The use of lightweight hardware which would then withstand lower temperatures for a longer period.
- The utilization of an electrical generator run by hot gases given off from a slower burning rate which would in turn produce electric power for a longer time.

Chem. Eng. J. **2013**, *225*, 809–817 14

Spherical agglomerates

- **Spherical crystallization**: the formation of agglomerates was promoted by the addition of a small amount of the bridging liquid, which was **immiscible** with the **antisolvent**.
- **Water** was environmentally benign and immiscible with both ethyl acetate and MTBE, water was selected as the bridging liquid.

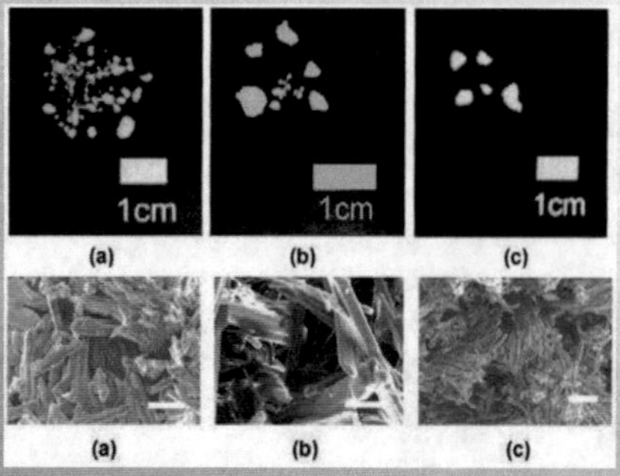

(a) 2.5 mL ethanol + 12.5 mL MTBE + 0.26 mL water
(b) 0.64 mL methanol + 11.6 mL ethyl acetate + 0.14 mL water
(c) 0.64 methanol + 11.6 mL MTBE + 0.08 mL water

	Methanol-MTBE-water	Methanol-ethyl acetate-water	Ethanol-MTBE-water
Yield (%)	87.3	24.8	52.3
Length mean diameter (cm)	0.297	0.293	0.146
Population density (1/cm^3)	20.6	10.9	1.85
Sphericity	0.86	0.82	0.71
Percentage friability index (%)	100	100	100
Angle of repose (deg.)	25.2	41.6	24.3

Methanol–MTBE–water system was the best!

- ✓ high yield
- ✓ high population density
- ✓ high sphericity
- ✓ high friability (i.e., greater strength)
- ✓ low angle of repose (i.e., good bulk flowability)

Chem. Eng. J. **2013**, *225*, 809–817 15

Conclusions

- This research had been successfully demonstrated that **co-crystallization** with crown ether could be utilized to minimize the drawbacks of ammonium nitrate:
  - ➢ polymorphic transition near room temperature
  - ➢ caking and smoke generation upon burning caused by the hygroscopic effect of ammonium nitrate
- The flowability of 1:1 co-crystals of AN-B18C6 could be further enhanced by **spherical crystallization**.

Chem. Eng. J. **2013**, *225*, 809–817 16

Acknowledgement

- My advisor – Prof. Tu Lee
- My lab partners
- My family
- National Science Council (Ministry of Science and Technology)
  - NSC 99-2119-M-008-016

17

Thanks for your attention !

18

gsk do more feel better live longer

Time To Maximum Rate Under Adiabatic Conditions Within 24hr ($TMR_{ad\ 24}$)

Christopher Morrison
Susan Shilcrat

Disclaimer:
This information is intended for educational purposes only by GlaxoSmithKline (hereinafter "GSK"). By offering this informational resource, GSK makes no representations about the suitability of this information for any purpose. The use of such information is entirely voluntary and reliance on it should only be undertaken after an independent review of it by qualified experts. In no event shall GSK be liable for any special, direct, indirect, consequential, exemplary, and/or any other damages whatsoever resulting from the use of any of this information.

Overview

- What is TMR_{ad} 24
- The Push for Change
  - 50K and 100K offset
  - Literature Review
- Calculation of TMR_{ad} 24
  - One DSC
  - One DSC + Cp
  - Three DSC + Cp
- When to Apply Each Technique

Introduction – What is TMRad 24?

- $TMR_{ad\ 24}$
  - Defined as the **T**ime to **M**aximum **R**ate given **ad**iabatic conditions such that there is **24** hours of time in order to correct the processing issue(s) prior to the reaction/decomposition reaction achieving its maximum rate of reaction
  - Based on kinetics of the reaction/decomposition reaction

Why Change? $TMR_{ad\ 24}$ vs. 100K Rule

- Even the 100K rule is unable to give an appropriate level of safety for some chemicals/mixtures when the onset temperature of decomposition is above 180°C.
- Neither the 50K nor the 100K rule take into account the kinetics of a decomposition reaction.
- $TMR_{ad\ 24}$ takes into consideration the kinetics of decomposition and in doing so may avoid the need for some ARCs for compounds with onset temperatures of decomposition below 180°C.
- Figure 11.8 from Francis Stoessel's "Thermal Safety of Chemical Processes" 2008.

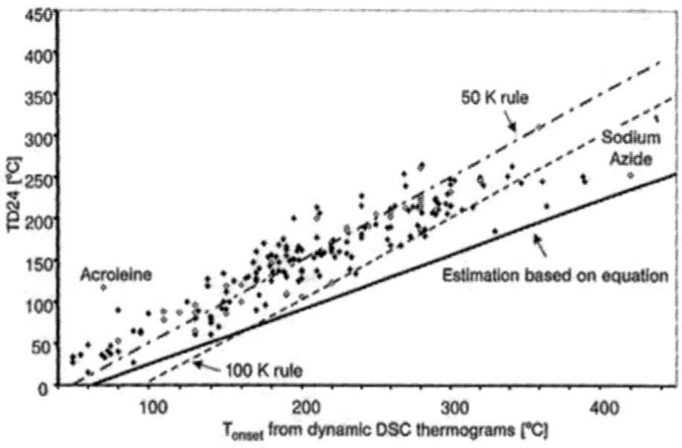

Figure 11.8 Comparison of experimental T_{D24} with the onset in DSC experiments. A prediction rule (lines) can be considered safe if all experimental values are located above the line representing the rule. White dots are pure compounds, black dots are reaction mixtures and distillation residues.

Literature Review of Energetic Materials

Source	Compound	Tonset °C	100K Rule	1 DSC TMRad 24	Tonset ARC	ARC 30K Rule	Lit. Ea (kJ/mol)	Lit. TMRad 24, °C
Bretherick's	4-Nitrobenzaldehyde	220	120	108	226	196	202	171
Bretherick's	2-Nitroaniline	280	180	150			171	223
Bretherick's	3-Nitroaniline	280	180	150			149	213
Bretherick's	4-Nitroaniline	280	180	150			160	217
Bretherick's	2-Nitrobenzoic acid	230	130	115			156	189
Bretherick's	2-Nitrophenol	260	160	136			135	182
Bretherick's	3-Nitrophenol						148	213
Bretherick's	4-Nitrophenol	240	140	122			117	185
Bretherick's	Phenyl azide	130	30	45			119	87
Bretherick's	3-Nitrobenzaldehyde	180	80	80	166	136	168	147
Bretherick's	2-Nitrobenzaldehyde	220	120	108	176	146	116	123
Bretherick's	3-Nitrobenzoic acid	320	220	178			151	245
Bretherick's	4-Nitrobenzoic acid	320	220	178			131	247
Bretherick's	2,3-Dichloroaniline						119	212
Bretherick's	2,4-Dichloroaniline						129	211
Bretherick's	2,5-Dichloroaniline						126	216
Bretherick's	2,6-Dichloroaniline						227	217
Bretherick's	3,4-Dichloroaniline						146	187
Bretherick's	1,2-Diphenylhydrazine	135	35	49			139	131
Bretherick's	4-Methoxy-1,3-phenylenediamine	260	160	136			114	207
Bretherick's	4-chloro-1,2-benzenediamine	160	60	66			140	100
Bretherick's	4-Chloroaniline	210	110	101			124	200
Bretherick's	4-Chloro-2-aminophenol						140	110
Bretherick's	Nitrobenzene	360	260	206			216	280
Bretherick's	2-Nitrotoluene						162	232
Bretherick's	4-Nitrotoluene	300	200	164			162	232

Calculating $TMR_{ad\ 24}$ Made Simple

- Step 1: Run a DSC on the sample at a rate of 2 K/min
- Step 2: Draw a baseline on the DSC data. (Accurate if done in excel since the graph can be greatly magnified.)
- Step 3: Locate the temperature when the heat produced by the sample is 10 W/kg (0.01mW/mg) above baseline this is the T_{onset}.
- Step 4: Calculate
  - Note T_{onset} is in °C

$$TMR_{ad}\,24 = 0.7\,T_{onset}\left(q(T)=10\frac{W}{kg}\right) - 46$$

- Step 5: Understanding the Assumptions:
  - Ea= 50 kJ/mol , low activation energy
  - Zero order kinetics
  - Cp= 1.0 kJ/kg/K

Pros and Cons to Simple (One DSC) $TMR_{ad\ 24}$

- Pros:
  - Fast and conservative
  - Requires only one DSC run
  - Only requires the temperature when the self heating rate, q(T), exceeds the equipment's detection limit of ~10 W/kg above baseline
  - Reduces the need to run an ARC at lower temperatures
  - Increases safety of processes with high onset temperatures (10 W/kg detection limit)

- Cons:
  - Does not take into account actual kinetics of the decomposition; assumes a **very conservative** low activation energy, zeroth-order kinetics, and a low specific heat

Calculating $TMR_{ad\ 24}$ One DSC

- Step 1: Run a DSC on the sample at a rate of 2 K/min
- Step 2: Draw a baseline on the DSC data. (Accurate if done in excel since the graph can be greatly magnified.)
- Step 3: Locate the temperature when the heat produced by the sample is 10 W/kg (0.01mW/mg) above baseline this is the T_{onset}.
- Step 4: Locate the temperature, T_{ref}, when the heat released, q_{ref}, is 20 W/kg above baseline (beginning of exotherm to avoid autocatalytic effects).
- Step 5: Extrapolate heat, q_T

$$q_T = q_{ref} \times \exp^{\left(\frac{E_A}{R}\left(\frac{1}{T_{ref}} - \frac{1}{T}\right)\right)}$$

- Step 6: Calculate TMR with known specific heat, Cp

$$TMR_{ad} = \frac{C_p R T^2}{q_T E_A}$$

- Step 7: Understanding the Assumptions:
  - Ea= 50 kJ/mol, low activation energy
  - Zero order kinetics

Pros and Cons to One DSC $TMR_{ad\ 24}$

- Pros:
  - Fast and conservative
  - Requires only one DSC run
  - Reduces the need to run an ARC
  - Allows for the adjustment of the specific heat, Cp, which results in a more accurate estimation of the $TMR_{ad\ 24}$
- Cons:
  - Requires the temperature, T_{onset}, when the self heating rate, q(T), exceeds the equipment's detection limit of ~10 W/kg above baseline
  - Requires the temperature, T_{ref}, when the self heating rate, q_{ref}, is at ~20^+ W/kg above baseline
  - Does not take into account actual kinetics of the decomposition; assumes a conservative low activation energy and zeroth-order kinetics

Calculating $TMR_{ad\ 24}$ Using Three DSC Experiments

- Step 1: Run three DSC on the sample at three different heating rates (1-8 K/min)
- Step 2: Draw a baseline on the DSC data.
- Step 3: Locate the temperature when the heat produced by the sample is 10 W/kg (0.01mW/mg) above baseline this is the T_{onset}. (Note that the DSC software is capable to do steps 2 and 3.)
- Step 4: Locate the temperature, T_{ref}, when the heat released, q_{ref}, is ~20 W/kg above baseline for a DSC run at 2-3 K/min.

Calculating $TMR_{ad\ 24}$ Using Three DSC Experiments

- Step 5: Enter heating rate and Onset Temperatures in the green boxes
  - Note: A correlation of 0.990 will result in an approximate TMR error of 8⁺ hrs
  - Note: A correlation of 0.999 will result in an approximate TMR error of 3 hrs

Heating rate (K/min)	LN Heating rate	Onset Temperature (°C)	Temperature (K)	1/T
1	0	213.05	486.2	0.002056767
3	1.098612289	228.48	501.63	0.001993501
5	1.609437912	236.89	510.04	0.001960631

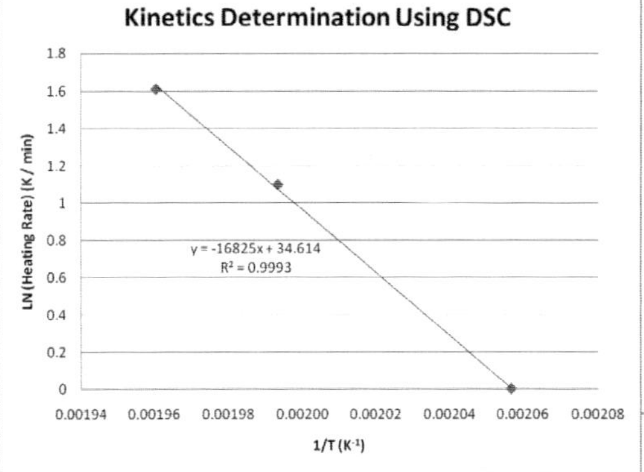

Intercept	34.61	LN(Pre-Exponent)
Pre-exponential factor	1.08E+15	A
Slope	-16824.97	-Ea/R (K)
Gas Constant	8.314	R (J/mol K)
Activation energy	139883	J/mol
Activation energy	139.88	kJ/mol

The activation energy is now known and can be applied to the decomposition reaction.

Calculating $TMR_{ad\ 24}$ Using Three DSC Experiments

- Step 6: Extrapolate the self heating rate, q_T

$$q_T = q_{ref} \times \exp^{\left(\frac{E_A}{R}\left(\frac{1}{T_{ref}} - \frac{1}{T}\right)\right)}$$

- Step 7: Calculate TMR with known specific heat, Cp, and activation energy, E_A

$$TMR_{ad} = \frac{C_p R T^2}{q_T E_A}$$

- Step 8: Understanding the Assumptions:
  - Zero order kinetics

Calculating $TMR_{ad\ 24}$ Using Three DSC Experiments

- Step 9: Do a sanity check of the data and values calculated
  - Run an isoage DSC or ARC at the temperature corresponding to a TMR_{ad} of ~8 hrs

  - Adjust TMR for Phi factor, Φ
    - Where:
    - M_c is the mass of the container
    - C_{pc} is the specific heat of the container
    - M_m is the mass of the material
    - C_{pm} is the specific heat of the material
    - TMR_{exp} is the experimental TMR
    - $TMR_{\Phi=1}$ is representative of the TMR on large scale

$$\Phi = \frac{(m_c C_{pc}) + (m_m C_{pm})}{(m_m C_{pm})}$$

$$TMR_{\Phi=1} = \frac{TMR_{exp}}{\Phi}$$

  - The TMR_{ad24} is considered conservative and safe as long as the experimental TMR_{ad8} is longer than the calculated TMR_{ad8}

Pros and Cons to Three DSC Determination of $TMR_{ad\ 24}$

- Pros:
  - Faster than an ARC and conservative
  - Reduces the need to run an ARC, can use DSC instead
  - Activation energy is known for the decomposition reaction
  - Allows for the adjustment of the specific heat, Cp, and activation energy, E_A, which results in a more accurate estimation of the $TMR_{ad\ 24}$

- Cons:
  - Requires the temperature, T_{onset}, when the self heating rate, q(T), exceeds the equipment's detection limit of ~10 W/kg above baseline
  - Requires the temperature, T_{ref}, when the self heating rate, q_{ref}, is at ~20+ W/kg above baseline for one of the DSC runs at 2-3 K/min

- Note: Activation energy can also be calculated from ARC exo files and the above mentioned process of using three DSCs could be used with one ARC

When to Apply Each Technique: Rules of Use

- Simplified TMR_{ad24} (One DSC)
  - Good for majority of decompositions
    - If autocatalytic may want to consider an isoage/isothermal ARC/DSC
  - If there is a melt that goes into an exotherm take T_{onset} from the beginning of the melt
- TMR_{ad24} C_p Adjusted (One DSC)
  - Allows for adjustment of Cp
  - Requires q_{ref} and T_{ref} in order to extrapolate q_T at T
  - If there is a melt that goes into an exotherm, q_{ref} and T_{ref} will not be accurate, so revert back to the simplified TMR_{ad24} calculation
- TMR_{ad24} E_A calculated and C_p Adjusted (3 DSC or 1 ARC and 1 DSC)
  - Calculation on E_A
  - Allows for adjustment of Cp
  - Requires q_{ref} and T_{ref} in order to extrapolate q_T at T
  - If there is a melt that goes into an exotherm, q_{ref} and T_{ref} will not be accurate, so revert back to the simplified TMR_{ad24} calculation
  - Run either an isothermal DSC or ARC to confirm that the TMR is conservatively accurate

Questions?

Food for Thought Regarding Heat Loss on Scale

Typical Heat Loss (Table 11.9 from Pharmaceutical Industry R&D to Manufacturing, 2011.)		
	Heat Loss (W/kg/K)	Time for 1°C Loss at 80°C
5000 L rxr	0.027	43 min
2500 L rxr	0.054	21 min
100 mL beaker	3.68	17 s
10 mL test tube	5.91	11 s
1 L Dewar	0.018	62 min

Onset Temperature from DSC (1 of 3)

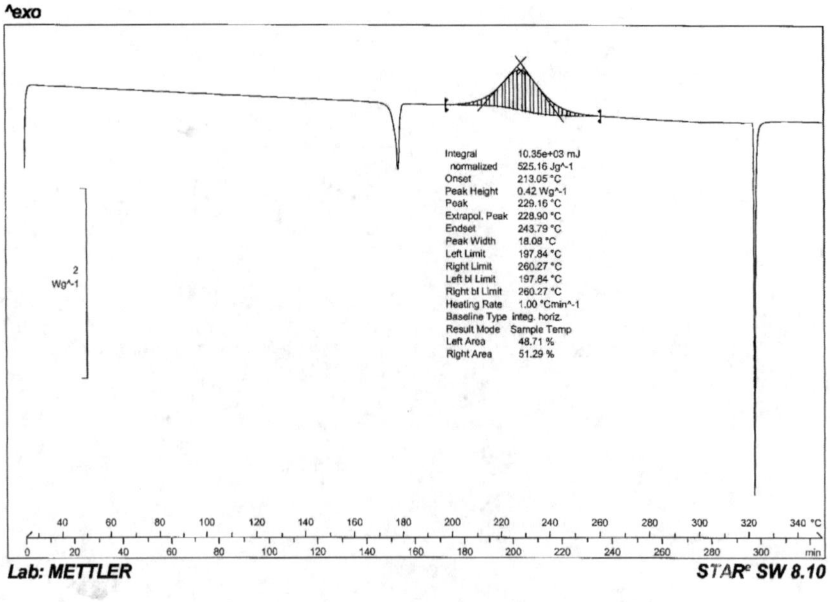

Onset Temperature from DSC (2 of 3)

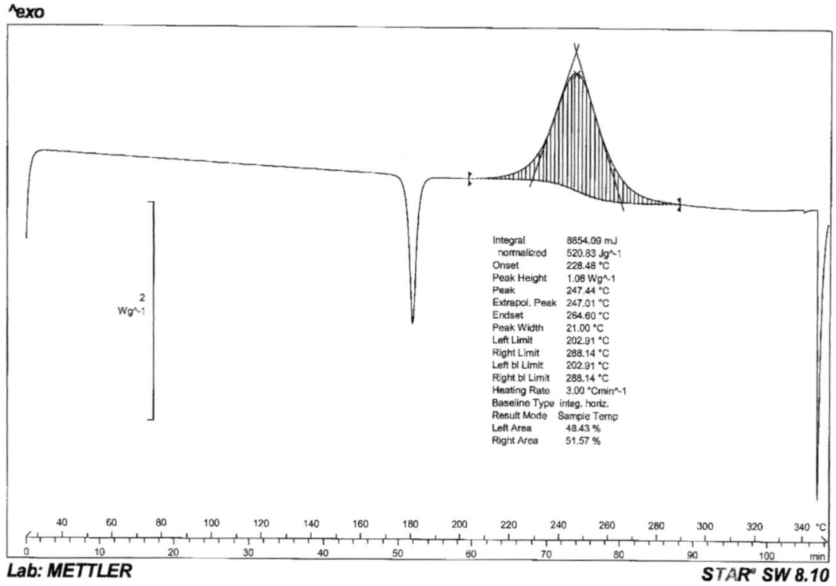

Onset Temperature from DSC (3 of 3)

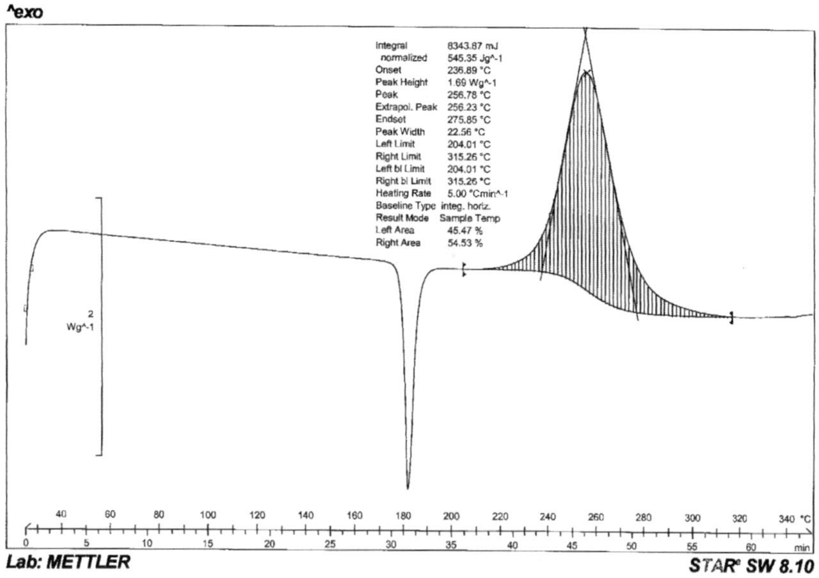

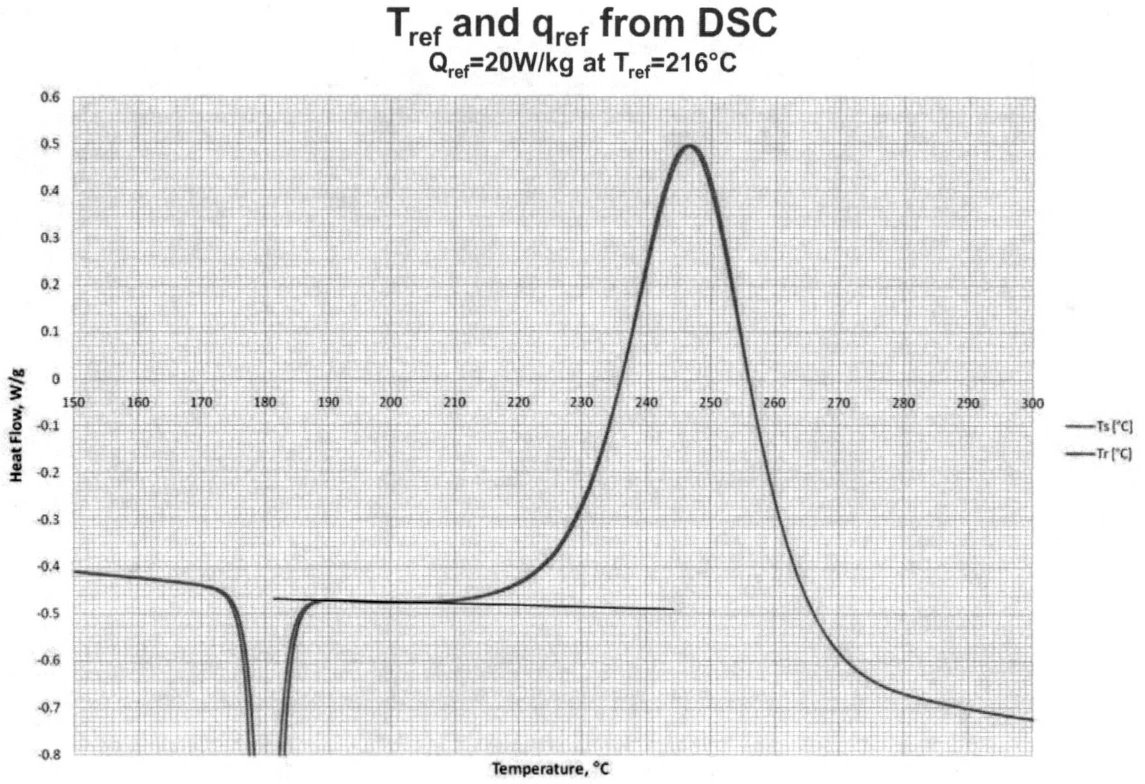

T_ref and q_ref from DSC
Q_ref=20W/kg at T_ref=216°C
Heat Flow, W/g
0.6
0.5
0.4
0.3
0.2
0.1
0
-0.1
-0.2
-0.3
-0.4
-0.5
-0.6
-0.7
-0.8
150
160
170
180
190
200
210
220
230
240
250
260
270
280
290
300
Ts [°C]
Tr [°C]
Temperature, °C

Abstract

Fouling is a complex phenomenon which commonly occurs in crude oil heat transfer equipment, reducing heat transfer, increasing pressure drops and in general resulting in reduced energy recovery efficiency and increased capital and operating costs of refineries.

Measuring with accuracy the fouling behaviour of an oil at or close to industrial conditions is not easy. A High Pressure Oil Rig (HIPOR), designed to overcome current limitations and operate in a regime close to refinery shell-and-tube exchangers, has been developed at Imperial College London under the direction of Prof. G.F. Hewitt. The apparatus includes a tubular and annulus flow sections with heat input supplied via Joule heating (Macchietto et al, 2011), and an experimental programme is under way.

The aim of this paper is two-fold: First, a predictive model of the HIPOR test section is briefly presented. The model is intended to assist in the analysis and interpretation of the experimental fouling data collected, and also in the development of the experimental programme itself. Second, some results are reported in respect of the calibration of the model to the equipment in clean (no-fouling) conditions, so as to establish a properly validated performance of the equipment (sensors and controllers, as well as thermo-hydraulic behaviour) in the absence of fouling and a reliable base-line for the assessment of differences due to fouling.

For the first objective, a dynamic, distributed thermo-hydraulic model of the tubular and annulus flow sections of HIPOR is presented. The model, an extension of Hexxcell's model for HiPOR, considers the two dimensional (axial and radial) heat transfer in different domains with variable thermal conductivity (heat generation), boundary conditions and other aspects (for example, heat losses) which are carefully determined. The calibration and validation of various model components through a variety of tests is described. The model predicts temperature profile and pressure drop of fluid in the test section as well as local conditions of the physical plant and fluid properties.

Instrumentation for temperature, flow rate and differential pressure measurement has been installed to record primary data for validation purposes. Heat input (heat flux) supplied via Joule heating is also carefully measured.

Tests were performed for varying inlet fluid temperatures and step changes in the heat input, enabling experimental results and model predictions to be compared. The simulation results (*temperature profile, wall temperature, differential pressure and temperature difference, outlet and inlet temperature*) show a consistent agreement with primary measurements from the rig at different process conditions, including transient responses. Furthermore, results also demonstrate the sensitivity of operating process parameters to differential pressure.

It is concluded that the model of the HIPOR rig at 'clean' oil conditions was successfully validated and could be a used as a reliable reference base to assist in the interpretation and validation of crude oil fouling data from HIPOR.

References

S. Macchietto, G.F. Hewitt, F. Coletti, B.D. Crittenden, D.R. Dugwell, A. Galindo, G. Jackson, R. Kandiyoti, S.G. Kazarian, P.F. Luckham, O.K. Matar, M. Millan-Agorio, E.A. Müller, W. Paterson, S.J. Pugh, S.M. Richardson, D.I. Wilson, "Fouling in crude oil preheat trains: a systematic solution to an old problem", *Heat Transfer Engineering*, **32**, 3, 197-215 (2011).

F. Coletti and S. Macchietto A dynamic, distributed model of shell-and-tube heat exchangers undergoing crude oil fouling. *Ind. Eng. Chem. Res.* **50** (8), pp 4515–4533 (2011).

AUTHOR INDEX

AUTHOR INDEX

AUTHOR INDEX

AUTHOR INDEX

AUTHOR INDEX